Cliffs Advanced Placement™

CliffsAP™
Chemistry

3RD EDITION

by

Gary S. Thorpe, M.S.

Consultant

Jerry Bobrow, Ph.D.

Hungry Minds™

HUNGRY MINDS, INC.

New York, NY ◆ Cleveland, OH ◆ Indianapolis, IN
Chicago, IL ◆ Foster City, CA ◆ San Francisco, CA

About the Author

Gary S. Thorpe has taught AP Chemistry and gifted programs for over twenty-five years. He currently teaches chemistry at Beverly Hills High School, Beverly Hills, California.

Author's Acknowledgments

I would like to thank my wife, Patti, and my two daughters, Krissi and Erin, for their patience and understanding while I was writing this book. Special thanks also goes to Dr. Norm Juster, Professor Emeritus of Chemistry, U.C.L.A., Dr. Jerry Bobrow of Bobrow Test Preparation Services, and Christopher Bushee for their input, proofreading, and suggestions.

Publisher's Acknowledgments

Editorial

Project Editor: Donna Wright

Acquisitions Editor: Sherry Gomoll

Technical Editor: Christopher Bushee

Production

Proofreader: Joel K. Draper

Hungry Minds Indianapolis Production Services

CliffsAP™ Chemistry, 3rd Edition

Published by
Hungry Minds, Inc.
909 Third Avenue
New York, NY 10022
www.hungryminds.com
www.cliffsnotes.com

Note: If you purchased this book without a cover, you should be aware that this book is stolen property. It was reported as "unsold and destroyed" to the publisher, and neither the author nor the publisher has received any payment for this "stripped book."

Library of Congress Control Number: 00-054070

ISBN: 0-7645-8684-X

Printed in the United States of America

10 9 8 7 6 5 4 3 2 1

3B/SX/QS/QR/IN

Distributed in the United States by Hungry Minds, Inc.

Distributed by CDG Books Canada Inc. for Canada; by Transworld Publishers Limited in the United Kingdom; by IDG Norge Books for Norway; by IDG Sweden Books for Sweden; by IDG Books Australia Publishing Corporation Pty. Ltd. for Australia and New Zealand; by TransQuest Publishers Pte Ltd. for Singapore, Malaysia, Thailand, Indonesia, and Hong Kong; by Gotop Information Inc. for Taiwan; by ICG Muse, Inc. for Japan; by Norma Comunicaciones S.A. for Colombia; by Intersoft for South Africa; by Eyrolles for France; by International Thomson Publishing for Germany, Austria and Switzerland; by Distribuidora Cuspide for Argentina; by LR International for Brazil; by Galileo Libros for Chile; by Ediciones ZETA S.C.R. Ltda. for Peru; by WS Computer Publishing Corporation, Inc., for the Philippines; by Contemporanea de Ediciones for Venezuela; by Express Computer Distributors for the Caribbean and West Indies; by Micronesia Media Distributor, Inc. for Micronesia; by Grupo Editorial Norma S.A. for Guatemala; by Chips Computadoras S.A. de C.V. for Mexico; by Editorial Norma de Panama S.A. for Panama; by American Bookshops for Finland. Authorized Sales Agent: Anthony Rudkin Associates for the Middle East and North Africa.

For general information on Hungry Minds' products and services please contact our Customer Care department; within the U.S. at 800-762-2974, outside the U.S. at 317-572-3993 or fax 317-572-4002.

For sales inquiries and resellers information, including discounts, premium and bulk quantity sales and foreign language translations please contact our Customer Care department at 800-434-3422, fax 317-572-4002 or write to Hungry Minds, Inc., Attn: Customer Care department, 10475 Crosspoint Boulevard, Indianapolis, IN 46256.

For information on licensing foreign or domestic rights, please contact our Sub-Rights Customer Care department at 650-653-7098.

For information on using Hungry Minds' products and services in the classroom or for ordering examination copies, please contact our Educational Sales department at 800-434-2086 or fax 317-572-4005.

Please contact our Public Relations department at 212-884-5163 for press review copies or 212-884-5000 for author interviews and other publicity information or fax 212-884-5400.

For authorization to photocopy items for corporate, personal, or educational use, please contact Copyright Clearance Center, 222 Rosewood Drive, Danvers, MA 01923, or fax 978-750-4470.

Hungry Minds™ is a trademark of Hungry Minds, Inc.

Table of Contents

PART III: AP CHEMISTRY LABORATORY EXPERIMENTS

PART IV: AP CHEMISTRY PRACTICE TEST

PART V: APPENDIXES

Preface

The AP chemistry exam is coming up! Your thorough understanding of months and months of college-level chemistry lectures, tests, quizzes, homework problems, lab write-ups, and notes are to be evaluated in a 3-hour examination. It's just you and the AP exam. In preparing to do the very best job possible, you have four options:

1. Read all of your textbook again.

2. Do all of your homework problems again.

3. Buy a test preparation guide that has every conceivable type of problem in it *and* that in many cases is thicker than your textbook *and* that you will never be able to finish *and* that does not explain how to do well on the essay portion of the exam *and* does not review all of the laboratory experiments required and tested.

4. Use *Cliffs Advanced Placement Chemistry,* 3rd Edition.

I'm glad you chose option 4. I have taught chemistry for over 25 years. I've put together for this book, in a reasonable number of pages, what I feel are the best examples of problems to help you prepare for the exam. With other AP exams to study for and other time commitments, you need a quick and neat book that you can finish in a few weeks and that covers just about everything you might expect to find on the exam. You have that book in your hands.

This guide is divided into five parts:

Part I: Introduction

Part I contains the following sections: Questions Commonly Asked About the AP Chemistry Exam, Strategies for Taking the AP Chemistry Exam, Methods for Writing the Essays, Mathematical Operations, and Mathematics Self-Test.

Part II: Specific Topics

Each chapter lists key vocabulary words, formulas, and equations and provides about ten completely worked-out multiple-choice questions and solutions for two or more free-response questions. Self-contained chapters cover gravimetrics, thermochemistry, gases, the structure of atoms, covalent bonding, ionic bonding, liquids and solids, solutions, kinetics, equilibrium, acids and bases, energy, organic chemistry, nuclear chemistry, and writing and predicting chemical reactions.

Part III: Laboratory Experiments

Background information, sample data, and accompanying exercises on analysis of the data for all 22 laboratory experiments recommended by the AP College Board. Also included are two sample laboratory write-ups. Provides a 'refresher' for the experiments that you have completed and that you will be tested on.

Part IV: AP Chemistry Practice Test

A complete exam consisting of both multiple-choice and free-response questions. All solutions are fully worked out.

Part V: Appendixes

Also provided for your convenience are four commonly used tables and charts that you will find useful as you work on problems. They are:

1. Appendix A: Commonly Used Abbreviations, Symbols, and Tables

2. Appendix B: Acid-Base Indicators

3. Appendix C: Flame Tests for Elements

4. Appendix D: Qualitative Analysis of Cations and Anions

This book is not a textbook. The last thing you need right now is another chemistry textbook. However, if you have forgotten concepts or if something is new to you, use this preparation guide with your textbook to prepare for the AP chemistry exam. Now turn to the Study Guide Checklist on page ix and check each item, in order, as you complete the task. When you have checked all the items, you will be ready for the AP chemistry exam. Good Luck!

Study Guide Checklist

❑ 1. Read the *Advanced Placement Course Description—Chemistry* (also commonly known as the "Acorn Book") produced by Educational Testing Service (ETS) and available from your AP chemistry teacher, testing office, or counseling center or directly from The College Board.

❑ 2. Read the Preface to this Cliffs preparation guide.

❑ 3. Read "Questions Commonly Asked about the AP Chemistry Exam."

❑ 4. Read "Strategies for Taking the AP Chemistry Exam."

❑ 5. Read "Methods for Writing the Essays."

❑ 6. Review "Mathematical Operations" and take the "Mathematics Self-Test."

❑ 7. Go through each of the chapters listed below, carefully reviewing each key term, each key concept, and all the worked-out examples of multiple-choice and free-response questions.

Key Terms. You should be familiar with the meanings of these words and with the principles underlying them. They can serve as foundations for your essays. The more of these key terms (provided that they are relevant) that you can use effectively in your explanations, the better your free-response answers will be.

Key Concepts. These are formulas, constants, and equations used in the chapter. You should be completely familiar with them.

Samples. Do not go on to a new sample until you thoroughly understand the example you are currently working on. If you do not understand a sample after considerable thought, ask your AP chemistry teacher or fellow classmates how they approached the problem. Be sure to work each sample with pencil and paper as you go through the book. Write out your answer to every free-response question. Practice makes perfect!

    ❑ Gravimetrics

    ❑ Thermochemistry

    ❑ The Gas Laws

    ❑ Electronic Structure of Atoms

    ❑ Covalent Bonding

    ❑ Ionic Bonding

    ❑ Liquids and Solids

    ❑ Solutions

    ❑ Kinetics

    ❑ Equilibrium

    ❑ Acids and Bases

    ❑ Energy and Spontaneity

❑ Reduction and Oxidation

❑ Organic Chemistry

❑ Nuclear Chemistry

❑ Writing and Predicting Chemical Reactions

❑ 8. Review the Laboratory Experiments—go through each of the 22 laboratory experiments and review the procedures and analyze the data.

❑ 9. Take the AP Chemistry Practice Test.

❑ 10. Analyze any remaining weaknesses that the Practice Test reveals.

❑ 11. Read "The Final Touches."

Format of the AP Chemistry Exam

Section I: Multiple-Choice Questions

90 minutes

75 questions   45% of total grade

Periodic table provided; no calculators allowed; no table of equations or constants provided.

Section II: Free-Response (Essay) Questions

Periodic table, a table of standard reduction potentials, and a table containing various equations and constants are provided.

90 minutes

6 questions   55% of total grade

Part A: 40 minutes; calculator allowed (no qwerty keyboards). Any programmable or graphing calculator may be used and you will not be required to erase the calculator memories before or after the examination. Questions require mathematical computations. It is essential that you show all steps in solving mathematical problems since partial credit is awarded in each problem for showing how the answer was obtained.

Question 1 (Required): 20%—Always on equilibrium: K_{sp}, K_a, K_b, K_c, or K_p

Question 2 or 3 (Choose either one): 20%—Only one of these problems will be scored. If you start both problems, be sure to cross out the one you do not want scored. Both questions require mathematical computations.

Part B: 50 minutes; calculator *not* allowed. Questions *do not* require mathematical computations.

Question 4 (Required): 15%—Write the formulas to show the reactants and the products for any five of eight chemical reactions. Each of the reactions occurs in aqueous solution unless otherwise indicated. Represent substances in solution as ions if the substance is extensively ionized. Omit formulas for any ions or molecules that are unchanged by the reaction. In all cases a reaction occurs. You need not balance the equations.

Question 5 (Required): 15%

Question 6 (Required): 15%

Questions 7 or 8 (Choose either one): 15% — Only one of the problems will be scored. If you start both problems, be sure to cross out the one you do not want scored.

Format and allotment of time may vary slightly from year to year.

Topics Covered by the AP Chemistry Exam*

The percentage after each major topic indicates the approximate proportion of questions on the examination that pertain to the specific topic. The examination is constructed using the percentages as guidelines for question distribution.

I. Structure of Matter (20%)

A. Atomic theory and atomic structure

1. Evidence for atomic theory

2. Atomic masses; determination by chemical and physical means

3. Atomic number and mass number; isotopes

4. Electron energy levels: atomic spectra, quantum numbers, atomic orbitals

5. Periodic relationships including, for example, atomic radii, ionization energies, electron affinities, oxidation states

B. Chemical bonding

1. Binding forces

a. Types: ionic, covalent, metallic, hydrogen bonding, van der Waals (including London dispersion forces)

b. Relationships to states, structure, and properties of matter

c. Polarity of bonds, electronegativities

2. Molecular models

a. Lewis structures

b. Valence bond: hybridization of orbitals, resonance, sigma and pi bonds

c. VSEPR

3. Geometry of molecules and ions, structural isomerism of simple organic molecules; relation of properties to structure

4. Nuclear chemistry: nuclear equations, half-lives, and radioactivity; chemical applications

II. States of Matter (20%)

A. Gases

1. Laws of ideal gases

a. Equation of state for an ideal gas

b. Partial Pressures

2. Kinetic-molecular theory

    a. Interpretation of ideal gas laws on the basis of this theory

    b. Avogadro's hypothesis and the mole concept

    c. Dependence of kinetic energy of molecules on temperature

    d. Deviations from ideal gas laws

B. Liquids and solids

  1. Liquids and solids from the kinetic-molecular viewpoint

  2. Phase diagrams of one-component systems

  3. Changes of state, including critical points and triple points

  4. Structure of solids; lattice energies

C. Solutions

  1. Types of solutions and factors affecting solubility

  2. Methods of expressing concentration (The use of normalities is not tested)

  3. Raoult's law and colligative properties (nonvolatile solutes); osmosis

  4. Non-ideal behavior (qualitative aspects)

III. Reactions (35–40%)

A. Reaction types

  1. Acid-base reactions; concepts of Arrhenius, Brønsted-Lowry, and Lewis; coordination complexes, amphoterism

  2. Precipitation reactions

  3. Oxidation-reduction reactions

    a. Oxidation number

    b. The role of the electron in oxidation-reduction

    c. Electrochemistry: electrolytic and galvanic cells; Faraday's laws; standard half-cell potentials; Nernst equation; prediction of the direction of redox reactions

B. Stoichiometry

  1. Ionic and molecular species present in chemical systems: net ionic equations

  2. Balancing of equations including those for redox equations

  3. Mass and volume relations with emphasis on the mole concept, including empirical formulas and limiting reactants

V. Laboratory (5–10%)

1. Determination of the formula of a compound

2. Determination of the percentage of water in a hydrate

3. Determination of molar mass by vapor density

4. Determination of molar mass by freezing-point depression

5. Determination of the molar volume of a gas

6. Standardization of a solution using a primary standard

7. Determination of concentration by acid-base titration, including a weak acid or weak base

8. Determination of concentration by oxidation-reduction titration

9. Determination of mass and mole relationship in a chemical reaction

10. Determination of the equilibrium constant for a chemical reaction

11. Determination of appropriate indicators for various acid-base titrations; pH determination

12. Determination of the rate of a reaction and its order

13. Determination of enthalpy change associated with a reaction

14. Separation and qualitative analysis of cations and anions

15. Synthesis of a coordination compound and its chemical analysis

16. Analytical gravimetric determination

17. Colorimetric or spectrophotometric analysis

18. Separation by chromatography

19. Preparation and properties of buffer solutions

20. Determination of electrochemical series

21. Measurements using electrochemical cells and electroplating

22. Synthesis, purification, and analysis of an organic compound

*2000 AP Course Description in Chemistry Published by The College Board

PART I

INTRODUCTION

Questions Commonly Asked About the AP Chemistry Exam

Q. What is the AP Chemistry Exam?

A. The AP chemistry exam is given once a year to high school students and tests their knowledge of concepts in first-year college-level chemistry. The student who passes the AP exam may receive one year of college credit for taking AP chemistry in high school. Passing is generally considered to be achieving a score of 3, 4, or 5. The test is administered each May. It has two sections.

- Section I, worth 45% of the total score, is 90 minutes long and consists of 75 multiple-choice questions. The total score for Section I is the number of correct answers minus $\frac{1}{4}$ for each wrong answer. If you leave a question unanswered, it does not count at all. A student generally needs to answer from 50% to 60% of the multiple-choice questions correctly to obtain a 3 on the exam. The multiple-choice questions fall into three categories:

    Calculations — These questions require you to quickly calculate mathematical solutions. Since you will not be allowed to use a calculator for the multiple-choice questions, the questions requiring calculations have been limited to simple arithmetic so that they can be done quickly, either mentally or with paper and pencil. Also, in some questions, the answer choices differ by several orders of magnitude so that the questions can be answered by estimation.

    Conceptual — These questions ask you to consider how theories, laws, or concepts are applied.

    Factual — These questions require you to quickly recall important chemical facts.

- Section II, worth 55% of the total score, is 90 minutes long and consists of four parts — one equilibrium problem, one mathematical essay, writing and predicting five chemical equations, and three nonmathematical essays.

Q. What are the Advantages of Taking AP Chemistry?

A.
- Students who pass the exam may, at the discretion of the college in which the student enrolls, be given full college credit for taking the class in high school.

- Taking the exam improves your chance of getting into the college of your choice. Studies show that students who successfully participate in AP programs in high school stand a much better chance of being accepted by selective colleges than students who do not.

- Taking the exam reduces the cost of a college education. In the many private colleges that charge upward of $500 a unit, a first-year college chemistry course could cost as much as $3,000! Taking the course during high school saves money.

- Taking the exam may reduce the number of years needed to earn a college degree.

- If you take the course and the exam while still in high school, you will not be faced with the college course being closed or overcrowded.

- For those of you who are not going on in a science career, passing the AP chemistry exam may fulfill the laboratory science requirement at the college, thus making more time available for you to take other courses.

- Taking AP chemistry greatly improves your chances of doing well in college chemistry. You will already have covered most of the topics during your high school AP chemistry program, and you will find yourself setting the curve in college!

Q. Do All Colleges Accept AP Exam Grades for College Credit?

A. Almost all of the colleges and universities in the United States and Canada, and many in Europe, take part in the AP program. The vast majority of the 2,900 U.S. colleges and universities that receive AP grades grant credit and/or advanced placement. Even colleges that receive only a few AP candidates and may not have specific AP policies are often willing to accommodate AP students who inquire about advanced-placement work.

To find out about a specific policy for the AP exam(s) you plan to take, write to the college's Director of Admissions. You should receive a written reply telling you how much credit and/or advanced placement you will receive for a given grade on an AP Exam, including any courses you will be allowed to enter.

The best source of specific and up-to-date information about an individual institution's policy is its catalog or Web site. Other sources of information include *The College Handbook™ with College Explorer CD-ROM* and *College Search™*. For more information on these and other products, log on to the College Board's online store at: http://cbweb2.collegeboard.org/shopping/.

Q. How is the AP Exam Graded and What Do the Scores Mean?

A. The AP exam is graded on a five-point scale:

5: Extremely well qualified. About 17% of the students who take the exam earn this grade.

4: Well qualified. Roughly 15% earn this grade.

3: Qualified. Generally, 25% earn this grade.

2: Possibly qualified. Generally considered "not passing." About 22% of the students who take the exam earn this grade.

1: Not qualified. About 21% earn this grade.

Of the roughly 49,000 students from 4,700 high schools who take the AP chemistry exam each year, the average grade is 2.85 with a standard deviation of 1.37. Approximately 1,500 colleges receive AP scores from students who pass the AP chemistry exam.

Section I, the multiple-choice section, is machine graded. Each question has five answers to choose from. Remember, there is a penalty for guessing: $\frac{1}{4}$ of a point is taken off for each wrong answer. A student generally needs to correctly answer 50% to 60% of the multiple-choice questions to obtain a 3 on the exam. Each answer in Section II, the free-response section, is read several times by different chemistry instructors who pay great attention to consistency in grading.

Q. Are There Old Exams Out There That I Could Look At?

A. Yes! Questions (and answers) from previous exams are available from The College Board. Request an order form by contacting: AP Services, P.O. Box 6671, Princeton, NJ 08541-6671; (609) 771-7300 or (888) 225-5427; Fax (609) 530-0482; TTY: (609) 882-4118; or e-mail: apexams@ets.org.

Q. What Materials Should I Take to the Exam?

A. Be sure to take your admission ticket, some form of photo and signature identification, your social security number, several sharpened No. 2 pencils, a good eraser, a watch, and a scientific calculator with fresh batteries. You may bring a programmable calculator (it will *not* be erased or cleared), but it must *not* have a typewriter-style (qwerty) keyboard. You may use the calculator only in Section II, Part A.

Q. When Will I Get My Score?

A. The exam itself is generally given in the second or third week of May. The scores are usually available during the second or third week of July.

Q. Should I Guess on the Test?

A. Except in certain special cases explained later in this book, you should *not* guess. There is a penalty for guessing on the multiple-choice section of the exam. As for the free-response section, it simply comes down to whether you know the material or not.

Q. Suppose I do Terribly on the Exam. May I Cancel the Test and/or the Scores?

A. You may cancel an AP grade permanently only if the request is received by June 15 of the year in which the exam was taken. There is no fee for this service, but a signature is required to process the cancellation. Once a grade is cancelled, it is permanently deleted from the records.

You may also request that one or more of your AP grades are not included in the report sent to colleges. There is a $5 fee for each score not included on the report.

Q. May I Write on the Test?

A. Yes. Because scratch paper is not provided, you'll need to write in the test booklet. Make your notes in the booklet near the questions so that if you have time at the end, you can go back to your notes to try to answer the question.

Q. How Do I Register or Get More Information?

A. For further information contact: AP Services, P.O. Box 6671, Princeton, NJ 08541-6671; (609) 771-7300 or (888) 225-5427; Fax (609) 530-0482; TTY: (609) 882-4118; or e-mail: apexams@ets.org.

Strategies for Taking the AP Chemistry Exam

Section I: The Multiple-Choice Section

The "Plus-Minus" System

Many students who take the AP chemistry exam do not get their best possible score on Section I because they spend too much time on difficult questions and fail to leave themselves enough time to answer the easy ones. Don't let this happen to you. Because every question within each section is worth the same amount, consider the following guidelines.

1. Note in your test booklet the starting time of Section I. Remember that you have just over 1 minute per question.

2. Go through the entire test and answer all the easy questions first. Generally, the first 25 or so questions are considered by most to be the easiest questions, with the level of difficulty increasing as you move through Section I. Most students correctly answer approximately 60% of the first 25 multiple-choice questions, 50% of the next 25 questions, and only 30% of the last 25 questions (the fact that most students do not have time to finish the multiple-choice questions is factored into the percentages).

3. When you come to a question that seems impossible to answer, mark a large minus sign (−) next to it in your test booklet. You are penalized for wrong answers, so do not guess at this point. Move on to the next question.

4. When you come to a question that seems solvable but appears too time-consuming, mark a large plus sign (+) next to that question in your test booklet. Do not guess; move on to the next question.

5. Your time allotment is just over 1 minute per question, so a "time consuming" question is one that you estimate will take you several minutes to answer. Don't waste time deciding whether a question gets a plus or a minus. Act quickly. The intent of this strategy is to save you valuable time.

   After you have worked all the easy questions, your booklet should look something like this:

       1.
   +2.
       3.
   −4.
       5.

   and so on

6. After doing all the problems you can do immediately (the easy ones), go back and work on your "+" problems.

7. If you finish working your "+" problems and still have time left, you can do either of two things:

> Attempt the "−" problems, but remember not to guess under any circumstance.

> Forget the "−" problems and go back over your completed work to be sure you didn't make any careless mistakes on the questions you thought were easy to answer.

You do not have to erase the pluses and minuses you made in your question booklet.

The Elimination Strategy

Take advantage of being able to mark in your test booklet. As you go through the "+" questions, eliminate choices from consideration by marking them out in your question booklet. Mark with question marks any choices you wish to consider as possible answers. See the following example:

A̸.
?**B.**
C̸.
D̸.
?**E.**

This technique will help you avoid reconsidering those choices that you have already eliminated and will thus save you time. It will also help you narrow down your possible answers.

If you are able to eliminate all but two possible answers, answers such as **B** and **E** in the previous example, you may want to guess. Under these conditions, you stand a better chance of raising your score by guessing than by leaving the answer sheet blank.

Section II: The Free-Response (Essay) Section

Many students waste valuable time by memorizing information that they feel they should know for the AP chemistry exam. Unlike the history exam, for which you need to have memorized hundreds of dates, battles, names, and treaties, the AP chemistry exam requires you to have memorized comparatively little. Rather, it is generally testing whether you can *apply* given information to new situations. You will be frequently asked to *explain, compare,* and *predict* in the essay questions.

Section II of the AP chemistry exam comes with

- a periodic table (see pages 340 and 341)
- an E°_{red} table (see page 346)
- a table of equations and constants (see pages 342–345)

Methods for Writing the Essays

The Restatement

In the second section of the AP exam, you should begin all questions by numbering your answer. You do not need to work the questions in order. However, the graders must be able to identify quickly which question you are answering. You may wish to underline any *key words* or *key concepts* in your answer. Do not underline too much, however, because doing so may obscure your reasons for underlining. In free-response questions that require specific calculations or the determination of products, you may also want to underline or draw a box around your final answer(s).

After you have written the problem number, restate the question in as few words as possible, but do not leave out any essential information. Often a diagram will help. By restating, you put the question in your own words and allow time for your mind to organize the way you intend to answer the questions. As a result, you eliminate a great deal of unnecessary language that clutters the basic idea. Even if you do not answer the question, a restatement may be worth 1 point.

If a question has several parts, such as (a), (b), (c), and (d), do *not* write all of the restatements together. Instead, write each restatement separately when you begin to answer that part. In this book, you will see many samples of the use of restatements.

Four Techniques for Answering Free-Response Questions

When you begin Section II, the essays, the last thing you want to do is start writing immediately. Take a minute and scan the questions. Find the questions that you know you will have the most success with, and put a star (*) next to them in your booklet. *You do not have to answer the questions in order;* however, you must number them clearly in your response book.

After you have identified the questions that you will eventually answer, the next step is to decide what format each question lends itself to. There are four basic formats. Let's do an actual essay question to demonstrate each format.

The Chart Format

In this format, you fill in a chart to answer the question. When you draw the chart, use the edge of your calculator case to make straight lines. Fill in the blanks with symbols, phrases, or incomplete sentences. The grid forces you to record all answers quickly and makes it unlikely that you will forget to give any part of the answer.

Essay 1

Given the molecules SF_6, XeF_4, PF_5, and ClF_3:

A. Draw a Lewis structure for each molecule.

B. Identify the geometry for each molecule.

C. Describe the hybridization of the central atom for each molecule.

D. Give the number of unshared pairs of electrons around the central atom.

Answer

1. Restatement: Given SF_6, XeF_4, PF_5, and ClF_3. For each, supply

A. Lewis structure

B. geometry

C. hybridization

D. unshared pairs of electrons around central atom

Characteristic	SF_6	XeF_4	PF_5	ClF_3
Lewis structure				
Geometry	Octahedral	Square Planar	Triangular Bypyramidal	T-shaped
Hybridization	sp^3d^2	sp^3d^2	sp^3d	sp^3d
Unshared pairs	0	2	0	2

The Bullet Format

The bullet format is also a very efficient technique because it, like the chart format, does not require complete sentences. In using this format, you essentially provide a list to answer the question. A • is a bullet, and each new concept receives one. Try to add your bullets in a logical sequence, and leave room to add more bullets. You may want to come back later and fill them in. Don't get discouraged if you do not have as many bullets as the samples contain — it takes practice. Reviewing the key terms at the beginning of each chapter may suggest additional points that you can incorporate.

Essay 2

> As one examines the periodic table, one discovers that the melting points of the alkali metals increase as one moves from cesium to lithium, whereas the melting points of the halogens increase from fluorine to iodine.
>
> **A.** Explain the phenomenon observed in the melting points of the alkali metals.
>
> **B.** Explain the phenomenon observed in the melting points of the halogens.
>
> **C.** Given the compounds CsI, NaCl, LiF, and KBr, predict the order of their melting points (from high to low) and explain your answer using chemical principles.

Answer

2. Given — melting points: alkali metals increase from Cs → Li
   halogens increase from F → I

   (a) Restatement: Explain alkali metal trend.

   - Observed melting point order: Li > Na > K > Rb > Cs

   - All elements are metals

   - All elements contain metallic bonds

   - Electrons are free to migrate in a "sea"

   - As one moves down the group, size (radius) of the atoms increases

   - As volume of atom increases, charge density decreases

   - Attractive force between atoms is directly proportional to melting point

   - Therefore, as attractive forces decrease moving down the group, melting point decreases

(b) Restatement: Explain halogen trend.

- Observed melting point order: I > Br > Cl > F
- All halogens are nonmetals
- Intramolecular forces = covalent bonding
- Intermolecular forces = dispersion (van der Waals) forces, which exist between molecules
- Dispersion forces result from "temporary" dipoles caused by polarization of electron clouds
- As one moves up the group, the electron clouds become smaller
- Smaller electron clouds result in higher charge density
- As one moves up the group, electron clouds are less readily polarized
- Less readily polarized clouds result in weaker dispersion forces holding molecules to other molecules
- Therefore, attractive forces between molecules decrease as one moves up the group, resulting in lower melting points

(c) Restatement: Predict melting point order (high to low) CsI, NaCl, LiF, and KBr and explain.

- LiF > NaCl > KBr > CsI
- All compounds contain a metal and a nonmetal
- Predicted order has ionic bonds
- Larger ionic radius results in lower charge density
- Lower charge density results in smaller attractive forces
- Smaller attractive forces result in lower melting point

The Outline Format

This technique is similar to the bullet format, but instead of bullets it uses the more traditional outline style that you may have used for years: roman numerals, letters, and so on. The advantages of this format are that it does not require full sentences and that it progresses in a logical sequence. The disadvantage is that it requires you to spend more time thinking about organization. Leave plenty of room here because you may want to come back later and add more points.

Essay 3

The boiling points and electrical conductivities of six aqueous solutions are as follows:

Solution	Boiling Point	Relative Electrical Conductivity
0.05 m $BaSO_4$	100.0254°C	0.03
0.05 m H_3BO_3	100.0387°C	0.78
0.05 m NaCl	100.0485°C	1.00
0.05 m $MgCl_2$	100.0689°C	2.00
0.05 m $FeCl_3$	100.0867°C	3.00
0.05 m $C_6H_{12}O_6$	100.0255°C	0.01

Discuss the relationship among the composition, the boiling point, and the electrical conductivity of each solution.

Answer

3. Given: Boiling point data and electrical conductivities of six aqueous solutions, all at 0.05 m.

   Restatement: Discuss any relationships between B.P. and electrical conductivities.

   I. $BaSO_4$ *(If you have a highlighter with you, highlight the main categories.)*

   A. $BaSO_4$ is an ionic compound.

   B. According to known solubility rules, $BaSO_4$ is not very soluble.

   1. If $BaSO_4$ were totally soluble, one would expect its B.P. to be very close to that of NaCl because $BaSO_4$ would be expected to dissociate into two ions (Ba^{2+} and SO_4^{2-}) just as NaCl would (Na^+ and Cl^-). The substantial difference between the B.P. of the NaCl solution and that of the $BaSO_4$ solution suggests that the dissociation of the latter is negligible.

   2. The electrical conductivity of $BaSO_4$ is closest to that of $C_6H_{12}O_6$, an organic molecule, which does not dissociate; this observation further supports the previous evidence of the weak-electrolyte properties of $BaSO_4$.

II. H_3BO_3

    A. H_3BO_3 is a weak acid.

    B. In the equation $\Delta t = i \cdot m \cdot K_b$, where Δt is the boiling-point elevation, m is the molality of the solution, and K_b is the boiling-point-elevation constant for water, i (the van't Hoff factor) would be expected to be 4 if H_3BO_3 were completely ionized. According to data provided, i is about 1.5. Therefore, H_3BO_3 must have a relatively low K_a.

III. NaCl, $MgCl_2$, and $FeCl_3$

    A. All three compounds are chlorides known to be completely soluble in water, so they are strong electrolytes and would increase electrical conductivities.

    B. The van't Hoff factor (i) would be expected to be 2 for NaCl, 3 for $MgCl_2$, and 4 for $FeCl_3$.

    C. Using the equation

$$\frac{\Delta t}{m \cdot K_b} = \frac{\text{B.P. of solution} - 100°\text{C}}{0.05 \frac{\text{mole solute}}{\text{kg}} \cdot 0.512°\text{C} \frac{\text{kg}}{\text{mole solute}}}$$

we find that the van't Hoff factors for these solutions are

Compound	Calculated i	Expected i
NaCl	1.9	2.0
$MgCl_2$	2.7	3.0
$FeCl_3$	3.4	4.0

which are in agreement.

    D. The electrical conductivity data support the rationale just provided: The greater the number of particles, which in this case are ions, the higher the B.P.

IV. $C_6H_{12}O_6$

    A. $C_6H_{12}O_6$, glucose, is an organic molecule. It would not be expected to dissociate into ions that would conduct electricity. The reported electrical conductivity for glucose supports this.

    B. Because $C_6H_{12}O_6$ does not dissociate, i is expected to be close to 1. The equation in III. C. gives i as exactly 1.

    C. The boiling-point-elevation constant of $0.512°\text{C} \cdot \text{kg/mole}$ would be expected to raise the B.P. $0.0256°\text{C}$ for a 0.05 m solution when $i = 1$. The data show that the boiling-point elevation is $0.0255°\text{C}$. This agrees with the theory. Therefore, $C_6H_{12}O_6$ does not dissociate. With few or no ions in solution, poor electrical conductivity is expected. This is supported by the evidence in the table.

The Free Style

This method is the one most commonly used, although in my opinion, it is the method of last resort. Free style often results in aimless, rambling, messy, incomplete answers. This method is simply writing paragraphs to explain the question. If you do adopt this method for an answer (and many questions lend themselves only to this method), you must organize the paragraphs before writing. Also review your list of key terms to see if there are any concepts you want to add to your answers. Note, however, that adding thoughts at a later time is difficult with this approach because they will be out of logical sequence. (Unlike the bullet and outline formats, free style doesn't leave you room to add more ideas where they belong.)

Essay 4

If one completely vaporizes a measured amount of a volatile liquid, the molecular weight of the liquid can be determined by measuring the volume, temperature, and pressure of the resulting gas. When one uses this procedure, one uses the ideal gas equation and assumes that the gas behaves ideally. However, if the sample is slightly above the boiling point of the liquid, the gas deviates from ideal behavior. Explain the postulates of the ideal gas equation, and explain why, when measurements are taken just above the boiling point, the calculated molecular weight of a liquid deviates from the true value.

Answer

4. Restatement: Explain ideal gas equation and why MW measurements taken just above boiling point deviate.

    The ideal gas equation, PV = nRT stems from three relationships known to be true for gases under ordinary conditions:

    1. The volume is directly proportional to the amount, $V \sim n$
    2. The volume is directly proportional to the absolute temperature, $V \sim T$
    3. The volume is inversely proportional to the pressure, $V \sim 1/P$

We obtain n, the symbol used for the moles of gas, by dividing the mass of the gas by the molecular weight. In effect, n = mass/molecular weight (n = m/MW). Substituting this relationship into the ideal gas law gives

$$PV = \frac{mRT}{MW}$$

Solving the equation for molecular weight yields

$$MW = \frac{mRT}{PV}$$

Real gas behavior deviates from the values obtained using the ideal gas equation because the ideal gas equation assumes that (1) the molecules do not occupy space and (2) there is no attractive force between the individual molecules. However, at low temperatures (just above the boiling point of the liquid), these factors become significant, and we must use an alternative equation, known as the van der Waals equation, that accounts for them.

At the lower temperatures, a greater attraction exists between the molecules, so the compressibility of the gas is significant. This causes the product of P · V to be smaller than predicted. Because P · V is found in the denominator in the foregoing equation, the calculated molecular weight would tend to be higher than the molecular weight actually is.

Mathematical Operations

Significant Figures

In order to receive full credit in Section II, the essay section, you must be able to express your answer with the correct number of significant figures (s.f.). There are slight penalties on the AP chemistry exam for not doing so. The "Golden Rule" for using significant figures is that *your answer cannot contain more significant figures than the least accurately measured quantity. Do not use conversion factors for determining significant figures.* Review the following rules for determining significant figures. Underlined numbers are significant.

- Any digit that is not zero is significant. $\underline{123}$ = 3 s.f.

- Zeros between significant figures (captive zeros) are significant. $\underline{80601}$ = 5 s.f.; $\underline{10.001}$ = 5 s.f.

- Zeros to the left of the first nonzero digit (leading zeros) are not significant. $0.00\underline{2}$ = 1 s.f.

- If a number is equal to or greater than 1, then all the zeros written to the right of the decimal point (trailing zeros) count as significant figures. $\underline{9.00}$ = 3 s.f. The number 100 has only one significant figure ($\underline{1}00$), but written as 100. (note the decimal point), it has three significant figures. $\underline{400}.$ = 3 s.f.

- For numbers less than 1, only zeros that are at the end of the number and zeros that are between nonzero digits are significant. $0.0\underline{70}$ = 2 s.f.

- For addition or subtraction, the limiting term is the one with the smallest number of decimal places, so count the decimal places. For multiplication and division, the limiting term is the number that has the least number of significant figures, so count the significant figures.

    $11.01 + 6.\mathbf{2} + 8.995 = 26.\mathbf{2}$ (one decimal place)

    $32.010 \times \underline{501} = 1.60 \times 10^4$ (three significant figures)

Logs and Antilogs

You will use your calculator in Section II to determine logs and antilogs. There are two types of log numbers that you will use on the AP exam: $\log_{10}$, or log, and natural log, or ln. Log base 10 of a number is that exponent to which 10 must be raised to give the original number. Therefore, the log of 10 is 1 because 10^1 is 10. The log of 100 is 2 because 10^2 is 100. The log of 0.001 is −3, and so on.

There are a few types of problems on the AP exam in which you may have to use a natural logarithm. The symbol for a natural logarithm is ln. The relationship between $\log_{10}$ and ln is given by the equation $\ln x = 2.303 \log_{10} x$.

Scientific Notation

Try to use scientific notation when writing your answers. For example, instead of writing 1,345,255, write 1.345255×10^6. Remember always to write one digit, a decimal point, the rest of the digits (making sure to use only the correct number of significant figures), and then times 10 to the proper power. An answer such as 0.000045 should be written 4.5×10^{-5} (2 s.f.). Also, don't forget that when you multiply exponents you add them and when you divide exponents you subtract them. Your chemistry textbook or math book probably has a section that covers significant figures, logs, antilogs, scientific notation, and the like. If your math background or algebra skills are weak, you must thoroughly review and polish these skills before attempting to do the problems in this book.

Accuracy

Absolute error = Experimental value − Accepted value

$$\% \, error = \frac{actual - measured}{actual} \times 100\%$$

Consider the following expression:

$$\frac{(32.56 \times 0.4303 \times 0.08700)}{4.3422} \times 100\%$$

The % error for each term is

Term	Calculation	% Error
32.56 ± 0.01	[(32.56 + 0.01) − 32.56/32.56] · 100%	0.0307
0.4303 ± 0.0001	[(0.4303 + 0.0001) − 0.4303/0.4303] · 100%	0.0232
0.08700 ± 0.00001	[(0.08700 + 0.00001) − 0.08700/0.08700] · 100%	0.0115
4.3422 ± 0.0001	[(4.3422 + 0.0001) − 4.3422/4.3422] · 100%	0.0023

Adding up the % errors: 0.0307 + 0.0232 + 0.0115 + 0.0023 = 0.0677% and

$$\frac{(32.56 \times 0.4303 \times 0.08700)}{4.3422} \times 100\% = 28.07147 \text{ with } 0.0677\% \text{ error}$$

reported as 28.07 ± 0.01

Use the term with the largest possible error (32.56 ± 0.01) for significant numbers.

Two types of errors may affect the accuracy of a measured value:

(a) determinate errors — errors that are instrumental, operative and involved in the methodology. These types of errors can be avoided or corrected.

(b) indeterminate errors — accidental and/or random. These types of errors cannot be estimated or predicted except through the use of probability theory and follow a Gaussian distribution.

Precision

Often an actual or accepted value is not known. If this is the case, the accuracy of the measurement cannot be reported since one does not know how close or far one is from the actual value. Instead, experiments are repeated several times and a measurement of how close together the values lie (precision) is done. It is the goal that experiments that give reproducible results will also give accurate results.

Absolute deviation = |Measured − Mean|

Average Deviation or Average Difference = Average of all the absolute deviations.

$$\text{Percent Deviation} = \frac{\text{Average Deviation}}{\text{Mean}} \times 100\%$$

Example: Given three masses of the same object: 1.51 g, 1.63 g, 1.48 g

$$\text{Mean or Average} = \frac{1.51 + 1.63 + 1.48}{3} = 1.54$$

Absolute Deviation of each value from mean:

$$|1.51 - 1.54| = 0.03$$

$$|1.63 - 1.54| = 0.09$$

$$|1.48 - 1.54| = 0.06$$

$$\text{Average Deviation} = \frac{0.03 + 0.09 + 0.06}{3} = 0.06$$

$$\text{Relative Deviation for Relative Difference} = \frac{\text{Average Deviation}}{\text{Mean}} \times 100\%$$

$$= \frac{0.06}{1.54} \times 100\% = 3.9\%$$

This says that the three measurements are within 3.9% of the average (and hopefully) true value of the object.

Rounding Off Numbers

318.04 = 318.0 (the 4 is smaller than 5)

318.06 = 318.1 (the 6 is greater than 5)

318.05 = 318.0 (the 0 before the 5 is an even number)

318.15 = 318.2 (the 1 before the 5 is an odd number)

Mathematics Self-Test

Try taking this short mathematics self-test. If you understand these math problems and get the answers correct, you're ready to go on. If you miss problems in this area, you need to back up and review those operations with which you are uncomfortable.

Determine the number of significant figures in the following numbers.

1. 100
2. 100.01
3. 0.010
4. 1234.100

Round the following numbers to the number of significant figures indicated and express in scientific notation.

5. 100.075 rounded to 3 significant figures
6. 140 rounded to 2 significant figures
7. 0.000787 rounded to 2 significant figures

Perform the following math operations, expressing your answers to the proper number of significant figures.

8. $(4.5 \times 10^{-3}) + (5.89 \times 10^{-4})$
9. $(5.768 \times 10^{9}) \times (6.78 \times 10^{-2})$
10. (5.661×10^{-9}) divided by (7.66×10^{-8})
11. $8.998 + 9.22 + 1.3 \times 10^{2} + 0.006$

Determine:

12. log of 98.71
13. log of 0.0043
14. ln of 3.99
15. ln of 0.0564
16. log (0.831/0.111)
17. ln $(1.5^{2}/3.0 \times 10^{-4})$

Evaluate:

18. $e^{7.82}$

Solve for x:

19. log $(12.0/x) = 3.0$
20. $40.1 = 5.13^{x}$

Answers to Mathematics Self-Test

1. 1 significant figure
2. 5 significant figures
3. 2 significant figures
4. 7 significant figures
5. 1.00×10^2
6. 1.4×10^2
7. 7.9×10^{-4}
8. 5.1×10^{-3}
9. 3.91×10^8
10. 7.39×10^{-2}
11. 1.5×10^2
12. 1.994
13. −2.4
14. 1.38
15. −2.88
16. 0.874
17. 8.9
18. 2.49×10^3
19. $x = 0.012$
20. $x = 2.26$

SPECIFIC TOPICS

Gravimetrics

Key Terms

Words that can be used as topics in essays:

accuracy

atomic theory

density

empirical formula

extensive property

fractional crystallization

heterogeneous

homogeneous

intensive property

isotopes

law of conservation of mass and energy

law of definite proportions
  (law of constant composition)

limiting reactant

mixture

molecular formula

percentage yield

precision

random error

systematic error

theoretical yield

uncertainty

Key Concepts

Equations and relationships that you need to know:

- $1 \text{ nm} = 1 \times 10^{-9} \text{ m} = 10 \text{ Å}$

  $1 \text{ cm}^3 = 1 \text{ mL}$

  $°F = 1.8(°C) + 32$

  $K = °C + 273.16$

  $\text{density} = \dfrac{\text{mass}}{\text{volume}}$

  $\text{Avogadro's number} = 6.02 \times 10^{23} = 1 \text{ mole}$

  $\text{number of moles} = \dfrac{\text{mass}}{\text{molecular weight}}$

  $\text{molecular weight of ideal gas} = \text{density} \times \text{molar volume}$

  $\%\,\text{yield} = \dfrac{\text{actual yield}}{\text{theoretical yield}} \times 100\%$

  $\%\,\text{composition} = \dfrac{\text{mass of element in compound}}{\text{mass of compound}} \times 100\%$

  $\%\,\text{error} = \dfrac{\text{observed value} - \text{expected value}}{\text{expected value}} \times 100\%$

Measurement Terms

SI (International System) Multipliers

Multiple	Prefix	Symbol
10^{12}	tera	T
10^9	giga	G
10^6	mega	M
10^3	kilo	k
10^2	hecto	h
10^1	deka	da
10^{-1}	deci	d
10^{-2}	centi	c
10^{-3}	milli	m
10^{-6}	micro	μ
10^{-9}	nano	n
10^{-12}	pico	p
10^{-15}	femto	f
10^{-18}	atto	a

SI Base Units

meter	m	
kilogram	kg	
second	s (sec)	
ampere	A	
kelvin	K	
mole	mol	
candela	cd	

SI Derived Units

becquerel	Bq	1 disintegration/sec
coulomb	C	$A \cdot sec$
farad	F	$A \cdot sec/V = A^2 \cdot sec^4 \cdot kg^{-1} \cdot m^{-2}$
gray	Gy	J/kg
henry	H	Wb/A
hertz	Hz	sec^{-1} (cycle/sec)
joule	J	$kg \cdot m^2 \cdot sec^{-2} = 10^7$ ergs

SI Derived Units (continued)

lumen	lm	cd · sr
lux	lx	lm/m^2
newton	N	$kg \cdot m \cdot sec^{-2}$
pascal	Pa	$N/m^2 = kg \cdot m^{-1} \cdot sec^{-2}$
ohm	Ω	$V/A = kg \cdot m^2 \cdot sec^{-3} \cdot A^{-2}$
siemens	S	$\Omega^{-1} = A \cdot V^{-1}$
tesla	T	Wb/m^2
volt	V	$J \cdot A^{-1} \cdot sec^{-1} = kg \cdot m^2 \cdot sec^{-3} \cdot A^{-1}$
watt	W	$J/sec = kg \cdot m^2 \cdot sec^{-3}$
weber	Wb	$V \cdot sec$

Non-SI Units

angstrom	Å	10^{-8} cm
atmosphere	atm	$101,325\ N/m^2$ or 760 mm Hg or 101.3 kPa or 760 torr
bar	bar	$10^5\ N/m^2$
calorie	cal	4.184 J
dyne	dyn	$10^{-5}\ N = 1\ g \cdot cm \cdot sec^{-2} = 2.39 \times 10^{-8}\ cal \cdot cm^{-1}$
erg	erg	10^{-7} J
inch	in	2.54 cm
millimeter of mercury	mm Hg	$133.282\ N/m^2$
pound	lb	0.453502 kg
torr	torr	$133.282\ N/m^2$

Samples: Multiple-Choice Questions

1. A popular Bourbon whiskey is listed as being "92 Proof." The liquor industry defines "Proof" as being twice the volume percentage of alcohol in a blend. Ethanol (drinking alcohol) has the structural formula CH_3CH_2OH (MW = 46 g/mol). The density of ethanol is 0.79 g/mL. How many liters of whiskey must one have in order to have 50. moles of carbon?

   A. 0.80 liters

   B. 1.6 liters

   C. 3.2 liters

   D. 4.0 liters

   E. 6.4 liters

Answer: C

Step 1: Write down an equals sign (=).

Step 2: To the right of the equals sign, write down the units you want the answer to be in. Examination of the problem reveals that you want your answers in "liters of whiskey," so you have

   = liters of whiskey

Step 3: Begin the problem with the item you are limited to. In this case, it is 50. moles of carbon — no more, no less. Place the 50. moles of carbon over 1.

$$\frac{50. \text{ moles of carbon}}{1} = \text{liters of whiskey}$$

Step 4: Get rid of the units "moles of carbon" by placing them in the denominator of the next factor. What do you know about moles of carbon? There are 2 moles of carbon in each mole of ethanol.

$$\frac{50. \text{ moles of carbon}}{1} \times \frac{1 \text{ mole ethanol}}{2 \text{ moles carbon}} = \text{liters of whiskey}$$

Step 5: Continue in this fashion, getting rid of unwanted units until you are left in the units you desire. At that point, stop and do the calculations.

$$\frac{50. \text{ moles carbon}}{1} \times \frac{1 \text{ mole ethanol}}{2 \text{ moles carbon}} \times \frac{46 \text{ grams ethanol}}{1 \text{ mole ethanol}}$$
$$\times \frac{1 \text{ mL ethanol}}{0.79 \text{ g ethanol}} \times \frac{200 \text{ mL whiskey}}{92 \text{ mL ethanol}} \times \frac{1 \text{ L whiskey}}{10^3 \text{ mL whiskey}} = 3.2 \text{ L whiskey}$$

2. A sample of a pure compound was found to contain 1.201 grams of carbon, 0.202 grams of hydrogen, and 7.090 grams of chlorine. What is the empirical formula for the compound?

   A. $CHCl_3$

   B. CH_2Cl

   C. CH_2Cl_2

   D. CH_3Cl

   E. $C_2H_2Cl_4$

Answer: C

First, change the grams of each element to moles. You end up with 0.100 mole of carbon, 0.200 mole of hydrogen, and 0.200 mole of chlorine. This represents a 1 carbon : 2 hydrogen : 2 chlorine molar ratio.

3. Balance the following equation using the lowest possible whole-number coefficients:

$$NH_3 + CuO \rightarrow Cu + N_2 + H_2O$$

The sum of the coefficients is

   A. 9

   B. 10

   C. 11

   D. 12

   E. 13

Answer: D

Step 1: Begin balancing equations by trying suitable coefficients that will give the same number of atoms of each element on both sides of the equation. Remember to change coefficients, not subscripts.

   $\underline{2}NH_3 \rightarrow \underline{1}N_2$

Step 2: Look for elements that appear only once on each side of the equation and with equal numbers of atoms on each side. The formulas containing these elements must have the same coefficients.

   $\_CuO \rightarrow \_Cu$

Step 3: Look for elements that appear only once on each side of the equation but in unequal numbers of atoms. Balance these elements.

$$\underline{2NH_3} \rightarrow \underline{3H_2O}$$

Step 4: Balance elements that appear in two or more formulas on the same side of the equation.

Step 5: Double check your balanced equation and be sure the coefficients are the lowest possible whole numbers.

$$2NH_3 + 3CuO \rightarrow 3Cu + N_2 + 3H_2O$$

$$2 + 3 + 3 + 1 + 3 = 12$$

(Be sure to include the unwritten 1 that is in front of N_2.)

4. When 0.600 mole of $BaCl_{2(aq)}$ is mixed with 0.250 mole of $K_3AsO_{4(aq)}$, what is the maximum number of moles of solid $Ba_3(AsO_4)_2$ that could be formed?

   A. 0.125 mole

   B. 0.200 mole

   C. 0.250 mole

   D. 0.375 mole

   E. 0.500 mole

Answer: A

Begin by writing a balanced equation:

$$3BaCl_{2\,(aq)} + 2K_5AsO_{4\,(aq)} \rightarrow Ba_3(AsO_4)_{2\,(s)} + 6KCl_{(aq)}$$

You may want to write the net-ionic equation:

$$3Ba^{2+}_{(aq)} + 2AsO_4^{-3}_{(aq)} \rightarrow Ba_3(AsO_4)_{2\,(s)}$$

Next, realize that this problem is a limiting-reactant problem. That is, one of the two reactants will run out first, and when that happens, the reaction will stop. You need to determine which one of the reactants will run out first. To do this, you need to be able to compare them on a 1:1 basis. But their coefficients are different, so you need to relate both reactants to a common product, say $Ba_3(AsO_4)_2$. Set the problem up like this:

$$\frac{0.600 \, \text{mole BaCl}_2}{1} \times \frac{1 \, \text{mole Ba}_3(AsO_4)_2}{3 \, \text{moles BaCl}_2} = 0.200 \, \text{mole Ba}_3(AsO_4)_2$$

$$\frac{0.250 \, \text{mole K}_3AsO_4}{1} \times \frac{1 \, \text{mole Ba}_3(AsO_4)_2}{2 \, \text{moles K}_3AsO_4} = 0.125 \, \text{mole Ba}_3(AsO_4)_2$$

Given the two amounts of starting materials, you discover that you can make a maximum of 0.125 mole of $Ba_3(AsO_4)_2$, because at that point you will have exhausted your supply of K_3AsO_4.

---

5. A test tube containing $CaCO_3$ is heated until *all* of the compound decomposes. If the test tube plus calcium carbonate originally weighed 30.08 grams and the loss of mass during the experiment was 4.400 grams, what was the mass of the empty test tube?

    A. 20.07 g

    B. 21.00 g

    C. 24.50 g

    D. 25.08 g

    E. 25.68 g

Answer: A

Begin by writing a balanced equation. Remember that all Group II carbonates decompose to yield the metallic oxide plus carbon dioxide gas.

$$CaCO_3(s) \rightarrow CaO(s) + CO_2(g)$$

According to your balanced equation, any loss of mass during the experiment would have to have come from the carbon dioxide gas leaving the test tube. 4.400 grams of CO_2 gas correspond to 0.1000 mole. Because all of the calcium carbonate decomposed, and the calcium carbonate and carbon dioxide gas are in a 1:1 molar ratio, you must originally have had 0.1000 mole of calcium carbonate, or 10.01 grams. The calcium carbonate and test tube weighed 30.08 grams, so if you get rid of the calcium carbonate, you are left with 20.07 grams for the empty test tube.

---

6. 32.0 grams of oxygen gas, 32.0 grams of methane gas, and 32.0 grams of sulfur dioxide gas are mixed. What is the mole fraction of the oxygen gas?

    A. 0.143

    B. 0.286

    C. 0.333

    D. 0.572

    E. 0.666

Answer: B

First change all the grams to moles:

32.0 grams of O_2 = 1.00 mole

32.0 grams of CH_4 = 2.00 moles

32.0 grams of SO_2 = 0.500 mole

Mole fraction of oxygen gas:

Mole fraction of oxygen gas: $\dfrac{1 \text{ mole } O_2}{3.50 \text{ total moles}}$ = 0.286 mole fraction

---

7. Element X is found in two forms: 90.0% is an isotope that has a mass of 20.0, and 10.0% is an isotope that has a mass of 22.0. What is the atomic mass of element X?

   A. 20.1

   B. 20.2

   C. 20.8

   D. 21.2

   E. 21.8

Answer: B

To solve this problem, multiply the percentage of each isotope by its atomic mass and add those products.

$(0.900 \times 20.0) + (0.100 \times 22.0) = 20.2$ atomic mass of element X

---

8. What is the formula of a compound formed by combining 50. grams of element X (atomic weight = 100.) and 32 grams of oxygen gas?

   A. XO_2

   B. XO_4

   C. X_4O

   D. X_2O

   E. XO

Answer: B

According to the information given, you have 0.50 mole of element X (50. g/100. g · mole^{-1} = 0.50 mole). For the oxygen, remember that you will use 16 g/mole for the atomic weight, giving you 2.0 moles of oxygen atoms. A 0.50:2.0 molar ratio is the same as a 1:4 molar ratio, so the answer is XO_4.

9. An oxide is known to have the formula X_2O_7 and to contain 76.8% X by mass. Which of the following would you use to determine the atomic mass of X?

A. $\dfrac{76.8}{\left(\dfrac{23.2}{16.0}\right) \times \left(\dfrac{7}{2}\right)}$

B. $\dfrac{76.8}{\left(\dfrac{16.0}{23.2}\right) \times \left(\dfrac{2}{7}\right)}$

C. $\dfrac{76.8}{\left(\dfrac{23.2}{16.0}\right) \times \left(\dfrac{2}{7}\right)}$

D. $\dfrac{76.8}{\left(\dfrac{16.0}{23.2}\right) \times \left(\dfrac{7}{2}\right)}$

E. $\dfrac{\left(\dfrac{16.0}{23.2}\right) \times \left(\dfrac{7}{2}\right)}{76.8}$

Answer: C

From the information provided, you know that the oxide X_2O_7 contains 76.8% X and 23.2% oxygen by weight. If you had 100.0 g of the oxide, you would have 76.8 g of X and 23.2 g of O (or 23.2/16.0 moles of O atoms). Because for each mole of O in the oxide you have 2/7 mole of X, you have in effect (23.2/16.0) × 2/7 mole of X. Since the units of atomic mass are g/mole, the setup is:

$$\frac{76.8}{\left(\dfrac{23.2}{16.0}\right) \times \left(\dfrac{2}{7}\right)}$$

Finding the solution is unnecessary for selecting an answer; however, here is the solution for your information.

$$\frac{76.8 \, g\,X}{\left(\dfrac{23.2 \, \cancel{g\,O}}{16.0 \, \cancel{g\,O}/mole\,O}\right) \times \left(\dfrac{2 \, mole\,X}{7 \, \cancel{mole\,O}}\right)} = 186 \, g/mol$$

(The element is rhenium.)

10. A freshman chemist analyzed a sample of copper(II) sulfate pentahydrate for water of hydration by weighing the hydrate, heating it to convert it to anhydrous copper(II) sulfate, and then weighing the anhydride. The % H_2O was determined to be 30%. The theoretical value is 33%. Which of the following choices is definitely NOT the cause of the error?

A. After the student weighed the hydrate, a piece of rust fell from the tongs into the crucible.

B. Moisture driven from the hydrate condensed on the inside of the crucible cover before the student weighed the anhydride.

C. All the weighings were made on a balance that was high by 10%.

D. The original sample contained some anhydrous copper(II) sulfate.

E. The original sample was wet.

Answer: E

30% H_2O in the hydrate sample represents

$$\frac{\text{mass of hydrate} - \text{mass of anhydride}}{\text{mass of hydrate}} \times 100\%$$

In a problem like this, I like to make up some *easy* fictitious numbers that I can use to fit the scenarios and see how the various changes affect the final outcome. Let's say the mass of the hydrate is 10 g and the mass of the anhydride is 7 g. This would translate as

$$\frac{10\ \text{g} - 7\ \text{g}}{10\ \text{g}} \times 100\% = 30\%\ H_2O$$

In examining choice A, the original mass of the hydrate would not change; however, because rust will not evaporate, the final mass of the anhydride would be higher than expected — let's say 8 g. Substituting this value into the formula for % water would give

$$\frac{10\ \text{g} - 8\ \text{g}}{10\ \text{g}} \times 100\% = 20\%\ H_2O$$

which is less than the theoretical value of 30%. This is in the direction of the student's experimental results, and since we are looking for the choice that is NOT the cause, we can rule out A as an answer.

In choice B, the mass of the hydrate would not change, but the mass of the anhydride would be higher than it should be. Let's estimate the anhydride at 8 g again.

$$\frac{10\ \text{g} - 8\ \text{g}}{10\ \text{g}} \times 100\% = 20\%\ H_2O$$

In choice C, because all masses are being measured on a *consistently* wrong balance, the faultiness does not matter in the final answer.

$$\frac{11.0 \text{ g} - 7.7 \text{ g}}{11.0 \text{ g}} \times 100\% = 30\% \text{ H}_2\text{O}$$

In choice D, the original mass of the hydrate would remain unchanged. However, the mass of the anhydride would be higher than expected because the sample would lose less water than if it had been a pure hydrate. This fits the scenario of

$$\frac{10 \text{ g} - 8 \text{ g}}{10 \text{ g}} \times 100\% = 20\% \text{ H}_2\text{O}$$

with the error being consistent with the direction of the student's results. Therefore, D is not the correct answer.

In choice E, the original sample is wet. The freshman chemist weighed out 10 g of the hydrate, but more weight is lost in the heating process than expected, making the final mass of the anhydride lower than expected, say 6 g. Using the equation for % H$_2$O shows

$$\frac{10 \text{ g} - 6 \text{ g}}{10 \text{ g}} \times 100\% = 40\% \text{ H}_2\text{O}$$

which is higher than the theoretical value of 30% and in line with the reasoning that this could NOT have caused the error.

---

11. When 100 grams of butane gas (C$_4$H$_{10}$, MW = 58.14) is burned in excess oxygen gas, the theoretical yield of H$_2$O is:

   A. $\dfrac{54.14 \times 18.02}{100 \times 5}$

   B. $\dfrac{5 \times 58.14}{100 \times 18.02}$

   C. $\dfrac{4 \times 18.02}{13/2 \times 100} \times 100\%$

   D. $\dfrac{5 \times 58.14 \times 18.02}{100}$

   E. $\dfrac{100 \times 5 \times 18.02}{58.14}$

Answer: E

Begin with a balanced equation:

$$\text{C}_4\text{H}_{10} + 13/2 \text{ O}_2 \rightarrow 4\text{CO}_2 + 5\text{H}_2\text{O}$$

Next, set up the equation in factor-label fashion:

$$\frac{100 \text{ g C}_4\text{H}_{10}}{1} \times \frac{1 \text{ mole C}_4\text{H}_{10}}{58.14 \text{ g C}_4\text{H}_{10}} \times \frac{5 \text{ mole H}_2\text{O}}{1 \text{ mole C}_4\text{H}_{10}} \times \frac{18.02 \text{ g H}_2\text{O}}{1 \text{ mole H}_2\text{O}} = \text{g H}_2\text{O}$$

12. Element Q occurs in compounds X, Y, and Z. The mass of element Q in 1 mole of each compound is as follows:

Compound	Grams of Q in Compound
X	38.00
Y	95.00
Z	133.00

Element Q is most likely:

   A. N

   B. O

   C. F

   D. Ir

   E. Cs

Answer: C

All of the numbers are multiples of 19.00 (fluorine). Use the law of multiple proportions.

13. Which one of the following represents an intensive property?

   A. temperature

   B. mass

   C. volume

   D. length

   E. heat capacity

Answer: A

The measured value of an intensive property does NOT depend on how much matter is being considered. The formula for heat capacity C is $C = m \cdot C_p$, where m = mass and C_p = specific heat.

14. Which of the following would have an answer with three significant figures?

    A. $103.1 + 0.0024 + 0.16$

    B. $(3.0 \times 10^4)(5.022 \times 10^{-3}) / (6.112 \times 10^2)$

    C. $(4.3 \times 10^5) / (4.225 + 56.0003 - 0.8700)$

    D. $(1.43 \times 10^3 + 3.1 \times 10^1) / (4.11 \times 10^{-6})$

    E. $(1.41 \times 10^2 + 1.012 \times 10^4) / (3.2 \times 10^{-1})$

Answer: D

$$(1.43 \times 10^3 + 3.1 \times 10^1) = 14.3 \times 10^2 + 0.31 \times 10^2 = 14.6 \times 10^2$$

$$= \frac{14.6 \times 10^2}{4.11 \times 10^{-6}} = 3 \text{ s.f.} \left(\underline{3.55} \times 10^8 \right)$$

Samples: Free-Response Questions

1. A student performed the following experiment in the laboratory: She suspended a clean piece of silver metal in an evacuated test tube. The empty test tube weighed 42.8973 grams. The silver weighed 1.7838 grams. Next, she introduced a stream of chlorine gas into the test tube and allowed it to react with the silver. After a few minutes, a white compound was found to have formed on the silver strip, coating it uniformly. She then opened the apparatus, weighed the coated strip, and found it to weigh 1.9342 grams. Finally, she washed the coated strip with distilled water, removing all of the white compound from the silver strip, and then dried the compound and the strip and reweighed. She discovered that the silver strip weighed 1.3258 grams.

(a) Show how she would determine

   (1) the number of moles of chlorine gas that reacted

   (2) the number of moles of silver that reacted

(b) Show how she could determine the simplest formula for the silver chloride.

(c) Show how her results would have been affected if

   (1) some of the white compound had been washed down the sink before it was dried and reweighed

   (2) the silver strip was not thoroughly dried when it was reweighed

Answer

Step 1: Do a restatement of the general experiment. In this case, I would draw a sketch of the apparatus before and after the reaction, labeling everything. This will get rid of all the words and enable you to visualize the experiment.

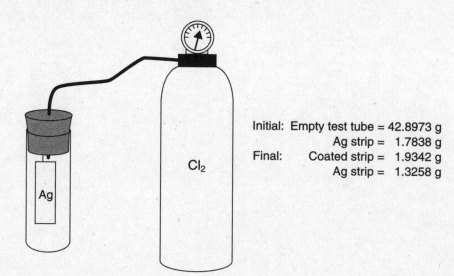

Initial: Empty test tube = 42.8973 g
              Ag strip = 1.7838 g
Final: Coated strip = 1.9342 g
              Ag strip = 1.3258 g

Step 2: Write a balanced chemical equation that describes the reaction.

silver + chlorine gas yields silver chloride.

$$2\,Ag\,(s) + Cl_2\,(g) \rightarrow 2\,AgCl\,(s)$$

Step 3: Begin to answer the questions asked. Remember to give a brief restatement for each question, to label each specific question so that the grader knows which question you are answering, and to underline the conclusion(s) where necessary.

1. **(a)** **(1)** Restatement: Number of moles of chlorine gas that reacted.

   mass of chlorine that reacted

   = (mass of silver strip + compound) − mass of original silver strip

   1.9342 g − 1.7838 g = 0.1504 g of chlorine atoms

   moles of chlorine atoms that reacted = mass of chlorine atoms/atomic mass of chlorine

   0.1504 g / 35.45 g · mole^{-1} = **0.004242 mole of chlorine atoms**

   (a) **(2)** Restatement: Moles of silver that reacted.

   $$\frac{(\text{mass of original silver strip} - \text{mass of dry strip after washing})}{\text{atomic mass of silver}}$$

   1.7838 g − 1.3258 g = 0.4580 g

   0.4580 g / 107.87 g · mole^{-1} = **0.004246 mole of silver atoms**

   (b) Restatement: Empirical formula for silver chloride.

   empirical formula = moles of silver atoms / moles of chlorine atoms

   $$\frac{0.004246 \text{ mole of silver}}{0.004242 \text{ mole of chlorine}} = 1.001 \rightarrow AgCl$$

   (c) **(1)** Restatement: Effect on the empirical formula if some of the white compound had been washed down the sink before the coated strip was dried and reweighed.

   The white product was silver chloride. Had she lost some before she weighed it, the mass of silver chloride would have been less than what it should have been. This would have made the number of grams of chlorine appear too low, which in turn would have made the number of moles of chlorine appear too low. Thus, in the ratio of moles of silver to moles of chlorine, the denominator would have been lower than expected and **the ratio would have been larger.** Because the mass of the compound does not enter into the calculations for the moles of silver that reacted, the moles of silver would not have been affected.

(c) (2) Restatement: Effect on the empirical formula if the silver strip had not been dried thoroughly after being washed free of the silver chloride.

Because the strip has been washed free of the compound (silver chloride), you assume that any silver missing from the strip went into the making of the silver chloride. If the strip had been wet when you weighed it, you would have been led to think that the strip was heavier than expected, and therefore that less silver had gone into the making of the silver chloride. Thinking that less silver had been involved in the reaction, you would have calculated fewer moles of silver. The calculated **ratio of moles of silver to moles of chlorine would have been less than expected.**

---

2. Three compounds, D, E, and F, all contain element G. The percent (by weight) of element G in each of the compounds was determined by analysis. The experimental data are presented in the following chart.

Compound	% by Weight of Element G	Molecular Weight
D	53.9	131.7
E	64.2	165.9
F	47.7	74.5

(a) Determine the mass of element G contained in 1.00 mole of each of compounds D, E, and F.

(b) What is the most likely value for the atomic weight of element G?

(c) Compound F contains carbon, hydrogen, and element G. When 2.19 g of compound F is completely burned in oxygen gas, 3.88 g of carbon dioxide gas and 0.80 g of water are produced. What is the most likely formula for compound F?

Answer

2. Given: Compounds D, E, and F with % (by weight of element G) and their respective MW's

(a) Restatement: Calculate the mass of element G in 1.00 mole of compounds D, E, and F.

0.539×131.7 g/mole = **71.0 g G/mole D**

0.642×165.9 g/mole = **107 g G/mole E**

0.477×74.5 g/mole = **35.5 g G/mole F**

(b) Restatement: Most likely atomic weight of G.

According to the law of multiple proportions, the ratios of the mass of element G to the masses of compounds D, E, and F must be small, whole numbers. The largest common denominator of 71.0, 106.5, and 35.5 is 35.5, so our best estimate is that the **atomic weight of G is 35.5** (chlorine).

(c) Restatement: Compound $F = C_xH_yG_z$ or $C_xH_yCl_z$?

$$C_xH_yCl_z + O_2(g) \rightarrow CO_2(g) + H_2O(\ell) + Cl_2(g)$$

$$2.19 \text{ g} + ? \text{ g} \rightarrow 3.88 \text{ g} + 0.80 \text{ g} + ? \text{ gCl}_2(g)$$

moles of carbon = moles of CO_2

$$\frac{3.88 \text{ g CO}_2}{44.01 \text{ g} \cdot \text{mole}^{-1}} = 0.0882$$

moles of hydrogen = $2 \times$ moles of H_2O

$$2 \times \frac{0.80 \text{ g H}_2\text{O}}{18.02 \text{ g} \cdot \text{mole}^{-1}} = 0.088$$

(2.19 grams of F)(1 mole F / 74.5 g F) = 0.0294 mole of compound F

This means that each mole of F contains 3 moles of C (0.0882 / 0.0294) and 3 moles of H (0.088), or 39 grams of CH. This leaves $74.5 - 39 = 36$ grams, corresponding to 1 mole of element G (Cl). Therefore, the empirical formula is **C_3H_3Cl.**

Thermochemistry

Key Terms

Words that can be used as topics in essays:

adiabatic

calorimeter

endothermic

enthalpy

entropy

exothermic

first law of thermodynamics

Gibbs free energy

heat

heat of dilution

heat of formation

heat of fusion

heat of hydration

heat of reaction

heat of solution

Hess's law

internal energy

kinetic energy

law of conservation of energy

potential energy

second law of thermodynamics

specific heat

standard state

state property (function)

surroundings

system (closed, isolated, open)

temperature

thermodynamics

third law of thermodynamics

work

Key Concepts

Equations and relationships that you need to know:

- Endothermic reaction has $+\Delta H$; $H_{products} > H_{reactants}$

  Exothermic reaction has $-\Delta H$; $H_{products} < H_{reactants}$

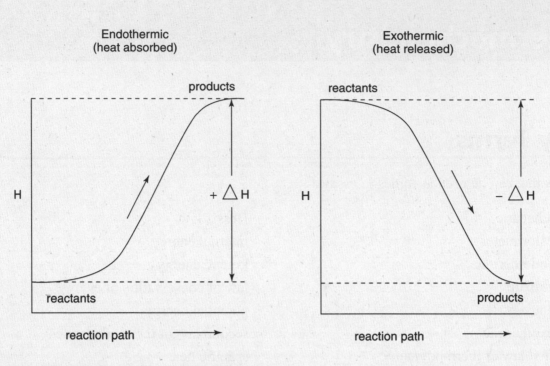

$$\Delta E = \Delta H - RT\Delta n = \Delta H - P\Delta V = q + w$$

$$\Delta H = \Delta E + \Delta(PV) = q - w + P\Delta V$$

$$w = -P\Delta V = -nRT \ln \frac{V_f}{V_i}$$

	Done on a system	*Done by a system*
Work, w	$+w$	$-w$
Heat, q	$+q$ (endothermic)	$-q$ (exothermic)

- "bomb" calorimeter

  $\Delta E = q_v$ (valid with constant volume: "bomb" calorimeter)

  $q_{reaction} = -(q_{water} + q_{bomb})$

  $q_{bomb} = C \cdot \Delta t$

  where C = calorimeter constant (heat capacity) in J/°C.

- "coffee cup" calorimeter

  $\Delta H = q_p$ (valid with constant pressure: "coffee cup" calorimeter)

  $q = m \cdot c \cdot \Delta t$

  $C_p = \dfrac{\Delta H}{\Delta t}$

$\Delta H_{reaction} = -q_{water}$

specific heat of water $= \dfrac{4.18J}{g \cdot {}^\circ C} = \dfrac{1cal}{g \cdot {}^\circ C}$

$\Delta H^\circ = \Sigma H^\circ_{f\,products} - \Sigma H^\circ_{f\,reactants} = q_p$

where q_p = heat flow, constant p

1 calorie = 4.184 Joules

101 Joules = 1 liter · atm

24.14 calories = 1 liter · atm

- law of Dulong and Petit

  molar mass · specific heat $\approx$ 25 J/mole · °C

- first law of thermodynamics

  $\Delta E = q + w = q_p - P\Delta V = \Delta H - P\Delta V$

  In any process, the total change in energy of the system, ΔE, is equal to the sum of the heat absorbed, q, and the work, w, done on the system.

- second law of thermodynamics

  $\Delta S_{univ} = \Delta S_{sys} + \Delta S_{surr} > 0$ spontaneous

  $\Delta S_{univ} = \Delta S_{sys} + \Delta S_{surr} < 0$ nonspontaneous

  $\Delta S_{univ} = \Delta S_{sys} + \Delta S_{surr} = 0$ equilibrium

  The entropy of the universe increases in a spontaneous process and remains unchanged in an equilibrium process.

- third law of thermodynamics

  $S^\circ = q_p\,/\,T$

  $\Delta S^\circ = \Sigma S^\circ_{products} - \Sigma S^\circ_{reactants}$

  The entropy of a perfect crystalline substance is zero at absolute zero.

- Hess's law: If reaction (1) has ΔH_1 and reaction (2) has ΔH_2 and reaction (1) + reaction (2) = reaction (3), then

  $$\Delta H_3 = \Delta H_1 + \Delta H_2$$

- Bond breaking: potential energy (enthalpy) of bond is increased; "strong" bonds $\rightarrow$ "weak" bonds; $\Delta H > 0$.

  Bond forming: potential energy (enthalpy) of bond is decreased; bond distance is decreased; "weak" bonds $\rightarrow$ "strong" bonds; $\Delta H < 0$.

  $\Delta H = \Sigma$ bond energy (reactants) $- \Sigma$ bond energy (products) = total energy input $-$ total energy released

- given: $aA + bB = cC + dD$

  $\Delta S_{\text{rxn (or sys)}} = [cS°(C) + dS°(D)] - [aS°(A) + bS°(B)]$

  $\Delta S_{\text{surr}} = \dfrac{-\Delta H}{T}$ which derives to $\Delta G = \Delta H - T\Delta S$

  $\Delta G° = \Sigma\Delta G°_{f\,\text{products}} - \Sigma\Delta G°_{f\,\text{reactants}}$

  $\Delta G° = -RT \ln K = -2.303\ RT \log K = -n\mathscr{F}E°$

  $\Delta G = \Delta G° + RT \ln Q = \Delta G° + 2.303\ RT \log Q$

Samples: Multiple-Choice Questions

1. Given the following information:

Reaction (1): $H_2(g) + \frac{1}{2}O_2(g) \rightarrow H_2O(\ell)$ $\Delta H° = -286 \text{ kJ}$

Reaction (2): $CO_2(g) \rightarrow C(s) + O_2(g)$ $\Delta H° = 394 \text{ kJ}$

Reaction (3): $2CO_2(g) + H_2O(\ell) \rightarrow C_2H_2(g) + \frac{5}{2}O_2(g)$ $\Delta H° = 1300 \text{ kJ}$

Find $\Delta H°$ for the reaction $C_2H_2(g) \rightarrow 2\,C(s) + H_2(g)$.

A. -226 kJ

B. -113 kJ

C. 113 kJ

D. 226 kJ

E. 452 kJ

Answer: A

Recognize that this is a Hess's law problem, which basically requires that you rearrange the three reactions listed if necessary (remembering to reverse the sign of $\Delta H°$ if you reverse the reaction) and then add up the reactions and the $\Delta H°$'s for the answer.

Looking at the first reaction, you notice that $H_2(g)$ is on the wrong side. This requires that you reverse the reaction, thus changing the sign of $\Delta H°$. In the second reaction, C is on the correct side of the equation, so the second equation should be left as is. In the third reaction, $C_2H_2(g)$ is on the wrong side of the equation, so the reaction and the sign of $\Delta H°$ must be reversed. At this point the set-up should look like

$H_2O(\ell) \rightarrow H_2(g) + 1/2\,O_2(g)$ $\Delta H° = +286 \text{ kJ}$

$CO_2(g) \rightarrow C(s) + O_2(g)$ $\Delta H° = +394 \text{ kJ}$

$C_2H_2(g) + 5/2\,O_2(g) \rightarrow 2CO_2(g) + H_2O(\ell)$ $\Delta H° = -1300 \text{ kJ}$

Before you add up the three reactions, note that there is no $CO_2(g)$ in the final reaction. The 2 $CO_2(g)$ in the third reaction must be able to cancel with two $CO_2(g)$'s in the second reaction. In order to get 2 $CO_2(g)$ in the second reaction, multiply everything in the second equation by 2, which yields

$\cancel{H_2O(\ell)} \rightarrow H_2(g) + \cancel{1/2\,O_2(g)}$ $\Delta H° = +286 \text{ kJ}$

$\cancel{2CO_2(g)} \rightarrow 2C(s) + \cancel{2O_2(g)}$ $\Delta H° = +788 \text{ kJ}$

$\underline{C_2H_2(g) + \cancel{5/2\,O_2(g)} \rightarrow \cancel{2CO_2(g)} + \cancel{H_2O(\ell)}}$ $\underline{\Delta H° = -1300 \text{ kJ}}$

$C_2H_2(g) \rightarrow H_2(g) + 2C(s)$ $\Delta H° = -226 \text{ kJ}$

2. A piece of metal weighing 500. grams is put into a boiling water bath. After 10 minutes, the metal is immediately placed in 250. grams of water at 40.°C. The maximum temperature that the system reaches is 50.°C. What is the specific heat of the metal? (The specific heat of water is 1.00 cal/g · °C.)

A. 0.010 cal/g · °C

B. 0.050 cal/g · °C

C. 0.10 cal/g · °C

D. 0.20 cal/g · °C

E. 0.50 cal/g · °C

Answer: C

Begin this problem by realizing that the heat gained by the water is equal to the heat lost by the metal.

heat lost or gained = specific heat of the substance × mass of the substance × Δt

Substituting the numbers into this concept gives

heat gained by water = heat lost by metal

$(1.00 \text{ cal/g} \cdot °C) \times (250. \text{ g H}_2\text{O}) \times (50.°C - 40.°C) =$

$-(x \text{ cal/g} \cdot °C) \times (500. \text{ g metal})(50.°C - 100.°C)$

$2500 = 25{,}000\, x$

$x = 0.10 \text{ cal/g} \cdot °C$

3. Given the following heat of reaction and the bond energies listed in the accompanying table, calculate the energy of the C=O bond. All numerical values are in kilocalories per mole, and all substances are in the gas phase.

$$CH_3CHO + H_2 \rightarrow CH_3CH_2OH \qquad \Delta H° = -17 \text{ kcal/mole}$$

Bond	O—H	C—H	C—C	C—O	H—H
Bond Energy (kcal/mole)	111	99	83	84	104

A. 79 kcal

B. 157 kcal

C. 173 kcal

D. 190 kcal

E. 277 kcal

Answer: C

Begin this problem by drawing a structural diagram of the reaction.

There are three steps you need to take to do this problem:

Step 1: Decide which bonds need to be broken on the reactant side of the reaction. Add up all the bond energies for the bonds that are broken. Call this subtotal ΔH_1. Assign ΔH_1 a positive value because energy is required when bonds are broken. In the example given, a $C\!=\!O$ and a H—H bond need to be broken. This becomes x kcal/mole + 104 kcal/mole, or $\Delta H_1 = 104 + x$ kcal/mole.

Step 2: Decide which bonds need to be formed on the product side of the reaction. Add up all of the bond energies that are formed. Call this subtotal ΔH_2. Assign ΔH_2 a negative value because energy is released when bonds are formed. In the example given, a C—H, a C—O, and a O—H bond need to be formed. This becomes 99 kcal/mole + 84 kcal/mole + 111 kcal/mole, or 294 kcal/mole. Remember to assign a negative sign, which makes $\Delta H_2 = -294$ kcal/mole.

Step 3: Apply Hess's law: $\Delta H° = \Delta H_1 + \Delta H_2$. You know that $\Delta H°$ is -17 kcal/mole, so Hess's law becomes

-17 kcal/mole = 104 kcal/mole + x kcal/mole $-$ 294 kcal/mole

$x = 173$ kcal/mole

which represents the bond energy of the $C\!=\!O$ bond.

---

4. Given the following heats of formation:

Substance	$\Delta H_f°$
acetic acid	−120 kcal/mole
carbon dioxide	−95 kcal/mole
water	−60 kcal/mole

Find $\Delta H°$ of combustion for acetic acid (CH_3COOH).

A. −430 kcal/mole

B. −190 kcal/mole

C. −45 kcal/mole

D. 45 kcal/mole

E. 190 kcal/mole

Answer: B

Begin this problem by realizing that $\Delta H°$ of combustion for acetic acid represents the amount of heat released when acetic acid burns in oxygen gas. The products are carbon dioxide and water. According to the following balanced equation,

$$CH_3COOH(\ell) + 2O_2(g) \rightarrow 2CO_2(g) + 2H_2O(\ell)$$

Because $\Delta H° = \Sigma H°_f \text{ products} - \Sigma H°_f \text{ reactants}$, you can substitute at this point to give

$\Delta H° = 2(-95 \text{ kcal/mole}) + 2(-60 \text{ kcal/mole}) - (-120 \text{ kcal/mole})$

$\Delta H° = -190 \text{ kcal/mole}$

---

5. For $H_2C = CH_2(g) + H_2(g) \rightarrow H_3C - CH_3(g)$, predict the enthalpy given the following bond dissociation energies:

H—C, 413 kJ/mole   H—H, 436 kJ/mole
C=C, 614 kJ/mole   C—C, 348 kJ/mole

A.   −656 kJ/mole

B.   −343 kJ/mole

C.   −289 kJ/mole

D.   −124 kJ/mole

E.   −102 kJ/mole

Answer: D

Begin this problem by drawing a structural diagram:

There are three steps you need to take to do this problem.

Step 1: Decide which bonds need to be broken on the reactant side of the reaction. Add up all the bond energies for the bonds that are broken. Call this subtotal ΔH_1, and assign it a positive value because when energy is required, bonds are broken. In the example given, a C=C and a H—H bond need to be broken. This becomes 614 kJ/mole + 436 kJ/mole = ΔH_1 = 1050 kJ/mole.

Step 2: Decide which bonds need to be formed on the product side of the reaction. Add up all the bond energies for the bonds that are formed. Call this subtotal ΔH_2. Assign ΔH_2 a negative value because when energy is released, bonds are formed. In the example given, two C—H bonds and a C—C bond need to be formed. This becomes (2 × 413 kJ/mole) + 348 kJ/mole, or 1174 kJ/mole. Remember to assign a negative sign, which makes $\Delta H_2 = -1174$ kJ/mole.

Step 3: Apply Hess's law: $\Delta H° = \Delta H_1 + \Delta H_2$

This becomes 1050 kJ/mole + (−1174 kJ/mole) = −124 kJ/mole.

---

6. According to the law of Dulong and Petit, the best prediction for the specific heat of technetium (Tc), MM = 100., is

   A. 0.10 J/g · °C

   B. 0.25 J/g · °C

   C. 0.50 J/g · °C

   D. 0.75 J/g · °C

   E. 1.0 J/g · °C

Answer: B

The law of Dulong and Petit states that

    molar mass × specific heat ≈ 25 J/mole · °C

You know that technetium has an atomic mass of 100., and substituting this into the law of Dulong and Petit gives you

$$100. \text{ g/mole} \times x \text{ J/g} \cdot °C \approx 25 \text{ J/mole} \cdot °C$$
$$x \approx 0.25 \text{ J/g} \cdot °C$$

---

7. How much heat is necessary to convert 10.0 grams of ice at −10.0°C to steam at 150°C? The specific heat capacity of ice is 0.500 cal/g · °C. The heat of fusion of ice is 76.4 cal/g. The specific heat capacity of water is 1.00 cal/g · °C. The heat of vaporization of water is 539 cal/g. The specific heat capacity of steam is 0.482 cal/g · °C.

   A. 2500 cal

   B. 4433 cal

   C. 7445 cal

   D. 8255 cal

   E. 9555 cal

Answer: C

First, the ice must be heated to its melting point, 0.0°C.

$$(\text{mass}) \times (\text{specific heat capacity}) \times (\Delta t)$$

$$\frac{10.0 \, \cancel{\text{g ice}}}{1} \times \frac{0.500 \, \text{cal}}{\cancel{\text{g} \cdot \text{°C}}} \times \frac{10.0 \, \cancel{\text{°C}}}{1} = 50.0 \, \text{cal}$$

Next, the ice must be melted.

$$(\text{mass}) \times (\Delta H_{\text{fus}})$$

$$\frac{10.0 \, \cancel{\text{g ice}}}{1} \times \frac{76.4 \, \text{cal}}{\cancel{\text{g}}} = 764 \, \text{cal}$$

Next, the water must be heated to its boiling point, 100.0°C.

$$\frac{10.0 \, \cancel{\text{g water}}}{1} \times \frac{1.00 \, \text{cal}}{\cancel{\text{g} \cdot \text{°C}}} \times \frac{100.0 \, \cancel{\text{°C}}}{1} = 1.00 \times 10^3 \, \text{cal}$$

Next, the water must be vaporized.

$$\frac{10.0 \, \cancel{\text{g water}}}{1} \times \frac{539 \, \text{cal}}{\cancel{\text{g}}} = 5390 \, \text{cal}$$

Next, the steam must be heated to 150.0°C

$$\frac{10.0 \, \cancel{\text{g stream}}}{1} \times \frac{0.482 \, \text{cal}}{\cancel{\text{g} \cdot \text{°C}}} \times \frac{50.0 \, \cancel{\text{°C}}}{1} = 241 \, \text{cal}$$

The last step is to add up all quantities of heat required.

50.0 cal + 764 cal + 1000 cal + 5390 cal + 241 cal = 7445 calories

---

8. Given these two standard enthalpies of formation:

Reaction (1) $S(s) + O_2(g) \rightleftharpoons SO_2(g)$ $\quad\quad\quad\quad \Delta H° = -295$ kJ/mole

Reaction (2) $S(s) + 3/2 \, O_2(g) \rightleftharpoons SO_3(g)$ $\quad\quad\quad \Delta H° = -395$ kJ/mole

What is the reaction heat for $2SO_2(g) + O_2(g) \rightleftharpoons 2SO_3(g)$ under the same conditions?

A. −1380 kJ/mole

B. −690. kJ/mole

C. −295 kJ/mole

D. −200. kJ/mole

E. − 100. kJ/mole

Answer: D

Examine the first reaction and realize that $SO_2(g)$ needs to be on the reactant side. Reverse the equation and change the sign of $\Delta H°$. When you examine the second reaction, you notice that $SO_3(g)$ is on the correct side, so there is no need to reverse this equation. At this point, your two reactions can be added together.

$$SO_2(g) \rightarrow S(s) + O_2(g) \qquad\qquad \Delta H° = 295 \text{ kJ/mole}$$
$$\underline{S(s) + 3/2O_2(g) \rightarrow SO_3(g) \qquad\qquad \Delta H° = -395 \text{ kJ/mole}}$$
$$SO_2(g) + 1/2O_2(g) \rightarrow SO_3(g) \qquad\qquad \Delta H° = -100 \text{ kJ/mole}$$

But before concluding that this is your answer, note that the question asks for $\Delta H°$ in terms of 2 moles of $SO_2(g)$. Doubling the $\Delta H°$ gives the answer, -200. kJ/mole.

---

9. In expanding from 3.00 to 6.00 liters at a constant pressure of 2.00 atmospheres, a gas absorbs 100.0 calories (24.14 calories = 1 liter · atm). The change in energy, ΔE, for the gas is

   A. −600. calories

   B. −100. calories

   C. −44.8 calories

   D. 44.8 calories

   E. 100. calories

Answer: C

The first law of thermodynamics states that $\Delta E = q + w$. Since $w = -P_{ext} \cdot \Delta V$, the equation can be stated as

$$\Delta E = \Delta H - P_{ext} \cdot \Delta V$$

$$\Delta E = 100.0 \text{ calories} - \left(2.00 \text{ ~~atmospheres~~} \cdot 3 \text{ ~~liters~~} 24.14 \text{ cal/}L \cdot atm\right) = -44.8 \text{ calories}$$

---

10. A gas which initially occupies a volume of 6.00 liters at 4.00 atm is allowed to expand to a volume of 14.00 liters at a pressure of 1.00 atm. Calculate the value of work, w, done by the gas on the surroundings

    A. −8.00 L · atm

    B. −7.00 L · atm

    C. 6.00 L · atm

    D. 7.00 L · atm

    E. 8.00 L · atm

Answer: A

$$w = -P\Delta V = -(1.00 \text{ atm})(8.00 \text{ liters}) = -8.00 \text{ L} \cdot \text{atm}$$

Because the gas was expanding, w is negative (work was being done by the system).

---

11. The molar heat of sublimation for molecular iodine is 62.30 kJ/mol at 25°C and 1.00 atm. Calculate the ΔH in (J $\cdot$ mole^{-1}) for the reaction

$$I_2(s) \rightarrow I_2(g) \qquad R = 8.314 \text{ J/mole} \cdot \text{K}$$

A. $\dfrac{62.30}{8.314 \cdot 298}$

B. $62.30 - (8.314)(298)$

C. $62.30 + (8.314)(298)$

D. $62.30(1000) + 1 \cdot (8.314)(298)$

E. none of the above are correct

Answer: D

Δn = moles of gaseous product − moles of gaseous reactants

$$= 1 - 0 = 1$$

Use the equation $\Delta H = \Delta E + \Delta n \cdot RT$

$$= \frac{62.30 \text{ kJ}}{\text{mole}} \times \frac{1000 \text{ J}}{1 \text{ kJ}} + \left[1 \cdot \left(8.314 \text{ J/mole} \cdot \text{K} \right) \cdot \left(298 \text{ K} \right) \right]$$

Samples: Free-Response Questions

1. To produce molten iron aboard ships, the Navy uses the thermite reaction, which consists of mixing iron(III) oxide (rust) with powdered aluminum, igniting it, and producing aluminum oxide as a by-product. ΔH_f° for aluminum oxide is −1669.8 kJ/mole, and that for iron(III) oxide is −822.2 kJ/mole.

    (a) Write a balanced equation for the reaction.

    (b) Calculate ΔH for the reaction.

    (c) The specific heat of aluminum oxide is 0.79 J/g · °C, and that of iron is 0.48 J/g · °C. If one were to start the reaction at room temperature, how hot would the aluminum oxide *and* the iron become?

    (d) If the temperature needed to melt iron is 1535°C, and the heat of fusion for iron is 270 J/g, confirm through calculations that the reaction will indeed produce molten iron.

Answer

1. (a) Restatement: Balanced equation.

    $$2Al(s) + Fe_2O_3(s) \rightarrow Al_2O_3(s) + 2Fe(s)$$

    (b) Restatement: Calculation of ΔH.

    $\Delta H = \Sigma \Delta H_{f \, products}^\circ - \Sigma \Delta H_{f \, reactants}^\circ$

    $\Delta H = [\Delta H_f \, Al_2O_3(s) + 2 \, \Delta H_f \, Fe(s)] - [2 \, \Delta H_f \, Al(s) + \Delta H_f \, Fe_2O_3(s)]$

    $\Delta H = [-1669.8 + 2(0)] - [2(0) + (-822.2)]$

    Remember that ΔH_f for elements is 0.

    $\Delta H = -847.6$ kJ

    (c) Restatement: How hot will the products become?

    Given: Specific heat of aluminum oxide = 0.79 J/g · °C

    Given: Specific heat of iron = 0.48 J/g · °C

    Going back to the balanced reaction (answer a), you note that 1 mole of Al_2O_3 is being produced for every 2 moles of iron — they are not being produced in a ratio of 1 gram to 1 gram. For each of the two products, determine how much energy (in joules) is required for a 1°C rise in temperature.

    For Al_2O_3: $\dfrac{1 \; \text{mole } Al_2O_3}{1} \times \dfrac{101.96 \; \text{g } Al_2O_3}{1 \; \text{mole } Al_2O_3} \times \dfrac{0.79 \; J}{g \cdot °C} = 81 \; J/°C$

    For Fe: $\dfrac{2 \; \text{moles Fe}}{1} \times \dfrac{55.85 \; \text{g Fe}}{1 \; \text{mole Fe}} \times \dfrac{0.48 \; J}{g \cdot °C} = 54 \; J/°C$

Together, the products are absorbing 81 J/°C + 54 J/°C = 135 joules for every l°C rise in temperature. Because the reaction is producing 847.6 kJ of energy (you know this from your answer to part b), the change in temperature, from the initial conditions of the reactants (presumably at room temperature) to those of the hot products, is found by dividing the heat of reaction by the energy absorbed per degree Celsius (J divided by J/°C = °C). Thus

$$\frac{847,600 \text{ J}}{135 \text{ J/}^\circ C} = 6.28 \times 10^3 \, ^\circ C \text{ change in temperature}$$

(d) Restatement: Confirmation that the heat produced from the thermite reaction is sufficient to melt iron.

Given: H_{fus} Fe = 270 J/g

Because ΔH_{fus} represents the amount of energy *absorbed* in melting iron, you must subtract this component from the ΔH you obtained in part (b). And because ΔH reflects the amount of heat produced when 2 moles of iron are produced, ΔH_{fus} for the reaction is

$$\frac{2 \text{ moles of Fe}}{1} \times \frac{55.85 \text{ g Fe}}{1 \text{ mole of Fe}} \times \frac{270 \text{ J}}{1 \text{ g Fe}} = 30,200 \text{ J}$$

The 30,200 J of energy was *absorbed* in melting iron, so you subtract it from the ΔH value obtained in part (b) to find how much energy was actually *released* in the reaction.

847,600 J – 30,200 J = 817,400 J

Finally, from part (c), you know that the amount of energy *released* divided by the energy absorbed per degree Celsius (J divided by J/°C = °C) equals the *change* in temperature. Thus you have

$$\frac{817,400 \text{ J}}{135 \text{ J/}^\circ C} = 6.05 \times 10^3 \, ^\circ C$$

which is above the melting point of iron, 1535°C.

---

2. You pack six aluminum cans of cola in a cooler filled with ice. Each aluminum can (empty) weighs 50.0 grams. When filled, each can contains 355 mL of cola. The density of the cola is 1.23 g/mL. The specific heat of aluminum is 0.902 J/g · °C, and that of the cola is 4.00 J/g · °C.

(a) If the cola is initially at 30°C and you wish to cool it to 10°C, how much heat must be absorbed by the ice?

(b) What is the minimum amount of ice (at 0°C) needed to cool the cola? ΔH_{fus} for ice is 6.00 kJ/mole.

Answer

2. Given: 6 aluminum cans (empty) at 50.0 g each

each can = 355 mL cola

density of cola = 1.23 g/mL

specific heat of aluminum = 0.902 J/g · °C

specific heat of cola = 4.00 J/g · °C

$t_i = 30°C$

$t_f = 10°C$

(a) Restatement: Amount of heat absorbed by ice.

quantity of heat released by six pack = heat released by cans + cola

$q_{\text{six-pack}} = q_{\text{cans}} + q_{\text{cola}}$

$q = \text{mass} \times \text{specific heat} \times \Delta t$

$$q_{\text{cans}} = \frac{6 \ \cancel{\text{cans}}}{1} \times \frac{50.0 \ \cancel{\text{g Al}}}{1 \ \cancel{\text{can}}} \times \frac{0.902 \ \text{J}}{\cancel{\text{g} \cdot °\text{C}}} \times \frac{\left(10 \ \cancel{°\text{C}} - 30 \ \cancel{°\text{C}}\right)}{1}$$

$$= -5412.00 \ \text{J}$$

$$q_{\text{cola}} = \frac{6 \ \cancel{\text{cans}}}{1} \times \frac{355 \ \cancel{\text{mL cola}}}{1 \ \cancel{\text{can}}} \times \frac{1.23 \ \cancel{\text{g cola}}}{1 \ \cancel{\text{mL cola}}} \times \frac{4.00 \ \text{J}}{\cancel{\text{g} \cdot °\text{C}}} \times \frac{\left(10 \ \cancel{°\text{C}} - 30 \ \cancel{°\text{C}}\right)}{1}$$

$$= -2.10 \times 10^5 \ \text{J}$$

$$q_{\text{six-pack}} = -5412 \ \text{J} + \left(-2.10 \times 10^5 \ \text{J}\right)$$

$$= -2.15 \times 10^5 \ \text{J}, \text{ or } -215 \ \text{kJ}$$

Therefore, the amount of heat absorbed by the ice is **215 kJ.**

(b) Restatement: Minimum amount of ice required to accomplish the necessary cooling.

For the system to reach 10°C, the heat absorbed *includes* the warming of the water.

Let x = mass of ice

$$215 \ \text{kJ} = x \left[\left(\frac{6.00 \ \text{kJ / mole}}{18.02 \ \text{g / mole}} \right) \right] + \left(4.184 \ \cancel{\text{J}}/\text{g} \cdot °\text{C} \times 10 \ °\text{C} \right)$$

$$x = \frac{215 \ \text{kJ}}{0.375 \ \text{kJ/g}} = 573 \ \textbf{g ice}$$

The Gas Laws

Key Terms

Words that can be used as topics in essays:

absolute zero

Avogadro's law

Boyle's law

Charles' law

combined gas law

Dalton's law

diffusion

effusion

Gay-Lussac's law

Graham's law of effusion

ideal gas

ideal gas law

kinetic theory

molar volume

mole fraction

partial pressure

real gas

root-mean-square velocity

STP

translational energy

van der Waals equation

Key Concepts

Equations and relationships that you need to know:

- 1 atm = 760 mm Hg = 760 torr = 101.3 kPa = 14.7 lb/in^2

  K = °C + 273.15

- Avogadro's law: $V_1 n_2 = V_2 n_1$

  Boyle's law: $P_1 V_1 = P_2 V_2$

  Charles' law: $V_1 T_2 = V_2 T_1$

  combined gas law: $\dfrac{P_1 V_1}{T_1} = \dfrac{P_2 V_2}{T_2}$

  Dalton's law of partial pressures: $P_{total} = P_1 + P_2 + P_3 + \ldots$

  derivation: $P_i = \dfrac{n_i}{n_{total}} \times P_{total}$

  density $= \dfrac{g}{V} = \dfrac{P \times MW}{R \cdot T}$

  Gay-Lussac's law: $P_1 T_2 = P_2 T_1$

ideal gas law: $PV = nRT$

$R = 0.0821$ liter $\cdot$ atm / mole $\cdot$ K

$\quad = 8.31$ liter $\cdot$ kPa / mole $\cdot$ K

$\quad = 8.31$ J / mole $\cdot$ K

$\quad = 8.31$ $V \cdot C$ / mole $\cdot$ K

$\quad = 8.31 \times 10^{-7}$ g $\cdot$ cm^2 / sec^2 $\cdot$ mole $\cdot$ K

(for calculating the average speed of molecules)

$\quad = 6.24 \times 10^4$ L $\cdot$ mm Hg / mole $\cdot$ K

$\quad = 1.99$ cal / mole $\cdot$ K

molecular weight $= MW = \dfrac{g \cdot R \cdot T}{P \cdot V}$

van der Waals (real gases): $(P + a/V^2)(V - b) = R \cdot T$

$$\text{or}$$

$$(P + n^2 a/V^2)(V - nb) = n \cdot R \cdot T$$

Here a corrects for force of attraction between gas molecules, and b corrects for particle volume.

- Graham's law of effusion:

$$\frac{r_1}{r_2} = \frac{\sqrt{d_2}}{\sqrt{d_1}} = \frac{\sqrt{MW_2}}{\sqrt{MW_1}} = \frac{t_2}{t_1} = \frac{u_1}{u_2}$$

where

$\quad r = $ rate of effusion

$\quad d = $ density

$MW = $ molecular weight

$\quad t = $ time

$\quad u = $ average speed

- **Kinetic Molecular Theory**

1. Gases are composed of tiny, invisible molecules that are widely separated from one another in empty space.

2. The molecules are in constant, continuous, random, and straight-line motion.

3. The molecules collide with one another, but the collisions are perfectly elastic (no net loss of energy).

4. The pressure of a gas is the result of collisions between the gas molecules and the walls of the container.

5. The average kinetic energy of all the molecules collectively is directly proportional to the absolute temperature of the gas. Equal numbers of molecules of any gas have the same average kinetic energy at the same temperature.

- $E_t = \dfrac{m \cdot u^2}{2} = cT$

$c = \dfrac{3R}{2N_A}$

$u^2 = \dfrac{3 \cdot R \cdot T}{m \cdot N_A} = \dfrac{3 \cdot R \cdot T}{\text{MW}}$

where

E_t = average kinetic energy of translation

m = mass (of particle)

u = velocity (average speed)

c = constant

N_A = Avogadro's number

R = 8.31×10^{-7} g $\cdot$ cm^2/sec^2 $\cdot$ mole $\cdot$ K

T = temperature in K

Samples: Multiple-Choice Questions

---

1. Which of the following would express the approximate density of carbon dioxide gas at 0°C and 2.00 atm pressure (in grams per liter)?

   A. 2 g/L

   B. 4 g/L

   C. 6 g/L

   D. 8 g/L

   E. None of the above

Answer: B

First, calculate the volume the CO_2 gas would occupy at 2.00 atm using the relationship

$$\frac{P_1 V_1}{T_1} = \frac{P_2 V_2}{T_2}$$

Since the temperature is remaining constant, we can use $P_1 V_1 = P_2 V_2$, where initial conditions are at STP and final conditions are at 0°C and 2.00 atm.

$$(1.00 \text{ atm})(22.4 \text{ liters}) = (2.00 \text{ atm})(V_2)$$
$$V_2 = 11.2 \text{ liters}$$

Since the amount of gas has not changed from the initial STP conditions (1 mole or 44.01 grams), the density of the gas at 2.00 atm and 0°C would be

$$\frac{44.01 \text{ grams}}{11.2 \text{ liters}} \approx 4 \text{ g/L}$$

Another approach to this problem would be to use the ideal gas law, $PV = nRT$.

$$\text{density} = \frac{g}{V} = \frac{P \cdot MW}{RT}$$

$$= \frac{200 \, \cancel{\text{atm}} \cdot 44.01 \, g \cdot \cancel{\text{mole}^{-1}}}{0.08211 \cdot \cancel{\text{atm}} \cdot \cancel{\text{mole}^{-1}} \cdot \cancel{\text{K}^{-1}} \cdot 273 \, \cancel{\text{K}}} \approx 4 \, g/L$$

---

2. The combustion of carbon monoxide yields carbon dioxide. The volume of oxygen gas needed to produce 22 grams of carbon dioxide at STP is

   A. 4.0 liters

   B. 5.6 liters

   C. 11 liters

   D. 22 liters

   E. 32 liters

Answer: B

Begin by writing down a balanced equation.

$$2CO(g) + O_2(g) \rightarrow 2\ CO_2(g)$$

Next, use the factor-label method to solve the problem.

$$\frac{22\ \cancel{g\ CO_2}}{1} \times \frac{1\ \cancel{mole\ CO_2}}{44\ \cancel{g\ CO_2}} \times \frac{1\ \cancel{mole\ O_2}}{2\ \cancel{moles\ CO_2}} \times \frac{22.4\ L\ O_2}{1\ \cancel{mole\ O_2}} = 5.6\ L\ O_2$$

Hint: Cancel the 22 and the 44 first, leaving 1/4 times 22.4 = 5.61 L.

---

3. If the average velocity of a methane molecule, CH_4 (MW = 16), is 5.00×10^4 cm/sec at 0°C, what is the average velocity of helium molecules at the same temperature and pressure conditions?

   A. 2.50×10^4 cm/sec

   B. 5.00×10^4 cm/sec

   C. 1.00×10^5 cm/sec

   D. 2.00×10^5 cm/sec

   E. 5.00×10^5 cm/sec

Answer: C

Graham's law of effusion:

$$\frac{u_1}{u_2} = \frac{\sqrt{MW_2}}{\sqrt{MW_1}}$$

$$\frac{5.00 \times 10^4}{x} = \frac{\sqrt{4}}{\sqrt{16}}$$

$$x = 1.00 \times 10^5\ cm/sec$$

---

4. When 2.00 grams of a certain volatile liquid is heated, the volume of the resulting vapor is 821 mL at a temperature of 127°C at standard pressure. The molecular weight of this substance is

   A. 20.0 g/mole

   B. 40.0 g/mole

   C. 80.0 g/mole

   D. 120. g/mole

   E. 160. g/mole

Answer: C

Begin this problem by listing the known facts.

$$m = 2.00 \text{ g} \qquad V = 0.821 \text{ liter} \qquad T = 400. \text{ K} \qquad P = 1.00 \text{ atm} \qquad MW = ?$$

You will need to use the ideal gas law to solve the problem: $PV = nRT$. Because moles can be calculated by dividing the mass of the sample by its molecular weight, the ideal gas law becomes

$$PV = \frac{m}{MW} R \cdot T$$

Solving for MW yields

$$MW = \frac{m \cdot R \cdot T}{P \cdot V} = \frac{(2.00 \text{ g})(0.0821 \, \text{liter} \cdot \text{atm})(400. \, \text{K})}{(1.00 \, \text{atm})(0.821 \, \text{liter}) \cdot \text{mole} \cdot \text{K}} = 80.0 \text{ g/mole}$$

---

5. A sample of zinc metal reacts completely with excess hydrochloric acid according to the following equation:

$$Zn_{(s)} + 2HCl_{(aq)} \rightarrow ZnCl_{2(aq)} + H_{2(g)}$$

8.00 liters of hydrogen gas at 720. mm Hg is collected over water at 40.°C (vapor pressure of water at 40.°C = 55 mm Hg). How much zinc was consumed by the reaction?

A. $\dfrac{(720/760) \cdot 8.00}{(0.0821) \cdot 313}$

B. $\dfrac{(760/720) \cdot 313}{(0.0821) \cdot 2}$

C. $\dfrac{(665/760) \cdot 8.00 \cdot (65.39)}{0.0821 \cdot 313}$

D. $\dfrac{(665/760) \cdot 8.00}{(65.39) \cdot (0.0821) \cdot 313}$

E. $\dfrac{8.00 \cdot 313 \cdot 65.39}{(665/760) \cdot (0.0821)}$

Answer: C

Begin by listing the information that is known.

$V = 8.00$ liters H_2

$P = 720.$ mm Hg $- 55$ mm Hg $= 665$ mm Hg (corrected for vapor pressure)

$T = 40.°C + 273 = 313 K$

Using the ideal gas law, $PV = nRT$, and realizing that one can determine grams from moles, the equation becomes

$$n H_2 = \frac{PV}{RT} = \frac{(665/760 \, \text{atm}) \cdot 8.00 \, \text{L} \, H_2}{(0.0821 \, \text{L atm} / \text{mole} \cdot \text{K}) \cdot 313 \, \text{K}}$$

Since for every mole of hydrogen produced, one mole of zinc is consumed, the last step would be to convert these moles to grams by multiplying by the molar mass of zinc.

$$\frac{\text{moles } H_2}{1} \cdot \frac{1 \text{ mole Zn}}{1 \text{ mole } H_2} \cdot \frac{65.39 \text{ g Zn}}{1 \text{ mole Zn}} = \frac{(665/760) \cdot 8.00 \cdot 65.39}{(0.0821) \cdot 313}$$

---

6. What is the partial pressure of helium when 8.0 grams of helium and 16 grams of oxygen are in a container with a total pressure of 5.00 atm?

    A. 0.25 atm

    B. 1.00 atm

    C. 1.50 atm

    D. 2.00 atm

    E. 4.00 atm

Answer: E

Use the formula

$$P_1 = \frac{n_1}{n_{\text{total}}} P_{\text{total}}$$

derived from Dalton's law of partial pressures. Find the number of moles of the two gases first.

$$\frac{8.0 \text{ g He}}{1} \times \frac{1 \text{ mole He}}{4.0 \text{ g He}} = 2.0 \text{ moles He}$$

$$\frac{16 \text{ g } O_2}{1} \times \frac{1 \text{ mole } O_2}{32 \text{ g } O_2} = 0.50 \text{ mole } O_2$$

$n_{\text{total}} = 2.0 \text{ moles} + 0.50 \text{ mole} = 2.5 \text{ moles}$

$$P_{\text{He}} = \frac{2.0 \text{ moles}}{2.5 \text{ moles}} \times 5.00 \text{ atm} = 4.00 \text{ atm}$$

---

7. For a substance that remains a gas under the conditions listed, deviation from the ideal gas law would be most pronounced at

    A. −100°C and 5.0 atm

    B. −100°C and 1.0 atm

    C. 0°C and 1.0 atm

    D. 100°C and 1.0 atm

    E. 100°C and 5.0 atm

Answer: A

The van der Waals constant *a* corrects for the attractive forces between gas molecules. The constant *b* corrects for particle volume. The attractive forces between gas molecules become pronounced when the molecules are closer together. Conditions which favor this are low temperatures (−100°C) and high pressures (5.0 atm).

8. 100 grams of $O_2(g)$ and 100 grams of He(g) are in separate containers of equal volume. Both gases are at 100°C. Which one of the following statements is true?

  A. Both gases would have the same pressure.

  B. The average kinetic energy of the O_2 molecules is greater than that of the He molecules.

  C. The average kinetic energy of the He molecules is greater than that of the O_2 molecules.

  D. There are equal numbers of He molecules and O_2 molecules.

  E. The pressure of the He(g) would be greater than that of the $O_2(g)$.

Answer: E

Oxygen gas weighs 32 grams per mole, whereas helium gas weighs only 4 grams per mole. One can see that there are roughly 3 moles of oxygen molecules and 25 moles of helium molecules. Gas pressure is proportional to the number of molecules and temperature, and inversely proportional to the size of the container. Since there are more helium molecules, you would expect a higher pressure in the helium container (with all other variables being held constant). As long as the temperatures of the two containers are the same, the average kinetic energies of the two gases are the same.

9. Which one of the manometers below represents a gas pressure of 750 mm Hg? (Atmospheric pressure is 760 mm Hg.)

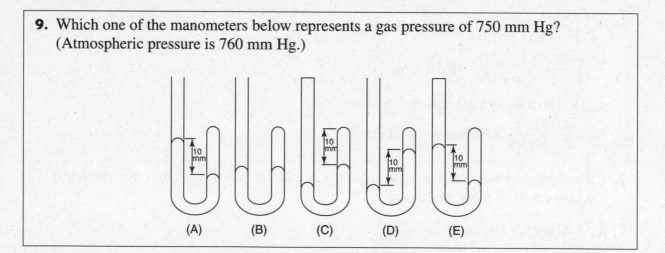

Answer: D

Manometer (D), the correct answer, shows the air pressure to be 10 mm Hg greater than that of the gas. Manometer (A) shows the gas pressure to be 10 mm Hg greater than air pressure. Manometer (B) shows the gas pressure to be equal to that of the air pressure. Manometer (C) is a closed manometer showing that the pressure of the gas is 10 mm Hg greater than the unknown pressure on the left side of the manometer. Manometer (E) is also closed and shows that the gas on the right side (770 mm Hg) is exerting a pressure 10 mm Hg greater than the gas on the left side.

Samples: Free-Response Questions

1. Assume that 185.00 grams of fluorine gas and 4.0 moles of xenon gas are contained in a flask at 0°C and 2.5 atm of pressure.

   (a) Calculate (1) the volume of the flask and (2) the partial pressure of each gas.

   (b) 23.00 grams of lithium metal is introduced into the flask, and a violent reaction occurs in which one of the reactants is entirely consumed. What weight of lithium fluoride is formed?

   (c) Calculate the partial pressures of any gas(es) present after the reaction in part (b) is complete and the temperature has been brought back to 0°C. (The volumes of solid reactants and products may be ignored.)

Answer

1. Given: 185.00 $F_2(g)$ 4.0 moles $Xe(g)$ $T = 273K$ $P = 2.5$ atm

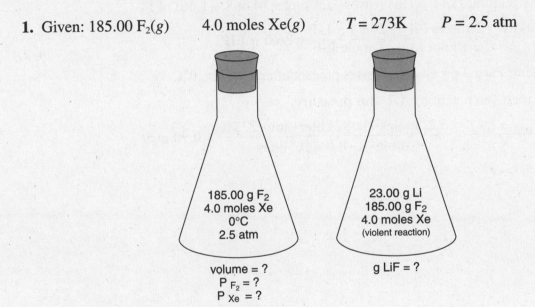

volume = ?
P_{F_2} = ?
P_{Xe} = ?

g LiF = ?

(a) (1) Restatement: Volume of flask.

$$PV = nRT$$

n = total moles = moles F_2 + moles xenon

$$\text{moles } F_2 = \frac{185.00 \text{ g } F_2}{1} \times \frac{1 \text{ mole } F_2}{38.00 \text{ g } F_2} = 4.868 \text{ moles } F_2$$

total moles of gas = 4.868 + 4.0 = 8.9 moles

Solve for the volume.

$$V = \frac{nRT}{P} = \frac{8.9 \text{ moles} \cdot (0.0821 \text{ liter} \cdot \text{atm}) \cdot 273 K}{\text{mole} \cdot K \cdot 2.5 \text{ atm}}$$

$$= 8.0 \times 10^1 \text{ liters}$$

(a) (2) Restatement: Find partial pressure of each gas.

$$P_{total} = P_{fluorine} + P_{xenon}$$

$$P_{flourine} = \frac{n_{xenon} \cdot R \cdot T}{V} = \frac{4.868 \; \cancel{moles} \cdot 0.0821 \; \cancel{liter} \cdot atm \cdot 273 \; \cancel{K}}{\cancel{mole} \cdot \cancel{K} \cdot 8.0 \times 10^1 \; \cancel{liters}} = \textbf{1.4 atm}$$

$$P_{xenon} = \frac{n_{xenon} \cdot R \cdot T}{V} = \frac{4.0 \; \cancel{moles} \cdot 0.0821 \; \cancel{liter} \cdot atm \cdot 273 \; \cancel{K}}{\cancel{mole} \cdot \cancel{K} \cdot 8.0 \times 10^1 \; \cancel{liters}} = \textbf{1.1 atm}$$

(b) Restatement: 23.00 g Li added to flask. Weight of LiF formed?

Xe is inert (no reaction with lithium).

The balanced equation for the reaction is thus $F_2(g) + 2Li(s) \rightarrow 2\,LiF(s)$.

$$\frac{23.00 \; \cancel{g\,Li}}{1} \times \frac{1 \; mole \; Li}{6.94 \; \cancel{g\,Li}} = 3.31 \; mole \; Li$$

Therefore, lithium is the limiting reagent. So 3.31 moles of Li is used up along with 3.31/2 or 1.66 moles of F_2. This leaves 3.21 moles (4.868 − 1.66) of F_2.

$$\frac{3.31 \; \cancel{moles\,Li}}{1} \times \frac{2 \; \cancel{moles\,LiF}}{2 \; \cancel{moles\,Li}} \times \frac{25.94 \; g \; LiF}{1 \; \cancel{mole\,LiF}} = \textbf{86.0 g LiF}$$

(c) Restatement: Partial pressures of gases present after reaction, 0°C

Xenon is inert (no reaction): **1.1 atm pressure**

$$P_{fluorine} = \frac{n_{fluorine} \cdot R \cdot T}{V} = \frac{3.21 \; \cancel{moles} \cdot 0.0821 \; \cancel{liter} \cdot atm \cdot 273 \; \cancel{K}}{\cancel{mole} \cdot \cancel{K} \cdot 8.0 \times 10^1 \; \cancel{liters}} = \textbf{0.90 atm}$$

2. Three students, Mason, Amir, and Kamyar, measured an empty Erlenmeyer flask and a tight-fitting stopper and found the mass to be 62.371 g. The students then carefully measured out 4.4 mL of a volatile liquid, poured the liquid into the flask, and heated the flask containing the liquid in a 101.1°C boiling water bath. As soon as all the liquid had vaporized, the students covered the flask with the stopper and set it aside to cool. Then, after a few minutes, they very quickly removed and replaced the stopper and reweighed the flask. The mass of the flask, stopper, and condensed vapor was 63.088 g. At the conclusion of the experiment, the students rinsed out the flask, filled it with water to the top, replaced the tight-fitting stopper (being sure there were no air bubbles), and obtained a volume of 261.9 mL. The barometric air pressure that day was 733 mm Hg.

   (a) Calculate the pressure of the vapor inside the flask (in atm) after the vapor had cooled and the students opened the flask momentarily.

   (b) What was the mass of the vapor in the flask?

   (c) Calculate the mass of 1 mole of the vapor.

   (d) Explain how each of the following errors in the laboratory procedure would affect the calculation of the molecular mass.

   (1) The volatile liquid contained nonvolatile impurities.

   (2) The students removed the flask from the water bath before all of the liquid had vaporized.

   (3) There was a hole in the stopper.

   (4) There were a few drops of water left on the flask from the water bath when the final mass was taken.

Answer

Begin this problem by numbering the problem and listing the critical information, getting rid of superfluous words, and clarifying the situation in your mind.

   2. Given: 62.371 g = Erlenmeyer flask + stopper

   4.4 mL of volatile liquid

   101.1°C boiling water bath = 374.1 K

   63.088 g = flask, stopper, condensed vapor

   261.9 mL = volume of stoppered flask

   733 mm Hg = barometric pressure

Then, number each section; restate the question in as few words as possible; list the necessary information; write down any generic formulas; substitute the data from the problem into the generic formulas; do the math; and underline or box your answer.

   (a) Restatement: Pressure inside of flask after vapor cooled and the flask was opened momentarily.

$$\frac{733 \; \text{mm Hg}}{1} \times \frac{1 \; \text{atm}}{760 \; \text{mm Hg}} = \textbf{0.964 atm}$$

The students opened the flask and thus made the pressure inside the flask equal to the pressure outside it.

(b) Restatement: Mass of vapor inside the flask.

63.088 g − 62.371 g = **0.717 g**

(c) Restatement: Mass of 1 mole of vapor.

$$PV = nRT \rightarrow PV = \frac{g \cdot R \cdot T}{MW}$$

$$MW = \frac{g \cdot R \cdot T}{P \cdot V}$$

$$= \frac{(0.717\,g) \cdot (0.0821\,\cancel{liter} \cdot \cancel{atm}) \cdot (374.1\,\cancel{K})}{mole \cdot \cancel{K}\,(0.964\,\cancel{atm}) \cdot (0.2619\,\cancel{liter})}$$

$$= \mathbf{87.2\ g/mole}$$

(d) Restatement: How could each of these mistakes affect the molecular mass of the liquid?

(1) The volatile liquid contained nonvolatile impurities.

The molecular mass would be **too high** because the nonvolatile impurities would contribute additional mass. The contribution to volume would be negligible.

(2) The students removed the flask from the water bath before all the liquid had vaporized.

The mass of the condensed vapor would be too high. The excess mass would be due to both the vapor and the mass of liquid that had not vaporized. Examine the equation for determining the molecular weight.

$$MW = \frac{g \cdot R \cdot T}{P \cdot V}$$

The value for g would be too high, so the calculated MW would also be **too high.**

(3) There was a hole in the stopper.

The mass of the vapor would be too small, because some of the vapor would have escaped through the hole in the stopper before condensing on the flask. The variable *g* would be smaller than expected, resulting in a molecular mass **lower than expected.**

(4) A few drops of water were left on the flask from the water bath when the final mass was taken.

The mass of the condensate would be too large, because it would include both the mass of the condensate and the mass of the water left on the flask. The variable *g* would be larger than expected, resulting in a molecular mass **higher than expected.**

Electronic Structure of Atoms

Key Terms

Words that can be used as topics in essays:

amplitude, ψ

atomic radii

atomic spectrum

Aufbau principle

Balmer series, $n_{lo} = 2$

Bohr model

continuous spectrum

de Broglie relation

degenerate orbital

diamagnetic

effective nuclear charge, Z_{eff}

electromagnetic radiation

electron affinity

electron configuration

electronegativity

electron spin

emission spectra

excited state

frequency, ν

ground state

Heisenberg uncertainty principle

Hund's rule

ionic radii

ionization energy

isoelectronic

line spectrum

Lyman series, $n_{lo} = 1$

nodal surface

orbital

orbital diagram

paramagnetic

Paschen series, $n_{lo} = 3$

Pauli exclusion principle

penetration effect

photon

probability distribution

quantization

quantum mechanics

quantum numbers

shielding

Schrödinger equation

valence electrons

wave function

wavelength, λ

wave mechanical model

wave-particle duality of nature

Key Concepts

Equations and relationships that you need to know:

- **Postulates of the Quantum Theory**

  1. Atoms and molecules can exist only in discrete states, characterized by definite amounts of energy. When an atom or molecule changes state, it absorbs or emits just enough energy to bring it to another state.

  2. When atoms or molecules absorb or emit light in moving from one energy state to another, the wavelength of the light is related to the energies of the two states as follows:

     $$E_{high} - E_{low} = \frac{h \cdot c}{\lambda}$$

  3. The allowed energy states of atoms and molecules are described by quantum numbers.

- h = Planck's constant $= 6.626 \times 10^{-34} \, \dfrac{J \cdot s}{particle}$

  $$= 6.626 \times 10^{-27} \, erg \cdot sec$$

  c = speed of light $= 2.998 \times 10^{10}$ cm/sec

  $$\Delta E = \frac{1.196 \times 10^5 \, kJ \cdot nm}{\lambda \cdot mole}$$

  $$E = \frac{-1312 kJ}{n^2 \cdot mole} = h \cdot v = \frac{2.18 \times 10^{-11} \, erg}{n^2} = mc^2$$

  $$= \left(\frac{Z_{eff}}{n^2}\right) 1312 \, kJ/mole$$

  $$Z_{eff} = Z - \sigma$$

  where Z_{eff} = effective nuclear charge, Z = actual nuclear charge, and σ = shielding or screening constant

  Rydberg-Ritz equation: $E_n = -R_H \left(\dfrac{1}{n^2}\right)$

  $$\Delta E = \frac{h \cdot c}{\lambda} = h \cdot v = R_H \left(\frac{1}{n_i^2} - \frac{1}{n_f^2}\right)$$

  n_i and n_f are quantum numbers

  $R_H = 2.18 \times 10^{-18} \, J = 109,737 \, cm^{-1}$

  $$m \cdot r \cdot v = \frac{n \cdot h}{2\pi}$$

  $$E_b - E_a = \frac{z^2 e^2}{2a_0} \left[\frac{1}{n_a^2} - \frac{1}{n_b^2}\right]$$

  where n = quantum energy level

  $\quad\quad E$ = energy (at states a and b)

  $\quad\quad$ e = charge on electron

  $\quad\quad a_0$ = Bohr radius

  $\quad\quad z$ = atomic radius

n = principal energy level

l = angular momentum = sublevel (s, p, d, and f)

m_l = magnetic quantum number = orientation of orbital

m_s = electron spin quantum number

- **Predicted Electron Configuration**

$1s^2\ 2s^2\ 2p^6\ 3s^2\ 3p^6\ 4s^2\ 3d^{10}\ 4p^6\ 5s^2\ 4d^{10}\ 5p^6\ 6s^2\ 4f^{14}\ 5d^{10}\ 6p^6\ 7s^2\ 5f^{14}\ 6d^{10}$

Common Exceptions	
chromium	$4s^1 3d^5$
copper	$4s^1 3d^{10}$
molybdenum	$5s^1 4d^{10}$
silver	$5s^1 4d^{10}$
gold	$6s^1 4f^{14} 5d^{10}$

- **Important Flame Test Colors***

Li^+	red	
Na^+	yellow	
K^+	violet	Group IA
Rb^+	purple	
Cs^+	blue	
Ca^{2+}	red	
Sr^{2+}	crimson	Group IIA
Ba^{2+}	green	

*For a complete list of flame test colors, turn to Appendix C.

- **Transition Metal Cations (in aqueous solution)**

Ag^+	colorless	Fe^{3+}	pale yellow
Cd^+	colorless	Hg^{2+}	colorless
Co^{2+}	pink	Mn^{2+}	pale pink
Cr^{3+}	purple	Ni^{2+}	green
Cu^{2+}	blue	Zn^{2+}	colorless
Fe^{2+}	pale green		

- **Polyatomic Anions (in aqueous solution)**

CrO_4^{2-}	yellow	$Cr_2O_7^{2-}$	orange

Samples: Multiple-Choice Questions

1. Which of the following setups would be used to calculate the wavelength (cm) of a photon emitted by a hydrogen atom when the electron moves from the $n = 5$ state to the $n = 2$ state? (The Rydberg constant is $R_H = 2.18 \times 10^{-18}$ J. Planck's constant is $h = 6.63 \times 10^{-34}$ J · sec. The speed of light = 3.00×10^{10} cm/sec.)

   A. $\left(2.18 \times 10^{-18}\right)\left(\dfrac{1}{5^2} - \dfrac{1}{2^2}\right)\left(6.63 \times 10^{-34}\right)$

   B. $\dfrac{\left(6.63 \times 10^{-34}\right)\left(3.00 \times 10^{10}\right)}{\left(2.18 \times 10^{-18}\right)\left(\dfrac{1}{5^2} - \dfrac{1}{2^2}\right)}$

   C. $\dfrac{\left(2.18 \times 10^{-18}\right)\left(3.00 \times 10^{10}\right)}{\left(6.63 \times 10^{-34}\right)\left(\dfrac{1}{5^2} - \dfrac{1}{2^2}\right)}$

   D. $\dfrac{\left(2.18 \times 10^{-18}\right) / \left(3.00 \times 10^{10}\right)}{\left(6.63 \times 10^{-34}\right) / \left(\dfrac{1}{5^2} - \dfrac{1}{2^2}\right)}$

   E. $\dfrac{\left(2.18 \times 10^{-18}\right)\left(3.00 \times 10^{10}\right)}{\left(6.63 \times 10^{-34}\right) / \left(\dfrac{1}{5^2} - \dfrac{1}{2^2}\right)}$

Answer: B

The following relationships are needed to solve this problem:

$$\Delta E = R_H \left(\frac{1}{n_i^2} - \frac{1}{n_f^2}\right) \text{ and } \lambda = \frac{h \cdot c}{\Delta E}$$

Combining these equations to solve for λ gives the equation

$$\lambda = \frac{h \cdot c}{R_H \left(\dfrac{1}{n_i^2} - \dfrac{1}{n_f^2}\right)} = \frac{\left(6.63 \times 10^{-34} \, \cancel{J} \cdot \cancel{sec}\right)\left(3.00 \times 10^{10} \, cm \cdot sec^{-1}\right)}{\left(2.18 \times 10^{-18} \, \cancel{J}\right)\left(\dfrac{1}{5^2} - \dfrac{1}{2^2}\right)}$$

2. A Co^{3+} ion has _____ unpaired electron(s) and is _____.

   A. 1, diamagnetic

   B. 3, paramagnetic

   C. 3, diamagnetic

   D. 4, paramagnetic

   E. 10, paramagnetic

Answer: D

The electron configuration for the Co^{3+} ion is $1s^2 2s^2 2p^6 3s^2 3p^6 3d^6$. If you missed this, review electron configurations of transitional metals. The Co^{3+} ion would have a total of 24 electrons: 10 pairs of electrons and four unpaired electrons in the $3d$ orbitals. Atoms in which one or more electrons are unpaired are paramagnetic.

---

3. Which of the following series of elements is listed in order of increasing atomic radius?

    A. Na, Mg, Al, Si

    B. C, N O, F

    C. O, S, Se, Te

    D. I, Br, Cl, F

    E. K, Kr, O, Au

Answer: C

Atomic radius increases as one moves down a column (or group).

---

4. A characteristic that is unique to the alkali metals is

    A. Their metallic character

    B. The increase in atomic radius with increasing atomic number

    C. The decrease in ionization energy with increasing atomic number

    D. The noble gas electron configuration of the singly charged positive ion

    E. None of these answer choices is correct.

Answer: D

The word *unique* in this question means that *only* the alkali metals possess this particular characteristic. Of the choices listed, D is the only property that is unique to the alkali metals.

---

5. Which of the following elements most readily shows the photoelectric effect?

    A. Noble gases

    B. Alkali metals

    C. Halogen elements

    D. Transition metals

    E. The chalcogen family

Answer: B

The photoelectric effect is the emission of electrons from the surface of a metal when light shines on it. Electrons are emitted, however, only when the frequency of that light is greater than a certain threshold value characteristic of the particular metal. The alkali metals, with only one electron in their valence shells, have the lowest threshold values.

---

6. The lithium ion and the hydride ion are isoelectronic. Which of the following statements is true of these two chemical species in the ground state?

    A. Li^+ is a better reducing agent than H^-.

    B. The H^- ion is several times larger than the Li^+ ion.

    C. It requires more energy to remove an electron from H^- than from Li^+.

    D. The chemical properties of the two ions must be the same because they have the same electronic structure.

    E. None of these is a true statement.

Answer: B

Both the lithium ion, Li^+, and the hydride ion, H^-, have the configuration $1s^2$. Both species have 2 electrons, but the lithium ion has 3 protons, which cause a greater "pull" on the 2 electrons than the 1 proton found in the hydride ion.

---

7. Which of the following configurations represents a neutral transition element?

    A. $1s^2\, 2s^2\, 2p^2$

    B. $1s^2\, 2s^2\, 2p^6\, 3s^2\, 3p^4$

    C. $1s^2\, 2s^2\, 3s^2$

    D. $1s^2\, 2s^2\, 2p^6\, 3s^2\, 3p^6\, 3d^8\, 4s^2$

    E. $1s^2\, 2s^2\, 2p^6\, 3s^2\, 3p^6\, 3d^{10}\, 4s^2\, 4p^6$

Answer: D

The transition elements are filling the *d* orbitals. When completely filled, the *d* orbitals hold a maximum of 10 electrons.

8. The four quantum numbers (n, l, m_l, and m_s) that describe the valence electron in the cesium atom are

    A. 6, 0, −1, +1/2

    B. 6, 1, 1, +1/2

    C. 6, 0, 0, +1/2

    D. 6, 1, 0, +1/2

    E. 6, 0, 1, −1/2

Answer: C

The valence electron for the cesium atom is in the 6s orbital. In assigning quantum numbers, n = principal energy level = 6. The quantum number l represents the angular momentum (type of orbital) with s orbitals = 0, p orbitals = 1, d orbitals = 2, and so forth. In this case, $l = 0$. The quantum number m_l is known as the magnetic quantum number and describes the orientation of the orbital in space. For s orbitals (as in this case), m_l always equals 0. For p orbitals, m_l can take on the values of −1, 0, and +1. For d orbitals, m_l can take on the values −2, −1, 0, +1, and +2. The quantum number m_s is known as the electron spin quantum number and can take only two values, +1/2 and −1/2, depending on the spin of the electron.

9. When subjected to the flame test, a solution that contains K^+ ions produces the color

    A. yellow

    B. violet

    C. crimson

    D. green

    E. orange

Answer: B

Refer to Key Concepts for colors of flame tests.

10. Which of the following results indicates exothermic change in atoms?

    A. The production of line emission spectra

    B. The appearance of dark lines in absorption spectra

    C. The formation of ions from a metal in the gaseous state

    D. The production of isotopic species of CO^+ and CO_2^+ when CO_2 is placed in a mass spectrometer

    E. The absence of all spectral lines in the region of excitation

Answer: A

When atoms absorb energy, electrons move from ground states to excited levels. When the electrons move back to their ground state, they release energy (exothermic) at particular wavelengths (or frequencies).

---

11. An energy value of 3.313×10^{-12} ergs is needed to break a chemical bond. What is the wavelength of energy needed to break the bond? (The speed of light = 3.00×10^{10} cm/sec; Planck's constant = 6.626×10^{-27} erg · sec.)

    A. 5.00×10^{-4} cm

    B. 1.00×10^{-5} cm

    C. 2.00×10^{-5} cm

    D. 6.00×10^{-5} cm

    E. 1.20×10^{-5} cm

Answer: D

You need to know two relationships to do this problem. First,

$$\upsilon = \frac{E}{h} = \frac{3.313 \times 10^{-12} \,\text{erg}}{6.626 \times 10^{-27} \,\text{erg} \cdot \text{sec}} = 5.000 \times 10^{14} \,\text{sec}^{-1}$$

The second relationship you need to know is

$$\lambda = \frac{c}{\upsilon} = \frac{3.00 \times 10^{10} \,\text{cm} \cdot \text{sec}}{\text{sec} \cdot 5.000 \times 10^{14}} = 6.00 \times 10^{-5} \,\text{cm}$$

An alternative approach would be to use the relationship

$$\lambda = \frac{h \cdot c}{E} = \frac{6.626 \times 10^{-27} \,\text{erg} \cdot \text{sec} \cdot 3.00 \times 10^{10} \,\text{cm} \cdot \text{sec}^{-1}}{3.313 \times 10^{-12} \,\text{erg}}$$

---

12. A characteristic of the structure of metallic atoms is that

    A.   they tend to share their electrons with other atoms

    B.   their atoms are smaller and more compact than those of nonmetallic elements

    C.   their outermost orbital of electrons is nearly complete, and they attract electrons from other atoms

    D.   the small numbers of electrons in their outermost orbital are weakly held and easily lost

    E.   they have heavier nuclei than nonmetallic atoms

Answer: D

Metals lose their electrons readily to become positively charged ions with charges of +1, +2, or +3.

Samples: Free-Response Questions

> **1.** The electon configuration of an element determines its chemical properties. For the elements sodium, magnesium, sulfur, chlorine, and argon, provide evidence that illustrates this statement and show how the evidence supports the statement.

Answer

1. Restatement: Electron configuration and chemical properties for Na, Mg, S, Cl, and Ar.

This question lends itself to the outline format.

I. Sodium, Na

    A. Electron configuration $1s^2\,2s^2\,2p^6\,3s^1$

    B. Lewis diagram Na ·

    C. Alkali metal

    D. Chemical properties

        1. Loses valence electron easily

        2. Very reactive

        3. Low ionization energy

        4. Reacts with nonmetals to form ionic solid

        5. When reacting with nonmetal, metal acts as a reducing agent

II. Magnesium, Mg

    A. Electron configuration $1s^2\,2s^2\,2p^6\,3s^2$

    B. Lewis diagram Mg :

    C. Alkaline earth metal

    D. Chemical properties

        1. Forms only divalent compounds that are stable and have high heats of formation

        2. Very reactive only at high temperatures

        3. Powerful reducing agent when heated

        4. Reacts with most acids

        5. Burns rapidly in air (O_2)

III. Sulfur, S

    A. Electron configuration $1s^2\,2s^2\,2p^6\,3s^2\,3p^4$

    B. Lewis diagram

:S̈:

    C. Nonmetal

    D. Chemical properties

        1. Combines with almost all elements except gold, iodine, platinum, and inert gases

        2. Burns in air to give sulfur dioxide

        3. Stable at room temperature

        4. Oxidizing agent

IV. Chlorine, Cl

    A. Electron configuration $1s^2\,2s^2\,2p^6\,3s^2\,3p^5$

    B. Lewis diagram

$$:\overset{\displaystyle\cdot}{\underset{\displaystyle\cdot}{\text{Cl}}}:$$

    C. Nonmetal

    D. Chemical properties

        1. Diatomic in natural state: Cl_2

        2. Strong oxidizing agent

V. Argon, Ar

    A. Electron configuration $1s^2\,2s^2\,2p^6\,3s^2\,3p^6$

    B. Lewis diagram

$$:\overset{\displaystyle\cdot\cdot}{\underset{\displaystyle\cdot\cdot}{\text{Ar}}}:$$

    C. Nonmetal

    D. Chemical properties

        1. Inert, will not react with any element

---

2. (a) Write the ground-state electron configuration for a phosphorus atom.

  (b) Write the four quantum numbers that describe all the valence electrons in the phosphorus atom.

  (c) Explain whether a phosphorus atom, in its ground state, is paramagnetic or diamagnetic.

  (d) Phosphorus can be found in such diverse compounds as PCl_3, PCl_5, $PCl_4{}^+$, $PCl_6{}^-$, and P_4. How can phosphorus, in its ground state, bond in so many different arrangements? Be specific in terms of hybridization, type of bonding, and geometry.

Answer

2. (a) Restatement: Electron configuration of P.

$1s^2\, 2s^2\, 2p^6\, 3s^2\, 3p^3$

(b) Restatement: Quantum numbers for valence electrons in P.

Here's a good place to use the chart format.

Electron Number	n	l	m_l	m_s
11	3	0	0	+1/2
12	3	0	0	−1/2
13	3	1	1	+1/2
14	3	1	0	+1/2
15	3	1	−1	+1/2

(c) Restatement: P paramagnetic or diamagnetic? Explain.

Phosphorus is paramagnetic, because a paramagnetic atom is defined as having magnetic properties caused by *unpaired* electrons. The unpaired electrons are found in the $3p$ orbitals, each of which is half-filled.

(d) Restatement: Explain how PCl_3, PCl_5, PCl_4^+, PCl_6^-, and P_4 exist in nature.

Here's another question where the chart format is appropriate.

	PCl_3	PCl_5	PCl_4^+	PCl_6^-	P_4
Type of bond	covalent	covalent	covalent	covalent	covalent
Lewis structure					
Geometry	pyramidal	triangular bipyramidal	tetrahedral	octahedral	tetrahedral
Hybridization	sp^3	sp^3d	sp^3	sp^3d^2	sp^3

Key Terms

Words that can be used as topics in essays:

antibonds	interatomic forces
atomic radius	intermolecular forces
bond energy	Lewis structures
bond length	localized electron model (LE model)
bond order	lone pair
bond polarity	molecular orbital theory
coordinate covalent bond or dative bond	molecular structure (geometry)
delocalization	network covalent solid
delocalized pi bonding	octet rule
diagonal relationships	paramagnetism
diamagnetism	pi bonds
dipole moment	polar covalent bond
electron affinity	polarity
electronegativity	resonance
expanded octet theory	sigma bonds
formal charge	valence bond model
hybridization	valence shell electron pair repulsion
hydrogen bonding	model (VSEPR model)

Key Concepts

Equations and relationships that you need to know:

- bond order $= \dfrac{\text{number of bonding electrons} - \text{number of antibonding electrons}}{2}$

- **Geometry and Hybridization Patterns**

Number of Atoms Bonded to Central Atom X	Number of Unshared Pairs on X	Hybridization	Geometry	Example*
2	0	sp	linear	$\underline{C}O_2$
2	1	sp^2	bent	$\underline{S}O_2$
2	2	sp^3	bent	$H_2\underline{O}$
2	3	sp^3d	linear	$\underline{Xe}F_2$
3	0	sp^2	trigonal planar	$\underline{B}Cl_3$
3	1	sp^3	trigonal pyramidal	$\underline{N}H_3$
3	2	sp^3d	T-shaped	$\underline{Cl}F_3$
4	0	sp^3	tetrahedral	$\underline{C}H_4$
4	1	sp^3d	distorted tetrahedral	$\underline{S}Cl_4$
4	2	sp^3d^2	square planar	$\underline{Xe}F_4$
5	0	sp^3d	triangular bipyramidal	$\underline{P}Cl_5$
5	1	sp^3d^2	square pyramidal	$\underline{Cl}F_5$
6	0	sp^3d^2	octahedral	$\underline{S}F_6$

*Underlined atom is central atom, X.

- **Character of Bonds**

Electronegativity Difference	Type of Bond	Example
$0 \rightarrow 0.2$	nonpolar covalent	Br_2, HI, CH_4
$0.2 \rightarrow 1.7$	polar covalent	NO, LiH
1.7 or greater	ionic	LiBr, CuF

- formal charge = $\left(\begin{array}{c}\text{total number of}\\\text{valence electrons}\\\text{in the free atom}\end{array}\right) - \left(\begin{array}{c}\text{total number of}\\\text{nonbonding}\\\text{electrons}\end{array}\right) - \dfrac{1}{2}\left(\begin{array}{c}\text{total number of}\\\text{bonding electrons}\end{array}\right)$

    single bond = 1 sigma bond

    double bond = 1 sigma bond, 1 pi bond

    triple bond = 1 sigma bond, 2 pi bonds

Homonuclear diatomic molecules of second-period elements B_2, C_2, and N_2:

$$\sigma 1s < \sigma * 1s < \sigma 2s < \sigma * 2s < \pi 2p_y = \pi 2p_z < \sigma 2p_x < \pi * 2p_y = \pi * 2p_z < \sigma * 2p_x$$

For O_2 and F_2:

$$\sigma 1s < \sigma * 1s < \sigma 2s < \sigma * 2s < \sigma 2p_x < \pi 2p_y = \pi 2p_z < \pi * 2p_y = \pi * 2p_z < \sigma * 2p_x$$

Samples: Multiple-Choice Questions

For Samples 1–5, use the following choices:

   A Trigonal planar

   B Tetrahedral

   C Pyramidal

   D Bent

   E Linear

---

1. What is the geometry of S_2Cl_2?

Answer: D

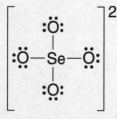

---

2. What is the geometry of OPN?

Answer: E

$$\ddot{\underset{..}{O}} = P = \ddot{\underset{}{N}}$$

---

3. What is the geometry of $SeO_4{}^{2-}$

Answer: B

$$\left[\begin{array}{c} :\ddot{O}: \\ :\ddot{O}-Se-\ddot{O}: \\ :\ddot{O}: \end{array} \right]^{2-}$$

---

4. What is the geometry of $SiO_3{}^{2-}$

Answer: A

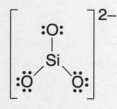

5. What is the geometry of the hydronium ion, H_3O^+?

Answer: C

$$\left[\begin{array}{c} H - \overset{\cdot\cdot}{O} - H \\ | \\ H \end{array} \right]^+$$

6. Which one of the following is a nonpolar molecule with one or more polar bonds?

    A. H—Br

    B. Cl—Be—Cl

    C. H—H

    D. H—O—H

    E. K—Cl

Answer: B

    A. $H - \overset{\cdot\cdot}{\underset{\cdot\cdot}{B}}r\!:$

    B. $:\!\overset{\cdot\cdot}{\underset{\cdot\cdot}{C}}l - Be - \overset{\cdot\cdot}{\underset{\cdot\cdot}{C}}l\!:$

    C. H—H

    D. $\overset{\displaystyle \cdot\overset{\cdot\cdot}{O}\cdot}{H \diagup \quad \diagdown H}$

    E. Compound is ionic, $K^+ :\!\overset{\cdot\cdot}{\underset{\cdot\cdot}{C}}l\!:^-$

The Cl atom is more electronegative than the Be atom, resulting in a polar bond. However, because the molecule is linear and the two ends are identical, the overall molecule is nonpolar.

7. How many total sigma bonds are in the benzene molecule, C_6H_6?

    A. 6

    B. 9

    C. 12

    D. 14

    E. 18

Answer: C

All single bonds are sigma (σ) in nature. Double bonds contain one sigma bond and one pi (π) bond. Triple bonds contain one sigma bond and two pi (π) bonds.

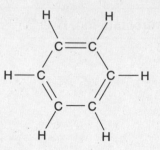

In the benzene molecule, there are 9 single bonds (9σ) and 3 double bonds (3σ and 3π) for a total of 12 sigma bonds.

---

8. Which one of the following does NOT exhibit resonance?

   A. SO_2

   B. SO_3

   C. HI

   D. CO_3^{2-}

   E. NO_3^-

Answer: C

There are no alternative ways of positioning electrons around the HI molecule. If you missed this question, refer to your textbook on the concept of resonance.

Note: Had you been given the choice, CH_3Br, you would not choose it due to a possible Baker-Nathan effect (no-bond resonance). In base, one result is protolysis:

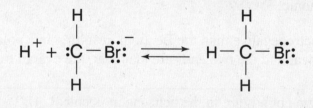

---

9. What type of hybridization would you expect to find in BCl_3?

   A. sp

   B. sp^2

   C. sp^3

   D. sp^3d^2

   E. No hybridization occurs in this molecule.

Answer: B

---

10. The electronegativity of carbon is 2.5, whereas that of oxygen is 3.5. What type of bond would you expect to find in carbon monoxide?

    A. Nonpolar covalent

    B. Polar covalent

    C. Covalent network

    D. Ionic

    E. Delta

Answer: B

Electronegativity differences less than 1.7 are classified as covalent. *Unequal* differences in *sharing* electrons are known as polar covalent.

---

11. Which one of the following contains a coordinate covalent bond?

    A. $N_2H_5^+$

    B. $BaCl_2$

    C. HCl

    D. H_2O

    E. NaCl

Answer: A

A coordinate covalent bond, also known as a dative bond, is a bond in which both electrons are furnished by one atom.

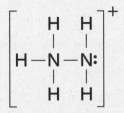

12. What is the formal number of pairs of unshared valence electrons in the NO_2^+ ion?

    A. 0

    B. 2

    C. 4

    D. 8

    E. 10

Answer: C

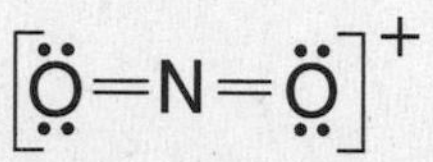

$$\left[\ddot{O} = N = \ddot{O} \right]^+$$

13. The bond energy of Br—Br is 192 kJ/mole, and that of Cl—Cl is 243 kJ/mole. What is the energy of the Cl—Br bond?

    A. 54.5 kJ/mole

    B. 109 kJ/mole

    C. 218 kJ/mole

    D. 435 kJ/mole

    E. 870 kJ/mole

Answer: C

If the polarity of the bond A—B is about the same as those of the nonpolar bonds A—A and B—B, then the bond energy of A—B can be taken as the average of the bond energies of A—A and B—B.

Samples: Free-Response Questions

> **1.** Given ClO_2^-, ClO_4^-, Cl_2O, ClO_3^-, and ClO_2.
>
> **(a)** Draw Lewis structures for all species.
>
> **(b)** Predict the bond angle for all species.
>
> **(c)** Predict the geometry for all species.
>
> **(d)** Predict the hybridization of the chlorine in all species.
>
> **(e)** Identify any of the species that might dimerize.
>
> **(f)** Identify any species that might be polar.

Answer

This problem is best done in the chart format. Note how the column headings record what is given and how the first entries in the rows serve as a restatement of what is wanted.

1.	ClO_2^-	ClO_4^-	Cl_2O	ClO_3^-	ClO_2
(a) Lewis structure					
(b) Bond angles	less than 109.5°	109.5°	less than 109.5°	less than 109.5°	less than 109.5°
(c) Geometry	bent	tetrahedral	bent	pyramidal	bent
(d) Hybridization	sp^3	sp^3	sp^3	sp^3	sp^3
(e) Dimerization	no	no	no	no	yes — unpaired electron will give
(f) Polarity	polar	nonpolar	polar	polar	polar

> **2.** As one moves down the halogen column, one notices that the boiling point increases. However, when examining the alkali metal family, one discovers that the melting point decreases as one moves down the column.
>
> **(a)** Account for the increase in boiling point of the halogens as one moves down the column.
>
> **(b)** Account for the decrease in melting point of the alkali metals as one moves down the column.
>
> **(c)** Rank Cs, Li, KCl, I_2, and F_2 in order of decreasing melting point, and explain your reasoning.

Answer

This answer might best be done in the bullet format.

2. (a) Restatement: Explain increase in B.P. of halogens as one moves down column.

- Halogens are nonmetals.

- Halogens are diatomic.

- Bonding found within halogen molecule is covalent — formed as a result of sharing of electrons.

- Forces found between halogen molecules are van der Waals forces, which are due to temporarily induced dipoles caused by polarization of electron clouds.

- Moving down the column, one would expect greater nuclear charge.

- Moving down the column, one would expect larger electron clouds due to higher energy levels being filled as well as greater atomic numbers and hence a greater number of electrons.

- Moving down halogen family, shielding effect and greater distance from nucleus would cause easier polarization of electron cloud.

- Therefore, greater polarization of electron cloud would cause greater attractive force (van der Waals force), resulting in higher boiling points.

- Furthermore, one must consider the effect of the molecular weight on the B.P. As the individual molecules become more and more massive, they need higher and higher temperatures to give them enough kinetic energy and velocity to escape from the surface and "boil."

(b) Restatement: Explain decrease in melting point of alkali metals as one moves down column.

- Alkali metal family are all metals.

- Metals have low electronegativity.

- Metals have low ionization energies.

- Metals exist in definite crystal arrangement — cations surrounded by "sea of electrons."

- As one moves down the alkali metal column, nuclear charge increases.

- As one moves down the alkali metal column, the electron cloud would be expected to get larger due to higher energy levels being filled.

- As one moves down the alkali metal family, the charge density would be expected to decrease due to significantly larger volume and more shielding.

- As one moves down the alkali metal family, one would expect the attractive forces holding the crystal structure together to decrease due to this last factor.

- Boiling point and melting point would be expected to be comparable because they both are functions of the strength of intermolecular attractive forces.

(c) Restatement: Rank Cs, Li, KCl, I_2, and F_2 in order of decreasing melting point. Explain.

- #1 KCl — highest melting point. Ionic bond present—formed by the transfer of electrons.

- #2 Li — alkali metal. Metallic bonds present (cations, mobile electrons). Low-density metal.

- #3 I_2 — solid at room temperature. Covalent bond present. Nonpolar.

- #4 Cs — liquid at near room temperature. Metallic bonds present; however, due to low charge density as explained above, attractive forces are very weak.

- #5 F_2 — gas at room temperature. Covalent bonds present. One would expect a smaller electron cloud than in I_2 due to reasons stated above.

Ionic Bonding

Key Terms

Words that can be used as topics in essays:

anions

atomic orbital model

bond energy

cation

chelates

complex ion

coordination number

Coulomb's law

crystal-field model

crystal-field splitting energy

dissociation constant, K_d

electronegativity

free radical

high-spin complex

inert complex

ionic radius

isoelectronic

labile complex

lattice energy

ligand

low-spin complex

noble-gas structure

orbital diagram

polarity

polydentate ligand

spectrochemical series

valence electron

Key Concepts

Equations and relationships that you need to know:

- $\% \text{ ionic character} = \dfrac{\text{measured dipole moment of } X^+Y^-}{\text{calculated dipole moment of } X^+Y^-} \cdot 100\%$

net dipole moment = charge × distance

Coulomb's law: $E = 2.31 \times 10^{-19} \, \text{J} \cdot \text{nm} \left(\dfrac{Q_1 Q_2}{r} \right)$

lattice energy $= k \left(\dfrac{Q_1 Q_2}{r} \right)$ $k = \text{constant}$

Q_1 and Q_2 — charges on the ions

r = shortest distance between centers of cations and anions

- ## Ligand Nomenclature

Br^-	bromo	I^-	iodo
$C_2O_4^{2-}$	oxalato	NH_3	amine
CH_3NH_2	methylamine	NO	nitrosyl
Cl^-	chloro	O^{2-}	oxo
CN^-	cyano	OH^-	hydroxo
CO	carbonyl	SO_4^{2-}	sulfato
CO_3^{2-}	carbonato	en	ethylenediamine
F^-	fluoro	EDTA	ethylenediamine-tetraaceto
H_2O	aquo		

- ## Coordination Numbers for Common Metal Ions

Ag^+	2	Co^{2+}	4, 6	Au^{3+}	4
Cu^+	2, 4	Cu^{2+}	4, 6	Co^{3+}	6
Au^+	2, 4	Fe^{2+}	6	Cr^{3+}	6
		Mn^{2+}	4, 6	Sc^{3+}	6
		Ni^{2+}	4, 6		
		Zn^{2+}	4, 6		

- ## Simple Cubic, Body-Centered Cubic, and Face-Centered Cubic Cells

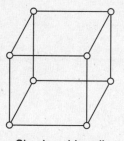

Simple cubic cell

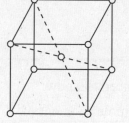

Body-centered cubic cell

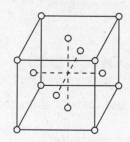

Face-centered cubic cell

- ## Common Polyatomic Ions

+1

ammonium	NH_4^+

+2

mercury (I)	Hg_2^{2+}

−1

acetate	$C_2H_3O_2^-$	hydrogen sulfite	
amide	NH_2^-	or bisulfite	HSO_3^-
azide	N_3^-	hydrosulfide	
benzoate	$C_7H_5O_2^-$	or bisulfide	HS^-
bitartrate	$HC_4H_4O_6^-$	hydroxide	OH^-
bromate	BrO_3^-	hypochlorite	ClO^-
carbonate		iodate	IO_3^-
or bicarbonate	HCO_3^-	monobasic phosphate	
chlorate	ClO_3^-	or dihydrogen	
chlorite	ClO_2^-	phosphate	$H_2PO_4^-$
cyanate	OCN^-	nitrate	NO_3^-
cyanide	CN^-	nitrite	NO_3^-
hydrazide	$N_2H_3^-$	perchlorate	ClO_4^-
hydrogen		permanganate	MnO_4^-
hydrogen sulfate		thiocyanate	SCN^-
or bisulfate	HSO_4^-	triiodide	I_3^-

−2

carbonate	CO_3^{2-}	metasilicate	SiO_3^{2-}
chromate	CrO_4^{2-}	oxalate	$C_2O_4^{2-}$
dibasic phosphate		peroxide	O_2^{2-}
or hydrogen		phthalate	$C_8H_4O_4^{2-}$
phosphate	HPO_4^{2-}	sulfate	SO_4^{2-}
dichromate	$Cr_2O_7^{2-}$	sulfite	SO_3^{2-}
disulfate		tartrate	$C_4H_4O_3^{2-}$
or pyrosulfate	$S_2O_7^{2-}$	tetraborate	$B_4O_7^{2-}$
manganate	MnO_3^{2-}	thiosulfate	$S_2O_3^{2-}$

−3

aluminate	AlO_3^{3-}	ferricyanide	$Fe(CN)_6^{3-}$
arsenate	AsO_4^{3-}	phosphate	
arsenite	AsO_3^{3-}	or tribasic	
borate	BO_3^{3-}	phosphate	PO_4^{3-}
citrate	$C_6H_5O_7^{3-}$	phosphite	PO_3^{3-}

−4

ferrocyanide	$Fe(CN)_6^{4-}$	silicate (ortho)	SiO_4^{4-}

- **Common Oxidation States of the Elements**

Actinium	+3	Indium	+3
Aluminum	**+3**	Iodine	+7, +5, +1, **−1**
Americium	+6, +5, +4, **+3**	Iridium	+4, +3
Antimony	+5, **+3**, −3	Iron	**+3, +2**
Argon	0	Krypton	+4, +2
Arsenic	+5, **+3**, −3	Lanthanum	**+3**
Astatine	**−1**	Lead	+4, **+2**
Barium	**+2**	Lithium	**+1**
Berkelium	+4, **+3**	Lutetium	**+3**
Beryllium	**+2**	Magnesium	**+2**
Bismuth	+5, **+3**	Manganese	**+7**, +6, **+4**, +3, **+2**
Boron	**+3**	Mercury	**+2**, +1
Bromine	+5, +3, +1, **−1**	Molybdenum	+6, +4, +3
Cadmium	**+2**	Neodymium	+3
Calcium	**+2**	Neon	0
Californium	+3	Neptunium	+6, **+5, +4**, +3
Carbon	**+4**, +2, −4	Nickel	**+2**
Cerium	**+3**, +4	Niobium	**+5, +4**
Cesium	**+1**	Nitrogen	**+5, +4**, +3, +2, +1, **−3**
Chlorine	+7, +6, +5, +4, +3, +1, **−1**	Osmium	+8, **+4**
Chromium	**+6**, +5, +4, **+3**, +2	Oxygen	+2, −½, −1, **−2**
Cobalt	**+3, +2**	Palladium	+4, **+2**
Copper	**+2**, + 1	Phosphorus	**+5**, +3, −3
Curium	+3	Platinum	**+4, +2**
Dysprosium	+3	Plutonium	+6, +5, **+4**, +3
Erbium	+3	Polonium	+2
Europium	**+3**, +2	Potassium	**+1**
Fluorine	−1	Praseodymium	**+3**, +4
Francium	+1	Promethium	+3
Gadolinium	+3	Protactinium	**+5**, +4
Gallium	+3	Radium	**+2**
Germanium	**+4**, −4	Radon	0
Gold	**+3**, +1	Rhenium	**+7**, +6, +4
Hafnium	**+4**	Rhodium	+4, **+3**, +2
Helium	0	Rubidium	**+1**
Holmium	+3	Ruthenium	+8, +6, +4, **+3**
Hydrogen	**+1, −l**	Samarium	**+3**, +2

(The most common or stable oxidation states are in bold.)

Scandium	+3		Thorium	+4
Selenium	+6, +4, −2		Thulium	+3, +2
Silicon	+4, −4		Tin	+4, +2
Silver	+1		Titanium	+4, +3, +2
Sodium	+1		Tungsten	+6, +4
Strontium	+2		Uranium	+6, +5, +4, +3
Sulfur	+6, +4, +2, −2		Vanadium	+5, +4, +3, +2
Tantalum	+5		Xenon	+6, +4, +2
Technetium	+7, +6, +4		Yitrium	+3
Tellurium	+6, +4, −2		Ytterbium	+3, +2
Terbium	+3, +4		Zinc	+2
Thallium	+3, +1		Zirconium	+4

Samples: Multiple-Choice Questions

Note to the student: The AP chemistry exam does not emphasize complex ions or coordination compounds. There is nothing on the AP exam that involves the concepts of crystal-field theory, low versus high spin, valence bond theory, or other related areas. If you understand the questions presented here, then you are basically "safe" in this area of the exam. Most high school AP chemistry programs do not focus much on this area of chemistry because of time constraints.

---

1. Which of the following is the electron configuration for Ni^{2+}?

   A. $1s^2\, 2s^2\, 2p^6\, 3s^2\, 3p^6\, 4s^2\, 3d^6$

   B. $1s^2\, 2s^2\, 2p^6\, 3s^2\, 3p^6\, 4s^2\, 3d^{10}$

   C. $1s^2\, 2s^2\, 2p^6\, 3s^2\, 3p^6\, 3d^8$

   D. $1s^2\, 2s^2\, 2p^6\, 3s^2\, 3p^6\, 4s^2\, 3d^8$

   E. $1s^2\, 2s^2\, 2p^6\, 3s^2\, 3p^6\, 4s^2\, 3d^4$

Answer: C

In forming ions, the transition metals lose their valence (outermost) shell electrons first, followed by their outer d electrons. *Note:* In order for transition metal ions to be colored, the d orbitals must be *partially* filled. In this case, the solution containing the Ni^{2+} ion would be colored (green).

---

2. As the atomic number of the elements increases down a column,

   A. the atomic radius decreases

   B. the atomic mass decreases

   C. the elements become less metallic

   D. ionization energy decreases

   E. the number of electrons in the outermost energy level increases

---

Answer: D

Because the distance between the electrons and the nucleus is increasing, the electrons are becoming further away from the nucleus, making it easier to remove them by overcoming the electrostatic force attracting them to the nucleus. Also, there are more electrons in the way, increasing interference (the electron shielding effect).

3. What is the oxidation number of platinum in $PtCl_6^{2-}$

    A. −4

    B. −2

    C. −1

    D. +4

    E. +6

Answer: D

Because chlorine is a halogen, it has an oxidation number of −1. And because there are 6 chlorines, there is a total charge of −6 for the chlorines. The overall charge is −2, so algebraically, $x + (-6) = -2$. Solving this yields $x = +4$. Thus, the charge (or oxidation number) of platinum is +4.

4. What type of bond would you expect in CsI?

    A. Ionic

    B. Covalent

    C. Hydrogen

    D. Metallic

    E. van der Waals

Answer: A

Generally, expect an ionic bond whenever you have a metal (Cs) bonded to a nonmetal (I). If you have a table of electronegativities to refer to, the electronegativity difference is greater than 1.7 for ionic compounds.

5. Arrange the following ions in order of increasing ionic radius: Mg^{2+}, F^-, and O^{2-}.

    A. O^{2-}, F^-, Mg^{2+}

    B. Mg^{2+}, O^{2-}, F^-

    C. Mg^{2+}, F^-, O^{2-}

    D. O^{2-}, Mg^{2+}, F^-

    E. F^-, O^{2-}, Mg^{2+}

Answer: C

Note that all the ions have 10 electrons; that is, they are all isoelectronic with neon. Because they all have the same number of electrons, the only factor that will determine their size will be the nuclear charge — the greater the nuclear charge, the smaller the radius. Therefore, magnesium, with a nuclear charge of +2, has the smallest radius among these ions.

---

6. Which of the following is the correct Lewis structure for the ionic compound $Ca(ClO_2)_2$?

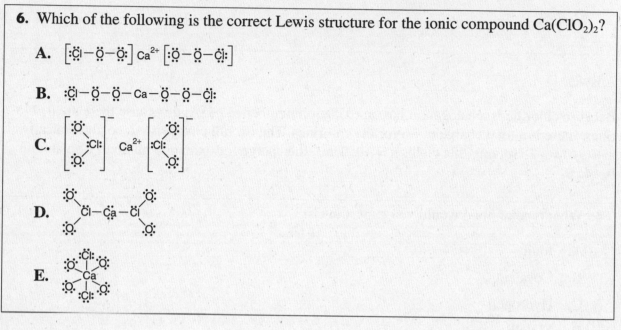

Answer: C

---

7. The compound expected when chlorine reacts with aluminum is

    A. Al_2Cl_3

    B. Al_3Cl_2

    C. $AlCl_3$

    D. Al_3Cl

    E. $AlCl_2$

Answer: C

The compound is ionic — a metal (Al) bonded to a nonmetal (Cl). All ionic compounds are solids at room temperature and pressure. Aluminum has 13 electrons. As an ion, it will lose 3 electrons to become isoelectronic with neon. Thus the aluminum ion will have the electronic configuration $1s^2 2s^2 2p^6$.

Chlorine as a free element is diatomic (Cl_2); however, as an ion, it will gain one electron to become isoelectronic with argon. The electronic configuration of the chloride ion is $1s^2 2s^2 2p^6 3s^2 3p^6$. The compound thus formed, aluminum chloride, has the formula $AlCl_3$.

8. What ions would you find in solution if potassium perchlorate was dissolved in water?

    A. KCl, O_2

    B. K^+, Cl^-, O^{2-}

    C. KCl, O^{2-}

    D. K^+, ClO_4^-

    E. K^+, Cl^-, O^{2-}

Answer: D

Potassium is a metal, and the polyatomic anion, ClO_4^- is a nonmetal; therefore, the compound is an ionic solid at room temperature. When the compound is dissolved in water, the ionic bond between the cation, K^+, and the polyatomic anion, ClO_4^-, is broken due to the polarity of the water molecule, resulting in the two aqueous ions, K^+ and ClO_4^-.

9. Which one of the following is correct?

    A. $KClO_3$ potassium perchlorate

    B. CuO copper oxide

    C. $Al_3(SO_3)_2$ aluminum sulfate

    D. $MgPO_4$ magnesium phosphate

    E. $Na_2Cr_2O_7$ sodium dichromate

Answer: E

In choice A, $KClO_3$ is potassium chlorate, not perchlorate. In choice B, CuO is copper (II) oxide, to distinguish it from copper (I) oxide, Cu_2O. In choice C, the formula for aluminum sulfate is $Al_2(SO_4)_3$. In D, the formula for magnesium phosphate is $Mg_3(PO_4)_2$. If you missed these, review inorganic nomenclature and refer to pages 94 and 95 of this book to become familiar with common ions and their charges.

Samples: Free-Response Questions

1. The first three ionization energies (I_1, I_2, and I_3) for beryllium and neon are given in the following table:

(kJ/mole)	I_1	I_2	I_3
Be	900	1757	14,840
Ne	2080	3963	6276

(a) Write the complete electron configuration for beryllium and for neon.

(b) Explain any trends or significant discrepancies found in the ionization energies for beryllium and neon.

(c) If chlorine gas is passed into separate containers of heated beryllium and heated neon, explain what compounds, if any, might be formed, and explain your answer in terms of the electron configurations of these two elements.

(d) An unknown element, X, has the following three ionization energies:

(kJ/mole)	I_1	I_2	I_3
X	419	3069	4600

On the basis of the ionization energies given, what is most likely to be the compound produced when chlorine reacts with element X?

Answer

The outline format might work well in part (b).

1. Given: First three ionization energies of Be and Ne.

   (a) Restatement: Electron configuration of Be and Ne.

   Be: $1s^2 2s^2$
   Ne: $1s^2 2s^2 2p^6$

   (b) Restatement: Significant trends/discrepancies in the first three ionization energies of Be and Ne.

   I. Note that in the case of both beryllium and neon, ionization energies increase as one moves from I_1 to I_2 to I_3.

   II. The general trend is for ionization energy to increase as one moves from left to right across the periodic table and to decrease as one moves down; this is the inverse of the trend one finds in examining the atomic radius.

   III. Note further that beryllium and neon are in the second period.

   IV. Beryllium

       A. There is generally not enough energy available in chemical reactions to remove inner electrons, as noted by the significantly higher third ionization energy.

       B. The Be^{2+} ion is a very stable species with a noble-gas configuration, so removing the third outermost electron from beryllium requires significantly greater energy.

   V. Neon

       A. Neon is an inert element with a full complement of 8 electrons in its valence shell.

       B. It is significantly more difficult to remove neon's most loosely held electron (I_1) than that of beryllium's I_1. This trend is also noted when examining I_2's and I_3's. Neon also has a greater nuclear charge than beryllium, which, if all factors are held constant, would result in a smaller atomic radius.

(c) Restatement: $Cl_2(g)$ passed into separate containers of Be and Ne. What compounds formed? Explanation.

The only compound formed would be **$BeCl_2$**. The Be atom readily loses 2 electrons to form the stable Be^{2+} ion. The third ionization energy is too high to form Be^{3+}. The electron affinity of neon is very low because it has a stable octet of electrons in its valence shell and the ionization energies of neon are too high.

(d) Restatement: Given the first three ionization energies of element X what compounds is it most likely to form with Cl_2?

The first ionization energy (I_1) of element X is relatively low when compared to I_2 and I_3. This means that X is probably a member of the Group I alkali metals. Thus, the formation of X^{2+} and X^{3+} would be difficult to achieve. Therefore, the formula is most likely to be **XCl**.

---

2. Bromine reacts with a metal (M) as follows:

$M(s) + Br_2(g) \rightarrow MBr_2(s)$

Explain how the heat of the reaction is affected by

  (a) The ionization energy for the metal M

  (b) The size of the atomic radius for the ion M^{2+}

Answer

This question is probably best answered in the outline format, because you will try to show a logical progression of concepts leading to two overall conclusions. Using the chart format would become too complicated. This question should take about ten minutes to answer.

2. Restatement: How are ionization energy and atomic radius affected by the heat of reaction for

$$M(s) + Br_2(g) \rightarrow MBr_2(s)$$

I. Ionization energy

   A. Definition — the amount of energy that a gaseous atom must absorb so that the outermost electron can be completely separated from the atom.

   B. The lower the ionization energy, the more metallic the element.

   C. With all other factors held constant, energy is required to form the M^{2+} ion (endothermic)

   D. With larger ionization energies, the heat of the reaction becomes more positive or more endothermic.

II. Atomic radius

   A. Definition — half the distance of closest approach between two nuclei in the ordinary form of an element.

   B. Positive ions are smaller than the metal atoms from which they are formed.

   C. Lattice energy — energy released when an ionic (metal M +; nonmetal Br) solid forms from its ions.

   D. The change that occurs from $M(s) \rightarrow M^{2+}$, which exists as the cation in the ionic solid MBr_2, results in a decrease in the radius.

   E. In the calculation of lattice energy,

   $$L.E. = \frac{Q_1 Q_2}{r}$$

   the lattice energy is inversely proportional to the radius, r.

   F. Because atomic distances (r) are decreasing in the reaction, lattice energy is increasing ($-\Delta H$), which has the effect of making the heat of reaction more negative, or more exothermic.

   G. Q_1 and Q_2 being opposite in sign further confirms the fact that bringing cations and anions together is an exothermic process.

Liquids and Solids

Key Terms

Words that can be used as topics in essays:

adhesion

alloys

amorphous solids

band model

boiling point

capillary action

closest packing

cohesion

condensation

coordination number

critical point

critical pressure

critical temperature

crystalline solid

cubic closest packing

deposition

dipole-dipole attraction

dipole-induced dipole

dipole force

dispersion force

dynamic equilibrium

electron sea model

freezing

heat of vaporization

heating curve

hexagonal closest packing

hydrogen bond

induced dipole

intermolecular forces

intramolecular forces

ion-dipole forces

lattice

London dispersion forces

melting

melting point

network solid

normal boiling point

normal melting point

phase diagram

polarization

sublimation

supercooled

superheated

surface tension

triple point

types of structural units

   ions

   macromolecules

   metals

   molecules

unit cell

van der Waals forces

vaporization

vapor pressure

viscosity

X-ray diffraction

Key Concepts

Equations and relationships that you need to know:

- $\Delta H_{\text{vaporization}} = H_{\text{vapor}} - H_{\text{liquid}}$

  $\Delta H_{\text{fusion}} = H_{\text{liquid}} - H_{\text{solid}}$

  $\Delta H_{\text{sublimation}} = H_{\text{fusion}} + H_{\text{vaporization}}$

- Raoult's law: $P_1 = X_1 P_1^{\circ}$

  where

  P_1 = vapor pressure of solvent over the solution

  X_1 = mole fraction of solvent

  P_1° = vapor pressure of pure solvent

- Clausius – Clapeyron equation:

  $\ln\left(P_{\text{vap}}\right) = -\dfrac{\Delta H_{\text{vap}}}{R}\left(\dfrac{1}{T}\right) + C$ where C is a constant that is characteristic of a liquid

  $\log\left(\dfrac{P_2}{P_1}\right) = \dfrac{\Delta H_{\text{vap}}}{2.303R}\left(\dfrac{T_2 - T_1}{T_2 T_1}\right)$

  $\ln\dfrac{P_1}{P_2} = \dfrac{\Delta H_{\text{vap}}}{R}\left(\dfrac{1}{T_2} - \dfrac{1}{T_1}\right)$

- Intermolecular forces
  - I. van der Waals
    - a. dipole-dipole   HBr—H$_2$S
    - b. dipole-induced dipole   NH$_3$—C$_6$H$_6$
    - c. dispersion   He—He
  - II. Ion-induced dipole   NO$_3^-$—I$_2$
  - III. Hydrogen bond   H$_2$O—H$_2$O

Samples: Multiple-Choice Questions

For Samples 1–5, use the following choices:

    A. Hydrogen bonding

    B. Metallic bonding

    C. Ionic bonding

    D. Dipole forces

    E. van der Waals forces (London dispersion forces)

---

1. What accounts for the intermolecular forces between CCl_4 molecules?

Answer: E

Because CCl_4 is a nonpolar molecule, the only forces present are dispersion forces.

---

2. What explains why the boiling point of acetic acid, CH_3COOH, is greater than the boiling point of dimethyl ether, CH_3-O-CH_3?

Acetic acid, CH_3COOH

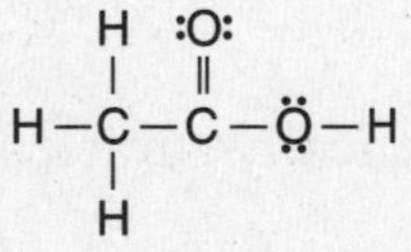

Dimethyl Ether, CH_3-O-CH_3

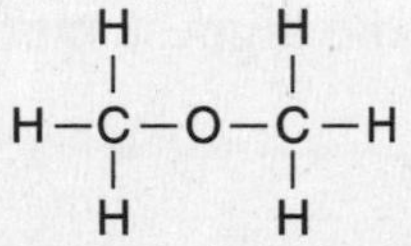

Answer: A

Note that for acetic acid there is a hydrogen attached to an oxygen atom (a prerequisite for H bonding) but that for dimethyl ether there is no H atom connected to an F, O, or N atom.

Note: Had a choice been acetone, CH_3COCH_3, it would not have been

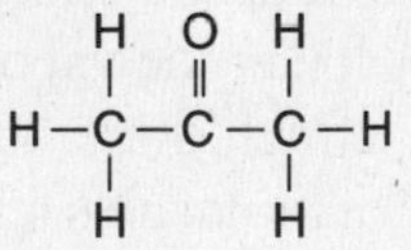

a good choice, since even though there is no hydrogen atom attached to an F, O, or N atom, acetone enol is in tautomeric equilibrium with the ketone tautomer, resulting in a small amount of hydrogen bonding, accounting for a higher than expected boiling point.

3. What holds solid sodium together?

Answer: B

Sodium is a metal. Metals are held together in a crystal lattice, which is a network of cations surrounded by a "sea" of mobile electrons.

4. What holds solid ICl together?

Answer: D

ICl is a polar molecule. Polar molecules have a net dipole — that is, a center of positive charge separated from a center of negative charge. Adjacent polar molecules line up so that the negative end of the dipole on one molecule is as close as possible to the positive end of its neighbor. Under these conditions, there is an electrostatic attraction between adjacent molecules. The key word here is *solid,* because if the question just asked for the force holding ICl (the molecule) together, the answer would be covalent bonding.

5. What holds calcium chloride together?

Answer: C

The metallic cations (Ca^{2+}) are electrostatically attracted to the nonmetallic anions (Cl^-).

6. Which of the following liquids has the highest vapor pressure at 25°C?

    A. Carbon tetrachloride, CCl_4

    B. Hydrogen peroxide, H_2O_2

    C. Water, H_2O

    D. Dichloromethane, CH_2Cl_2

    E. Trichloromethane, $CHCl_3$

Answer: D

You can rule out choice B, hydrogen peroxide, and choice C, water, because the very strong hydrogen bonds between their molecules lower the vapor pressure (the ease at which the liquid evaporates). Although answer A, carbon tetrachloride, the only nonpolar molecule in the list, has only dispersion forces present between molecules, choice D, dichloromethane, has the lowest molecular weight and consequently the lowest amount of dispersion forces.

7. Which of the following statements is true of the critical temperature of a pure substance?

   A. The critical temperature is the temperature above which the liquid phase of a pure substance can exist.

   B. The critical temperature is the temperature above which the liquid phase of a pure substance cannot exist.

   C. The critical temperature is the temperature below which the liquid phase of a pure substance cannot exist.

   D. The critical temperature is the temperature at which all three phases can coexist.

   E. The critical temperature is the temperature at which the pure substance reaches, but cannot go beyond, the critical pressure.

Answer: B

This is the definition of critical temperature.

8. A certain metal crystallizes in a face-centered cube measuring 4.00×10^2 picometers on each edge. What is the radius of the atom? (1 picometer (pm) $= 1 \times 10^{-12}$ meter)

   A. 141 pm

   B. 173 pm

   C. 200. pm

   D. 282 pm

   E. 565 pm

Answer: A

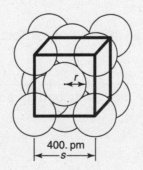

400. pm
s

The formula which relates the radius of an atom (r) to the length of the side (s) of the unit cell for a face-centered cubic cell is $4r = s\sqrt{2}$.

$$r = \frac{400.\ \text{pm}\sqrt{2}}{4} = 100.(1.414) = 141\ \text{pm}$$

9. The molecules butane and 2-methylpropane are structural isomers. Which of the following characteristics would be the same for both isomers, assuming constant temperature where necessary?

    A. Boiling point

    B. Vapor pressure

    C. Melting point

    D. Solubility

    E. Gas density

Answer: E

Choices A, B, C, and D involve the strength of intermolecular forces. Because the two compounds differ in structure, there would be intermolecular differences between the two compounds. However, because both isomers have the same molecular mass, they would have the same gas density.

10. Which of the following choices represents intermolecular forces listed in order from strongest to weakest?

    A. dipole attractions, dispersion forces, hydrogen bonds

    B. hydrogen bonds, dispersion forces, dipole attractions

    C. dipole attractions, hydrogen bonds, dispersion forces

    D. hydrogen bonds, dipole attractions, dispersion forces

    E. dispersion forces, hydrogen bonds, dipole attractions

Answer: D

Hydrogen bonds are the strongest of the intermolecular forces listed; dispersion forces are the weakest.

11. Arrange the following in order of increasing boiling point: $NaCl$, CO_2, CH_3OH, CH_3Cl.

    A. CH_3Cl, CO_2, CH_3OH, $NaCl$

    B. CO_2, CH_3Cl, CH_3OH, $NaCl$

    C. CO_2, CH_3OH, CH_3Cl, $NaCl$

    D. $NaCl$, CH_3OH, CH_3Cl, CO_2

    E. CH_3OH, CO_2, CH_3Cl, $NaCl$

Answer: B

CO_2 — nonpolar, dispersion forces only
CH_3Cl — polar molecule
CH_3OH — polar molecule, hydrogen bonds
$NaCl$ — ionic compound

---

12. An imaginary metal crystallizes in a cubic lattice. The unit cell edge length is 100. picometers (1 picometer = 1×10^{-12} meter). The density of this metal is 200. g/cm^3. The atomic mass of the metal is 60.2 g/mol. How many of these metal atoms are there within a unit cell?

   A. 1.00

   B. 2.00

   C. 4.00

   D. 6.00

   E. 12.0

Answer: B

First, calculate the mass of one cell, 100. pm on an edge:

$$\frac{200.\ g}{1\ cm^3} \times \left(\frac{100\ cm}{1\ m}\right)^3 \times \left(\frac{1\ m}{10^{12}\ pm}\right)^3 \times \left(\frac{100.\ pm}{1}\right)^3 = 2.00 \times 10^{-22}\ g/cell$$

Next, calculate the mass of one metal atom:

$$\frac{60.2\ g}{1\ mole} \times \frac{1\ mole}{6.02 \times 10^{23}\ atoms} = 10.0 \times 10^{-23} = 1.00 \times 10^{-22}\ g/atom$$

Finally, calculate the number of metal atoms in one cell:

$$\frac{2.00 \times 10^{-22}\ g \cdot cell^{-1}}{1.00 \times 10^{-22}\ g \cdot atom^{-1}} = 2.00\ atoms/cell$$

Samples: Free-Response Questions

> **1.** Explain each of the following in terms of (1) inter- and intra-atomic or molecular forces and (2) structure.
>
> **(a)** ICl has a boiling point of 97°C, whereas NaCl has a boiling point of 1400°C.
>
> **(b)** KI(*s*) is very soluble in water, whereas I_2 (*s*) has a solubility of only 0.03 gram per 100 grams of water.
>
> **(c)** Solid Ag conducts an electric current, whereas solid $AgNO_3$ does not.
>
> **(d)** PCl_3 has a measurable dipole moment, whereas PCl_5 does not.

Answer

The bullet format will work well here.

1. Restatement: Explain each of the following:

(a) ICl has a significantly lower B.P. than NaC1.

- ICl is a covalently bonded, molecular solid; NaCl is an ionic solid.
- There are dipole forces between ICl molecules but electrostatic forces between Na^+ and Cl^- ions.
- Dipole forces in ICl are much weaker than the ionic bonds in NaCl.
- I and Cl are similar in electronegativity — generates only partial δ^+ and δ^- around molecule.
- Na and Cl differ greatly in electronegativity — greater electrostatic force.
- When heated slightly, ICl boils because energy supplied (heat) overcomes weak dipole forces.

(b) KI is water soluble; I_2 is not.

- KI is an ionic solid, held together by ionic bonds.
- I_2 is a molecular solid, held together by covalent bonds.
- KI dissociates into K^+ and I^- ions.
- I_2 slightly dissolves in water, maintaining its covalent bond.
- Solubility rule: Like dissolves like. H_2O is polar; KI is polar; I_2 is not polar.

(c) Ag conducts; $AgNO_3$ does not.

- Ag is a metal.
- $AgNO_3$ is an ionic solid.
- Ag structure consists of Ag^+ cations surrounded by mobile or "free" electrons.
- $AgNO_3$ structure consists of Ag^+ cations electrostatically attracted to NO_3^- polyatomic anions — no free or mobile electrons.

(d) PCl_3 has a dipole; PCl_5 does not.

- PCl_3 Lewis diagram:

- PCl_3 — note the lone pair of unshared electrons.
- PCl_3 is pyramidal, and all pyramidal structures are polar.
- PCl_5 Lewis diagram:

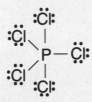

- PCl_5 — no unshared electrons on P
- PCl_5 is trigonal bipyramidal and thus perfectly symmetrical, so there is no polarity; all dipoles cancel.

---

2. Solids can be classified into four categories: ionic, metallic, covalent network, and molecular. For each of the four categories, identify the basic structural unit; describe the nature of the force both within the unit and between units; cite the basic properties of each type of solid; give two examples of each type of solid; and describe a laboratory means of identifying each type of solid.

Answer

This question lends itself to the chart format. Here the column headings express what is given, and the first entry in each row serves as a restatement of what is wanted.

Characteristic	Ionic	Metallic	Covalent Network	Molecular
Structural Unit	ions	cations surrounded by mobile "sea" of electrons	atoms	polar or nonpolar molecules
Force Within Unit (Intra)	covalent bond within polyatomic ion	—	—	covalent bond

(continued)

(continued)

Characteristic	Ionic	Metallic	Covalent Network	Molecular
Force Between Units (Inter)	ionic bond electrostatic attraction	metallic bond	covalent bond	dipole-dipole dispersion (London) H bonds dipole-induced dipole
Basic Properties				
Melting Point	high	variable	very high	nonpolar — low polar — high
Conduction of electricity	in water solution or molten state	always conducts	does not conduct	does not conduct
Solubility	solubility in water varies	not soluble	not soluble	nonpolar — insoluble in water polar — some degree of solubility in water solubility in organic solvents varies
Hardness	hard, brittle	variable malleable ductile	very hard	soft
Conduction of heat	poor	good	poor, except diamond	poor
Examples	NaCl $CaCl_2$	Cu Au Fe	SiO_2 C (diamond)	H_2O – polar I_2 – nonpolar CO_2 – nonpolar
Lab Test	conducts in pure state when molten or in H_2O or ionizing solvents	always conducts	extremely hard; nonconductor	low M.P. nonconductor

Key Terms

Words that can be used as topics in essays:

boiling-point elevation	nonelectrolyte
colligative properties	normality
colloids	osmosis
electrolyte	osmotic pressure
fractional crystallization	Raoult's law
fractional distillation	saturated
freezing-point depression	solubility
heat of solution	solute
Henry's law	solution
ideal solution	solvent
ion pairing	supersaturated
isotonic solutions	Tyndall effect
mass percent	unsaturated
molality	van't Hoff factor
molarity	vapor pressure
mole fraction	volatility
net ionic equation	

Key Concepts

Equations and relationships that you need to know:

- Solubility Rules

Group	Solubility	Exceptions
Nitrates NO_3^-	All soluble	$Bi(NO_3)_3$, $Hg(NO_3)_2$
Chlorates ClO_3^-	All soluble	$Bi(ClO_3)_3$, $Hg(ClO_3)_2$
Perchlorates ClO_4^-	All soluble	$KClO_4^*$
Acetates $C_2H_3O_2^-$	All soluble	$AgC_2H_3O_2$
Alkali metal compounds Li, Na, K, Rb, Cs	All soluble	Some complex (ternary) alkali metal salts
Ammonium compounds NH_4^+	All soluble	none
Chlorides Cl^- Bromides Br^- Iodides I^-	All soluble	Those that contain Ag^+, Hg_2^{2+}, Pb^{2+}, and $PbCl_2^*$, $CrCl_3$, BiI_3, CuI, AuI, AuI_3, PtI_2, $CrBr_2^*$, $CuBr_2^*$, $PtBr_4^*$, HgI_2^*
Sulfates SO_4^{2-}	Most are soluble	$SrSO_4$, $Sr_2(SO_4)_3$, $BaSO_4$, $PbSO_4$, $HgSO_4$, Hg_2SO_4, $CaSO_4^*$, $Ag_2SO_4^*$, $Cr_2(SO_4)_3^*$, $Fe_2(SO_4)_3^*$
Carbonates CO_3^{2-}	Group I, $(NH_4)_2CO_3$, $CrCO_3$	All others
Phosphates PO_4^{3-}	Group I, $(NH_4)_3PO_4$	All others
Sulfites SO_3^{2-}	Group I, $(NH_4)_2SO_3$	All others
Hydroxides OH^-	Group I, $Sr(OH)_2$, $Ba(OH)_2$, $AuOH$	All others, $Ca(OH)_2^*$
Sulfides S^{2-}	Group I, $(NH_4)_2S$, SrS	All others, CaS^*
Oxides O^{2-}	CaO, SrO, BaO	All others, Na_2O and K_2O decompose in H_2O

*slightly soluble

Solubility Rules (Abbreviated)

1. All common salts of the Group 1 (IA) elements and ammonium ion are soluble.

2. All common acetates and nitrates are soluble.

3. All binary compounds of Group 17 (VIIA) elements (other than F) with metals are soluble except those of silver, mercury (I), and lead.

4. All sulfates are soluble except those of barium, strontium, lead, calcium, silver, and mercury(I).

5. Except for those in Rule 1, carbonates, hydroxides, oxides, sulfides, and phosphates are insoluble.

- mass percent solute $= \dfrac{\text{mass solute}}{\text{total mass solution}} \times 100\%$

molality $= \dfrac{\text{moles solute}}{\text{kg solvent}}$

molarity $= \dfrac{\text{moles solute}}{\text{liter solution}}$

In dilute aqueous solutions, molarity $\approx$ molality.

$M_1V_1 = M_2V_2$

- extent of solubility

    1. nature of solute-solvent interactions

    2. temperature

    3. pressure of gaseous solute

- Henry's law: $C = kP$

where

P = partial pressure of gas solute over the solution (atm)

C = concentration of dissolved gas (mole/liter)

k = Henry's constant (dependent on temperature, mole / L $\cdot$ atm)

- Raoult's law: $P_1 = X_1P_1^{\circ}$

where

P_1 = vapor pressure of solvent over the solution

P_1° = vapor pressure of pure solvent at same temperature

X_1 = mole fraction of solvent

- $\pi = \dfrac{nRT}{V} = iMRT$

  where

  π = osmotic pressure (atm)

  i = van't Hoff factor; should be 1 for all nonelectrolytes

  $= \dfrac{\text{actual number of particles in solution after dissociation}}{\text{number of formula units initially dissolved in solution}}$

  $MW = \dfrac{gRT}{\pi V}$

- $\Delta T_b = i \cdot k_b \cdot m$ for water, $k_b = 0.52°C/m$

  $\Delta T_f = i \cdot k_f \cdot m$ for water, $k_f = 1.86°C/m$

Samples: Multiple-Choice Questions

> **1.** A 10.0% sucrose solution has a density of 2.00 g/mL. What is the mass of sucrose dissolved in 1.00 liter of this solution?
>
> **A.** 1.00×10^2 g
>
> **B.** 2.00×10^2 g
>
> **C.** 5.00×10^2 g
>
> **D.** 1.00×10^3 g
>
> **E.** 1.00×10^4 g

Answer: B

This problem can be easily solved using the factor-label method:

$$\frac{1.00 \text{ liter solution}}{1} \times \frac{1000 \text{ mL solution}}{1 \text{ liter solution}} \times \frac{2.00 \text{ g solution}}{1 \text{ mL( ) solution}} \times \frac{10.0 \text{ g sucrose}}{100.0 \text{ g solution}}$$

$= 2.00 \times 10^2$ g sucrose

> **2.** How many milliliters of a 50.0% (by mass) HNO_3 solution, with a density of 2.00 grams per milliliter, are required to make 500. mL of a 2.00 M HNO_3 solution?
>
> **A.** 50.0 mL
>
> **B.** 63.0 mL
>
> **C.** 100. mL
>
> **D.** 200. mL
>
> **E.** 250. mL

Answer: B

This problem can be easily solved using the factor-label method:

$$\frac{500. \text{ mL (2.00 M sol'n)}}{1} \times \frac{1 \text{ liter (2.00 M sol'n)}}{1000 \text{ mL (2.00 M sol'n)}} \times \frac{2.00 \text{ moles } HNO_3}{1 \text{ liter (2.00 M sol'n)}}$$

$$\times \frac{63.0 \text{ g } HNO_3}{1 \text{ mole } HNO_3} \times \frac{100. \text{ g 50.0\% sol'n}}{50.0 \text{ g } HNO_3} \times \frac{1 \text{ mL 50.0\% sol'n}}{2.00 \text{ g 50.0\% sol'n}}$$

$= 63.0$ mL of a 50.0% sol'n

3. What is the normality of a solution that contains 9.80 g of H_2SO_4 in 200. mL of solution, assuming 100% ionization, $H_2SO_{4(aq)} \rightarrow 2H^+{}_{(aq)} + SO_4{}^{2-}{}_{(aq)}$?

   A. 0.500 N

   B. 1.00 N

   C. 1.50 N

   D. 2.00 N

   E. 2.50 N

Answer: B

This problem can be solved using the factor-label method:

$$\frac{9.80 \text{ g } H_2SO_4}{200. \text{ mL sol'n}} \times \frac{1000. \text{ mL sol'n}}{1 \text{ liter sol'n}} \times \frac{1 \text{ mole } H_2SO_4}{98.1 \text{ g } H_2SO_4}$$

$$\times \frac{2 \text{ g-eq } H_2SO_4}{1 \text{ mole } H_2SO_4} = 1.00 \text{ g-eq/liter} = 1.00 \text{ N}$$

4. Calculate the number of gram-equivalents of solute in 0.500 liter of a 3.00 N solution.

   A 1.00 g-eq

   B 1.50 g-eq

   C 2.00 g-eq

   D 3.00 g-eq

   E 6.00 g-eq

Answer: B

This problem can be solved using the factor-label method:

$$\frac{0.500 \text{ liter sol'n}}{1} \times \frac{3.00 \text{ g-eq}}{1 \text{ liter sol'n}} = 1.50 \text{ g-eq}$$

5. What is the percentage (by mass) of NaCl (FW = 58.50) in a 10.0-molal solution?

   A. $\dfrac{10.0 \times 58.50}{1585}$

   B. $\dfrac{10.0 \times 58.50}{1000.00}$

   C. $\dfrac{2 \times 58.50 \times 10.0}{1000.00}$

   D. $\dfrac{10.0 \times 58.50}{100.00}$

   E. $\dfrac{100 \times 58.50}{1000.00}$

Answer: A

This problem can be solved using the factor-label method:

$$\frac{10.0 \ \cancel{\text{moles NaCl}}}{1000. \ \text{g H}_2\text{O}} \times \frac{58.50 \ \text{g NaCl}}{1 \ \cancel{\text{mole NaCl}}} = \frac{585 \ \text{g NaCl}}{1000. \ \text{g H}_2\text{O}}$$

The question is asking for the parts of NaCl per *total* solution (solute + solvent).

$$\frac{585 \ \text{g NaCl}}{1000. \ \text{g H}_2\text{O} + 585 \ \text{g NaCl}}$$

---

6. When 5.92 grams of a nonvolatile, nonionizing compound is dissolved in 186 grams of water, the freezing point (at normal pressure) of the resulting solution is −0.592°C. What is the molecular weight of the compound?

 A. 10.0 g/mol

 B. 100. g/mol

 C. 110. g/mol

 D. 200. g/mol

 E. 210. g/mol

Answer: B

This problem can be solved using the factor-label method:

$$\frac{0.592 \ \cancel{°\text{C}}}{1} \times \frac{1 \ \text{mole solute}}{1.86 \ \cancel{°\text{C}} \cdot \cancel{\text{kg H}_2\text{O}}} \times \frac{186 \ \cancel{\text{g H}_2\text{O}}}{5.92 \ \text{g solute}} \times \frac{1 \ \cancel{\text{kg H}_2\text{O}}}{1000 \ \cancel{\text{g H}_2\text{O}}} = 0.0100 \ \text{mol/g} = 100. \ \text{g/mol}$$

---

7. Calculate the number of grams of glycerol, $C_3H_5(OH)_3$ (MW = 92.1 g/mol), that must be dissolved in 520. grams of water to raise the boiling point to 102.00°C.

 A. 5.65 g

 B. 92.0 g

 C. 184 g

 D. 194 g

 E. 204 g

Answer: C

This problem can be solved using the factor-label method:

$$\frac{520. \ \cancel{\text{g H}_2\text{O}}}{1} \times \frac{1 \ \cancel{\text{kg H}_2\text{O}}}{1000 \ \cancel{\text{g H}_2\text{O}}} \times \frac{1 \ \cancel{\text{mole C}_3\text{H}_5(\text{OH})_3}}{0.52 \ \cancel{°\text{C}} \cdot \cancel{\text{kg H}_2\text{O}}} \times \frac{92.1 \ \text{g C}_3\text{H}_5(\text{OH})_3}{1 \ \cancel{\text{mole C}_3\text{H}_5(\text{OH})_3}} \times \frac{2.00 \ \cancel{°\text{C}}}{1}$$

$$= 184 \ \text{g C}_3\text{H}_5(\text{OH})_3$$

8. In order to determine the molecular weight of a particular protein, 0.010 g of the protein was dissolved in water to make 2.93 mL of solution. The osmotic pressure was determined to be 0.821 torr at 20.0°C. What is the molecular weight of the protein?

A. 3.8×10^3 g/mole

B. 7.6×10^3 g/mole

C. 3.8×10^4 g/mole

D. 7.6×10^4 g/mole

E. None of the above

Answer: D

Begin with the equation

$$MW = gRT/\pi V$$

$$= \frac{0.010\,g}{1} \times \frac{0.0821\,\text{liter} \cdot \text{atm}}{\text{mole} \cdot \text{K}} \times \frac{293\,\text{K}}{1} \times \frac{1}{0.821\,\text{torr}} \times \frac{1}{293\,\text{mL}} \times \frac{760\,\text{torr}}{1\,\text{atm}} \times \frac{1000\,\text{mL}}{1\,\text{liter}}$$

$$= 7.6 \times 10^4 \text{ g/mol}$$

9. A solution of NH_3 dissolved in water is 10.0 *m*. What is the mole fraction of water in the solution?

A. 1.00/1.18

B. 1.00/2.18

C. 0.18/1.00

D. 0.18/10.0

E. 1.18

Answer: A

This problem can be solved using the factor-label method:

$$\frac{10.0 \text{ moles } NH_3}{1 \text{ kg } H_2O} \times \frac{1 \text{ kg } H_2O}{1000 \text{ g } H_2O} \times \frac{18.02 \text{ g } H_2O}{1 \text{ mole } H_2O} = \frac{0.180 \text{ mole } NH_3}{1.00 \text{ mole } H_2O}$$

total number of moles = 0.180 mole NH_3 + 1.00 mole H_2O = 1.18 moles sol'n

mole fraction of water $= \dfrac{1.00 \text{ mole } H_2O}{1.18 \text{ mole sol'n}}$

10. At 37°C and 1.00 atm of pressure, nitrogen dissolves in the blood at a solubility of 6.0×10^{-4} M. If a diver breathes compressed air where nitrogen gas constitutes 80. mole % of the gas mixture, and the total pressure at this depth is 3.0 atm, what is the concentration of nitrogen in her blood?

   A. 1.4×10^{-4} M

   B. 6.0×10^{-4} M

   C. 1.0×10^{-3} M

   D. 1.4×10^{-3} M

   E. 6.0×10^{-3} M

Answer: D

Determine k by using $C = kP$ (Henry's law).

$$k = \frac{\text{concentration N}_2}{\text{pressure N}_2} = \frac{6.0 \times 10^{-4}\ \text{M}}{1.00\ \text{atm}} = 6.0 \times 10^{-4}\ \text{M} \cdot \text{atm}^{-1}$$

To solve the problem

$$P = 0.80 \times 3.0\ \text{atm} = 2.4\ \text{atm}$$

$$C = kP = \frac{6.0 \times 10^{-4}\ \text{moles}}{\text{liter} \cdot \text{atm}} \times \frac{2.4\ \text{atm}}{1} = 1.4 \times 10^{-3}\ \text{M}$$

11. The vapor pressure of an ideal solution is 450. mm Hg. If the vapor pressure of the pure solvent is 1000. mm Hg, what is the mole fraction of the nonvolatile solute?

   A. 0.450

   B. 0.500

   C. 0.550

   D. 0.950

   E. None of the above

Answer: C

$$P_1 = X_1 P_1^\circ$$

$$X_1 = \frac{P_1}{P_1^\circ} = \frac{450.\ \text{mm Hg}}{1000.\ \text{mm Hg}} = 0.450$$

The mole fraction of the solute is

$$1.000 - X_1 = 1.000 - 0.450 = 0.550$$

Samples: Free-Response Questions

> **1.** An unknown hydrocarbon is burned in the presence of oxygen in order to determine its empirical formula. Another sample of the hydrocarbon is subjected to colligative property tests in order to determine its molecular mass.
>
> **(a)** Calculate the empirical formula of the hydrocarbon, if upon combustion at STP, 9.01 grams of liquid H_2O and 11.2 liters of CO_2 gas are produced.
>
> **(b)** Determine the mass of the oxygen gas that is used.
>
> **(c)** The hydrocarbon dissolves readily in CCl_4. A solution prepared by mixing 135 grams of CCl_4 and 4.36 grams of the hydrocarbon has a boiling point of 78.7°C. The molal boiling-point-elevation constant of CCl_4 is 5.02°C/molal, and its normal boiling point is 76.8°C. Calculate the molecular weight of the hydrocarbon.
>
> **(d)** Determine the molecular formula of the hydrocarbon.

Answer

1. Given: Unknown hydrocarbon is combusted in O_2 and subjected to colligative tests.

(a) Given: 9.01 g H_2O + 11.2 liters CO_2 produced.

Restatement: Find empirical formula of hydrocarbon.

$$\frac{9.01 \text{ g } H_2O}{18.02 \text{ g/mole}} = 0.500 \text{ mole } H_2O$$

$$\frac{0.500 \text{ mole } H_2O}{1} \times \frac{2 \text{ mole H}}{1 \text{ mole } H_2O} = 1.00 \text{ mole H}$$

$$\frac{11.2 \text{ liters } CO_2}{22.4 \text{ liters/mole}} = 0.500 \text{ mole } CO_2$$

$$\frac{0.500 \text{ mole } CO_2}{1} \times \frac{1 \text{ mole C}}{1 \text{ mole } CO_2} = 0.500 \text{ mole C}$$

empirical formula $= C_{0.5}H_1 \rightarrow \mathbf{CH_2}$

(b) Restatement: Calculate mass of O_2 required for complete combustion.

$$2 CH_2(g) + 3O_2(g) \rightarrow 2 CO_2(g) + 2 H_2O(\ell)$$

$$\frac{0.500 \text{ mole } H_2O}{1} \times \frac{3 \text{ moles } O_2}{2 \text{ moles } H_2O} \times \frac{32.00 \text{ g } O_2}{1 \text{ mole } O_2} = \mathbf{24.0 \text{ g } O_2}$$

(c) Given:

- The unknown hydrocarbon dissolves in CCl_4.
- 135 grams of CCl_4 + 4.36 grams of hydrocarbon.
- B.P. of CCl_4 = 76.8°C
- New B.P. = 78.7°C

- $K_b = 5.02°C/m$

Restatement: Calculate MW of hydrocarbon.

$$\Delta T = K_b \cdot m$$

$$\Delta T = \frac{(K_b)(g/MW)}{kg\ solvent}$$

$$MW = \frac{(K_b)(grams\ solute)}{(\Delta T)(kg\ solvent)}$$

$$MW = \frac{(5.02°C \cdot kg\ solvent / moles\ solute)(4.36\ g)}{(78.7°C - 76.8°C)(0.135\ kg\ solvent)} = 85.2\ g/mol$$

(d) Restatement: Determine molecular formula of hydrocarbon.

$$\frac{85\ g/mol(molecular\ mass)}{14.03\ g/mol(empirical\ mass)} = 6.1$$

Thus, the molecular mass is 6 times greater than the empirical mass.

molecular formula = C_6H_{12}

---

2. An experiment is to be performed to determine the formula mass of a solute (KNO_3) through boiling-point elevation.

(a) What data are needed to calculate the formula mass of the solute? Create appropriate data that can be used in part (c).

(b) What procedures are needed to obtain these data?

(c) List the calculations necessary to determine the formula mass; use your data to calculate the formula mass.

(d) Calculate the % error in your determination of the formula mass of KNO_3 and account for possible error(s).

Answer

2. (a) Restatement: Data needed to determine FM through B.P. elevation.

- $K_b = 0.52°C \cdot m^{-1}$

- Boiling point of water: 100.0°C

- Boiling point of KNO_3 solution: 102.2°C

- Changes in temperature between water and KNO_3 solution: 2.2°C

- Grams of solute: 10.0 g

- Grams of solvent: 50.0 g

(b) Restatement: Procedures needed.

- Measure 50.0 grams of distilled water into a 125-mL Erlenmeyer flask.

- Heat the water to the boiling point, and record the temperature of the water to the nearest 0.5°C.

- Do not let thermometer touch sides or bottom of flask.

- Be sure temperature is constant when reading.

- Prepare a solution of 10.0 grams of KNO_3 in 50.0 grams of distilled water. May have to add more water to the Erlenmeyer due to loss by evaporation.

- Determine B.P. of this solution.

(c) Restatement: Formula mass of solute.

- Change in boiling point

  change in B.P. = B.P. of solution − B.P. of solvent

  = 102.2°C − 100.0°C = 2.2°C

- Molal concentration of solution

  $$\frac{\text{change in B.P. temperature}}{i\left(0.52°C\right)} \times \frac{1 \text{ mole}}{1 \text{ kg}}$$

  where $i = 2$ because two moles of ions are formed for each mole of KNO_3 used.

  $$\frac{2.2°\cancel{C}}{2\left(0.52°\cancel{C}\right)} \times \frac{1 \text{ mole}}{1 \text{ kg}} = 2.1 \text{ moles/kg}$$

- Formula mass of KNO_3

  $$\frac{\text{grams of solute}}{\text{grams of solvent}} \times \frac{1.0 \times 10^3 \text{ g water}}{\text{moles solute}}$$

  $$\frac{10.0 \text{ g } KNO_3}{50.0 \text{ \cancel{g } H_2O}} \times \frac{1.0 \times 10^3 \text{ \cancel{g } H_2O}}{2.1 \text{ moles } KNO_3} = \textbf{95 g/mole}$$

(d) Restatement: Calculate the % error and possible sources.

KNO_3 formula mass (theoretical) = 101 g/mole

KNO_3 formula mass (from data) = 95 g/mole

$$\% \text{ error} = \frac{\text{difference between theoretical and experimental values}}{\text{theoretical value}} \times 100\%$$

$$\% \text{ error} = \frac{101 - 95}{101} \times 100\% = \textbf{6\%}$$

Sources of Error: 1. Experimental error in measurement; 2. Measuring tools not reliable or not sensitive enough; 3. Activity of ions as proposed by Debye and Huckel which states an "effective" concentration called activity which takes into account interionic attractions resulting in a decrease in the magnitude of colligative properties, especially for concentrated solutions.

Key Terms

Words that can be used as topics in essays:

activated complex

activation energy

adsorption

Arrhenius equation

catalyst (heterogeneous
    or homogeneous)

chain reaction

collision model

elementary reaction (step)

enzyme

first order

half-life

integrated rate law

kinetics

molecular orientations

molecular steps (uni, bi, ter)

order of reaction

overall order

rate constant

rate-determining step

rate expression

rate of reaction

reaction mechanism

second order

steric factor

Key Concepts

Equations and relationships that you need to know:

Reaction Order	*0*	*1*	*2*
Rate Law	k	$k[A]$	$k[A]^2$
Integrated Rate Law	$[A] = -kt + [A]_0$	$\ln [A] = -kt + \ln [A]_0$	$\dfrac{1}{[A]} - \dfrac{1}{[A]_0} = kt$
Relationship Between Concentration and Time	$[A]_0 - [A] = kt$	$\log_{10} \dfrac{[A]_0}{[A]} = \dfrac{kt}{2.30}$	$\dfrac{1}{[A]} - \dfrac{1}{[A]_0} = kt$
Half-life	$\dfrac{[A]_0}{2k}$	$0.693/k$	$1/k[A]_0$
Linear Plot	$[A]$ vs. t	$\log [A]$ vs. t	$1/[A]$ vs. t
Slope	$-k$	$-k$	k

- Zero order: $m = 0$, rate is independent of the concentration of the reactant. Doubling the concentration of the reactant does not affect the rate.

  First order: $m = 1$, rate is directly proportional to the concentration of the reactant. Doubling the concentration of the reactant doubles the rate.

  Second order: $m = 2$, rate is proportional to the square of the concentration of the reactant. Doubling the concentration of the reactant increases the rate by a factor of 4.

- rate $= \dfrac{\Delta \text{ concentration}}{\Delta \text{ time}}$

- Arrhenius equation

  $k = Ae^{-E_a/RT}$

  where

  k = rate constant

  A = Arrhenius constant

  e = base of natural logarithm

  E_a = activation energy

  R = universal gas constant

  T = temperature (K)

  $\ln k = \dfrac{-E_a}{RT} + \ln A \quad \ln \dfrac{k_1}{k_2} = \dfrac{E_a}{R}\left(\dfrac{T_1 - T_2}{T_1 T_2}\right)$

  $\log k = \log A - \dfrac{E_a}{2.30RT}$

  $\text{slope} = \dfrac{-E_a}{2.30R}$

  $\log \dfrac{k_2}{k_1} = \dfrac{E_a}{2.30R} \cdot \dfrac{(T_2 - T_1)}{(T_1 T_2)}$

- $\Delta E = \Sigma E_{\text{products}} - \Sigma E_{\text{reactants}}$

- collision theory: rate $= f \cdot Z$

  where

  Z = total number of collisions

  f = fraction of total number of collisions that occur at sufficiently high energy for reaction

  $Z = Z_0[A]^n[B]^m$

  where

  Z_0 = collision frequency when all reactants are at unit concentration

- $\Delta H = E_a - E_a'$

  where

  E_a = forward reaction activation energy

  E_a' = reverse reaction activation energy

Samples: Multiple-Choice Questions

1. Acetaldehyde, CH_3CHO, decomposes into methane gas and carbon monoxide gas. This is a second-order reaction. The rate of decomposition at 140°C is 0.10 mole/liter · sec when the concentration of acetaldehyde is 0.010 mole/liter. What is the rate of the reaction when the concentration of acetaldehyde is 0.50 mole/liter?

   A. 0.50 mole/liter · sec

   B. 1.0 mole/liter · sec

   C. 1.5 mole/liter · sec

   D. 2.0 mole/liter · sec

   E. 2.5 mole/liter · sec

Answer: E

Begin this problem by writing a balanced equation representing the reaction.

$$CH_3CHO(g) \rightarrow CH_4(g) + CO(g)$$

Next, write a rate expression.

$$rate = k(conc.\ CH_3CHO)^2$$

Because you know the rate and the concentration of CH_3CHO, solve for k, the rate-specific constant.

$$k = \frac{rate}{(conc.\ CH_3CHO)^2} \rightarrow \frac{0.10\ mole/liter \cdot sec}{(0.01\ mole/liter)^2} = 10.\ liters/mole \cdot sec$$

Finally, substitute the rate-specific constant and the new concentration into the rate expression.

$$rate = \frac{10.\ \text{liter}}{1\ \text{mole} \cdot sec} \times \left(\frac{0.50\ mole}{1\ liter}\right)^2 = 2.5\ moles/liter \cdot sec$$

2. The rate of the chemical reaction between substances A and B is found to follow the rate law

   $$rate = k[A]^2[B]$$

   where k is the rate constant. The concentration of A is reduced to half its original value. To make the reaction proceed at 50% of its original rate, the concentration of B should be

   A. decreased by ¼

   B. halved

   C. kept constant

   D. doubled

   E. increased by a factor of 4

Answer: D

Let X be what needs to be done to [B].

$$\frac{\text{rate}_{\text{new}}}{\text{rate}_{\text{old}}} = \frac{1}{2} = \frac{k\left([A]/2\right)^2 \cdot X[B]}{k[A]^2 \cdot [B]} = \frac{X}{4}$$

$$X = 2$$

---

3. Which of the following changes will decrease the rate of collisions between gaseous molecules of type A and B in a closed container?

A. Decrease the volume of the container.

B. Increase the temperature of the system.

C. Add A molecules.

D. Take away B molecules.

E. Add an accelerating catalyst.

Answer: D

With all other factors held constant, decreasing the number of molecules decreases the chance of collision. Adding an accelerating catalyst has no effect on the *rate* of collisions. It lowers the activation energy, thereby increasing the chance for *effective* molecular collisions. Furthermore, it increases the rate of production.

---

4. For a certain decomposition reaction, the rate is 0.50 mole/liter · sec when the concentration of the reactant is 0.10 M. If the reaction is second order, what will the new rate be when the concentration of the reactant is increased to 0.40 M?

A. 0.50 mole/liter · sec

B. 1.0 mole/liter · sec

C. 8.0 mole/liter · sec

D. 16 mole/liter · sec

E. 20. mole/liter · sec

Answer: C

The concentration of the reactant is increased by a factor of 4, from 0.10 M to 0.40 M. If the reaction is second order, the rate will then increase by a factor of $4^2 = 16$.

$$\frac{16}{1} \times \frac{0.50 \text{ mole}}{1 \text{ liter} \cdot \text{sec}} = \frac{8.0 \text{ moles}}{1 \text{ liter} \cdot \text{sec}}$$

5. The rate-determining step of a several-step reaction mechanism has been determined to be

$$3X(g) + 2Y(g) \rightarrow 4Z(g)$$

When 3.0 moles of gas X and 2.0 moles of gas Y are placed in a 5.0-liter vessel, the initial rate of the reaction is found to be 0.45 mole/liter · min. What is the rate constant for the reaction?

A. $\dfrac{0.45}{\left(\dfrac{3.0}{5.0}\right)^3 \left(\dfrac{2.0}{5.0}\right)^2}$

B. $\dfrac{0.45}{(3.0)(2.0)}$

C. $\dfrac{0.45}{\left(\dfrac{3.0}{5.0}\right)^2 \left(\dfrac{2.0}{5.0}\right)^3}$

D. $\dfrac{0.45}{\left(\dfrac{3.0}{5.0}\right)\left(\dfrac{2.0}{5.0}\right)}$

E. $\dfrac{(3.0)^3 (2.0)^3}{0.45}$

Answer: A

Given a reaction mechanism, the order with respect to each reactant is its coefficient in the chemical equation for that step. The slowest step is the rate-determining step, so

$$\text{rate} = k[\text{X}]^3[\text{Y}]^2$$

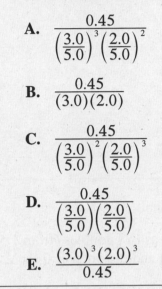

$$k = \frac{\text{rate}}{[\text{X}]^3[\text{Y}]^2} = \frac{0.45 \text{ mole/liter} \cdot \text{min}}{\left(\dfrac{3.0 \text{ moles}}{5.0 \text{ liters}}\right)^3 \left(\dfrac{2.0 \text{ moles}}{5.0 \text{ liters}}\right)^2}$$

6. The reaction $2X + Y \rightarrow 3Z$ was studied and the following data were obtained:

Experiment	X	Y	Rate (mole / liter · sec)
1	3.0	1.5	1.8
2	1.5	3.0	0.45
3	1.5	1.5	0.45

What is the proper rate expression?

A. rate = $k[X][Y]$

B. rate = $k[Y]^2$

C. rate = $k[X]$

D. rate = $k[X]^2[Y]$

E. rate = $k[X]^2$

Answer: E

Examine experiments 2 and 3, wherein [X] is held constant. Note that as [Y] doubles (from 1.5 to 3.0), the rate does not change. Hence, the rate is independent of [Y] and the order is 0 for Y.

Now examine experiments 1 and 3, wherein [Y] is held constant. Note that as [X] doubles, the rate is increased by a factor of 4. In this case, the rate is proportional to the square of the concentration of the reactant. This is a second-order reactant.

Combining these reactant orders in a rate equation gives

$$\text{rate} = k[X]^2[Y]^0 = k[X]^2$$

For Samples 7 and 8, refer to the following diagram.

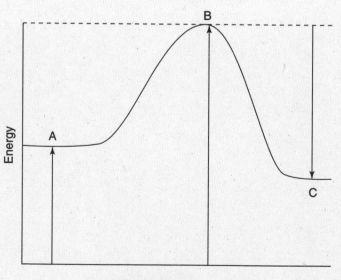

Path of Reaction →

---

7. The activation energy is represented by

    A. *A*

    B. *B*

    C. *C*

    D. *B − A*

    E. *B − C*

Answer: D

The activation energy is the amount of energy that the reactants must absorb from the system in order to react. In the reaction diagram, the reactants begin at *A*. The reactants must absorb the energy from *A* to *B* in order to form the activated complex. The energy necessary to achieve this activated complex is the distance from *A* to *B* in the diagram and is mathematically the difference $(B − A)$.

---

8. The enthalpy of the reaction is represented by

    A. B − (C − A)

    B. B

    C. C − A

    D. B − C

    E. A − (B − C)

Answer: C

The enthalpy of the reaction, ΔH, is the difference between the enthalpies of the products and the enthalpies of the reactants.

$$\Delta H = H_{products} − H_{reactants}$$

The products are represented at point *C* and the reactants are represented at point *A*, so the change in enthalpy is $C − A$.

9. Given the following reaction mechanism, express the rate of the overall reaction.

$2\,A + 3\,C \rightarrow 4\,D + 5\,E$ slow
$A + 2\,D \rightleftharpoons 2\,A + 2\,E$ fast
$2\,E + C \rightarrow 3\,D + 2\,C$ fast

A $k \cdot 2[A]^2 3[C]^3$

B $k \cdot [D]^4[E]^5 / [A]^2[C]^3$

C $k \cdot [D]^4[E]^5$

D $k \cdot [A]^3[C]^2$

E $k \cdot [A]^2[C]^3$

Answer: E

The rate is determined by the slowest step. Only when given the complete mechanisms *in steps* can you identify the slowest step and use the coefficients of the reactants as the orders in the rate equation.

10. Referring to example 9, express the overall reaction.

A. $3\,C + 2\,D \rightarrow 2\,A + 4\,E$

B. $3\,A + 3\,C \rightarrow 3\,D + 4\,E$

C. $2\,A + 3\,C \rightarrow A + 4\,E$

D. $A + 2\,C \rightarrow 5\,D + 5\,E$

E. $2\,A + C \rightarrow D + 2\,E$

Answer: D

To find the overall equation when given the reaction mechanism, simply add the reactions.

$2\,A + 3\,C \rightarrow 4\,D + 5\,E$ slow
$A + 2\,D \rightleftharpoons 2\,A + 2\,E$ fast
$2\,E + C \rightarrow 3\,D + 2\,C$ fast

There is a total of 3 A's on the left and 2 A's on the right: balance of 1 A on the left. There is a total of 4 C's on the left and 2 C's on the right: balance of 2 C's on the left. There is a total of 2 D's on the left and 7 D's on the right: balance of 5 D's on the right. There is a total of 2 E's on the left and 7 E's on the right: balance of 5 E's on the right.

$A + 2\,C \rightarrow 5\,D + 5\,E$

Samples: Free-Response Questions

> **1.** The reaction
>
> $$2 NO_2(g) + Cl_2(g) \rightarrow 2 NO_2Cl(g)$$
>
> was studied at 20°C and the following data were obtained:
>
> | Experiment | Initial $[NO_2]$ (mole · liter^{-1}) | Initial $[Cl_2]$ (mole · liter^{-1}) | Initial Rate of Increase of $[NO_2Cl]$ (mole · liter^{-1} · sec^{-1}) |
> |---|---|---|---|
> | 1 | 0.100 | 0.005 | 1.35×10^{-7} |
> | 2 | 0.100 | 0.010 | 2.70×10^{-7} |
> | 3 | 0.200 | 0.010 | 5.40×10^{-7} |
>
> **(a)** Write the rate law for the reaction.
>
> **(b)** What is the overall order for the reaction? Explain.
>
> **(c)** Calculate the rate-specific constant, including units.
>
> **(d)** In Experiment 3, what is the initial rate of decrease of $[Cl_2]$?
>
> **(e)** Propose a mechanism for the reaction that is consistent with the rate law expression you found in part (a).

Answer

1. Given: $2 NO_2(g) + CO_2(g) \rightarrow 2 NO_2Cl(g)$

(a) Restatement: Rate law.

rate $= k[NO_2]^n[Cl_2]^m$

Expt. 1: rate $= 1.35 \times 10^{-7}$ mole/liter · sec

$= k(0.100 \text{ M})^n(0.0050 \text{ M})^m$

Expt. 2: rate $= 2.70 \times 10^{-7}$ mole/liter · sec

$= k(0.100 \text{ M})^n(0.010 \text{ M})^m$

Expt. 3: rate $= 5.40 \times 10^{-7}$ mole/liter · sec

$= k(0.200 \text{ M})^n(0.010 \text{ M})^m$

$$\frac{\text{rate } 2}{\text{rate } 1} = \frac{2.70 \times 10^{-7} \text{ mole/liter} \cdot \text{sec}}{1.35 \times 10^{-7} \text{ mole/liter} \cdot \text{sec}}$$

$$= \frac{\cancel{k}(\cancel{0.100 \text{M}})^n(0.010 \text{M})^m}{\cancel{k}(\cancel{0.100 \text{M}})^n(0.0050 \text{M})^m}$$

$$= 2.00 = (2.0)^m \quad m = 1$$

$$\frac{\text{rate } 3}{\text{rate } 2} = \frac{5.40 \times 10^{-7}\,\text{mole/liter} \cdot \text{sec}}{2.70 \times 10^{-7}\,\text{mole/liter} \cdot \text{sec}}$$

$$= \frac{k(0.200\text{ M})^n \cancel{(0.010\text{ M})^m}}{k(0.100\text{ M})^n \cancel{(0.010\text{ M})^m}}$$

$$= 2.00 = (2.00)^n \quad n = 1$$

$$\text{rate} = k[NO_2]^1[Cl_2]^1$$

(b) Restatement: Overall order. Explain.

overall order $= m + n = 1 + 1 = 2$

The rate is proportional to the product of the concentrations of the two reactants:

$$2\,NO_2(g) + Cl_2(g) \rightarrow 2\,NO_2Cl(g)$$

$$\text{rate} = \frac{-\Delta[NO_2]}{\Delta t} = k[NO_2][Cl_2]$$

or

$$\text{rate} = \frac{-2\Delta[Cl_2]}{\Delta t} = k[NO_2][Cl_2]$$

(c) Restatement: Rate-specific constant k.

rate $= k[NO_2][Cl_2]$

$k = $ rate $/ [NO_2][Cl_2]$

$$= \frac{1.35 \times 10^{-7}\,\text{mole} \cdot \text{liter}^{-1} \cdot \text{sec}^{-1}}{(0.100\,\text{mole} \cdot \text{liter}^{-1})(0.005\,\text{mole} \cdot \text{liter}^{-1})}$$

$$= 2.7 \times 10^{-4}\,\text{liter/mole} \cdot \text{sec}$$

(d) Restatement: In experiment 3, initial rate of decrease of $[Cl_2]$.

$$2\,NO_2(g) + Cl_2(g) \rightarrow 2\,NO_2Cl(g)$$

$$\frac{-d[Cl_2]}{dt} = \frac{1}{2}\left(\frac{d[NO_2Cl]}{dt}\right)$$

$$= -(5.40 \times 10^{-7})/2$$

$$= -2.7 \times 10^{-7}\,\text{mole} \cdot \text{liter}^{-1} \cdot \text{sec}^{-1}$$

(e) Restatement: Possible mechanism.

The proposed mechanism must satisfy two requirements:

(1) The sum of the steps must give a balanced equation.

(2) The mechanism must agree with the experimentally determined rate law.

$$NO_2(g) + Cl_2(g) \xrightarrow{k_1} NO_2Cl(g) + Cl(g) \quad \text{slow}$$

$$Cl(g) + NO_2(g) \xrightarrow{k_2} NO_2Cl(g) \quad \text{fast}$$

Requirement 1:

$$NO_2 + Cl_2 \rightarrow NO_2Cl + \cancel{Cl}$$

$$\underline{\cancel{Cl} + NO_2 \rightarrow NO_2Cl}$$

$$2NO_2 + Cl_2 \rightarrow 2NO_2Cl$$

Requirement 2:

$$NO_2(g) + Cl_2(g) \rightarrow NO_2Cl + Cl$$

is the rate-determining step. This step is bimolecular.

$$rate = k_1[NO_2][Cl_2]$$

as found in part (a).

Meeting these two requirements does not prove that this *is* the mechanism for the reaction — only that it *could be*.

---

2. Hydrogen peroxide, H_2O_2, decomposes by first-order decomposition and has a rate constant of 0.015/min at 200°C. Starting with a 0.500 M solution of H_2O_2, calculate:

(a) The molarity of H_2O_2 after 10.00 min

(b) The time it will take for the concentration of H_2O_2 to go from 0.500 M to 0.150 M

(c) The half-life

Answer

2. Given: H_2O_2, first-order decomposition

$$k = 0.015/min \text{ at } 200°C$$

$$[H_2O_2]_0 = 0.500 \text{ M}$$

(a) Restatement: Calculate $[H_2O_2]$ after 10.00 min.

For first-order reactions: $rate = kX \rightarrow -dX/dt = kX$

Integrating from X_0 to X gives

$$-\ln\left(\frac{X}{X_0}\right) = kt$$

$$\log\left(\frac{X_0}{X}\right) = \frac{kt}{2.30}$$

$$\log\left(\frac{0.500M}{X}\right) = 0.065$$

$$\frac{0.500}{X} = 10^{0.065}$$

$$\frac{0.500}{X} = 1.2$$

$$X = \frac{0.500}{1.2} = 0.42 \text{ M}$$

(b) Restatement: Time for H_2O_2 to go from 0.500 M to 0.150 M.

For first-order reactions:

$$\log\left(\frac{X_0}{X}\right) = \frac{kt}{2.30}$$

$$t = \frac{2.30}{k}\log\left(\frac{X_0}{X}\right)$$

$$= \left(\frac{2.30}{0.015/\text{min}}\right)\log\left(\frac{0.500\,\text{M}}{0.150\,\text{M}}\right)$$

$$= \frac{2.30}{0.015/\text{min}} \times 0.523$$

$$= 8.0 \times 10^1\,\text{min}$$

(c) Restatement: Calculate the half-life.

For first-order reactions:

$$t_{1/2} = \frac{0.693}{k} = \frac{0.693}{0.015/\text{min}} = 46\,\text{min}$$

Equilibrium

Key Terms

Words that can be used as topics in essays:

5% rule

buffer

common ion effect

equilibrium expression

equivalence point

Henderson-Hasselbalch
equation

heterogeneous equilibria

homogeneous equilibria

indicator

ion product, P

K_a

K_b

K_c

K_{eq}

K_p

K_{sp}

K_w

law of mass action

Le Chatelier's principle

limiting reactant

method of successive approximation

net ionic equation

percent dissociation

pH

pK_a

pK_b

pOH

reaction quotient, Q

reciprocal rule

rule of multiple equilibria

solubility

spectator ions

strong acid

strong base

van't Hoff equation

weak acid

weak base

Key Concepts

Equations and relationships that you need to know:

- $\Delta G° = -RT \ln K = -0.0191\ T \log K_p$

  $\Delta G° \ll 0$, K is large; $\Delta G° \gg 0$, K is very small

- $pK_a = -\log K_a$

  $$HA_{(aq)} + H_2O_{(1)} \rightleftharpoons H_3O^+{}_{(aq)} + A^-{}_{(aq)}$$

  acid　　　base　　conjugate　　conjugate
  　　　　　　　　　acid　　　　　base

  $$K_a = \frac{[H_3O^+][A^-]}{[HA]} \rightleftharpoons \frac{[H^+][A^-]}{[HA]}$$

  $$B_{(aq)} + H_2O_{(1)} \rightleftharpoons BH^+{}_{(aq)} + OH^-{}_{(aq)}$$

  acid　　　base　　conjugate　　conjugate
  　　　　　　　　　acid　　　　　base

  $$K_b = \frac{[BH^+][OH^-]}{[B]}$$

- $K_a \times K_b = 1.0 \times 10^{-14} = K_w = [H^+][OH^-]$

  $K = K_1 \times K_2 \times \ldots$

- Q = ion product (or reaction quotient) = calculated as K_{sp} except using initial concentrations

  $Q > K_{sp}$, supersaturated solution, precipitate forms until such point that $Q = K_{sp}$

  $Q < K_{sp}$, unsaturated solution, no precipitate

  $Q = K_{sp}$, saturated solution, no precipitate forms

- Henderson-Hasselbalch: $pH = pK_a + \log \dfrac{[A^-]}{[HA]} = pK_a + \log \dfrac{[base]}{[acid]}$

- van't Hoff equation: $\ln \dfrac{K_2}{K_1} = -\dfrac{\Delta H°}{R}\left[\dfrac{1}{T_2} - \dfrac{1}{T_1}\right]$

- Clausius-Clapeyron equation: $\ln \dfrac{P_2°}{P_1°} = \dfrac{\Delta H°_{vap}}{R}\left[\dfrac{1}{T_2} - \dfrac{1}{T_1}\right]$

- $K_p = K_c(RT)^{\Delta n}$

  where Δn = moles gaseous products − moles gaseous reactants

Samples: Multiple-Choice Questions

1. 6.0 moles of chlorine gas are placed in a 3.0-liter flask at 1250 K. At this temperature, the chlorine molecules begin to dissociate into chlorine atoms. What is the value for K_c if 50.% of the chlorine molecules dissociate when equilibrium has been achieved?

   A. 1.0

   B. 3.0

   C. 4.0

   D. 6.0

   E. 12.0

Answer: C

Begin by writing the balanced equation in equilibrium.

$$Cl_2(g) \leftrightarrow 2\ Cl(g)$$

Next, write an equilibrium expression.

$$K_c = \frac{[Cl]^2}{[Cl_2]}$$

Then create a chart that outlines the initial and final concentrations for the various species.

Species	Initial Concentration	Final Concentration
Cl_2	$\dfrac{6.0 \text{ moles}}{3.0 \text{ liters}} = 2.0 \text{ M}$	$6.0 \text{ moles} - (0.5)(6.0) = 3.0 \text{ moles}$ $\dfrac{3.0 \text{ moles}}{3.0 \text{ liters}} = 1.0 \text{ M}$
Cl	$\dfrac{0 \text{ moles}}{3.0 \text{ liters}} = 0 \text{ M}$	$\dfrac{3.0 \text{ moles } Cl_2 \text{ dissociated}}{1} \times \dfrac{2 \text{ moles Cl}}{1 \text{ mole } Cl_2}$ $= 6.0 \text{ moles Cl at equilibrium}$ $\dfrac{6.0 \text{ moles Cl}}{3.0 \text{ liters}} = 2.0 \text{ M}$

Finally, substitute the concentrations (at equilibrium) into the equilibrium expression.

$$K_c = \frac{[Cl]^2}{[Cl_2]} = \frac{(2.0)^2}{1.0} = 4.0$$

2. 6.00 moles of nitrogen gas and 6.00 moles of oxygen gas are placed in a 2.00-liter flask at 500°C and the mixture is allowed to reach equilibrium. What is the concentration, in moles per liter, of nitrogen monoxide at equilibrium if the equilibrium constant is found to be 4.00?

- **A.** 3.00 M
- **B.** 6.00 M
- **C.** 8.00 M
- **D.** 10.0 M
- **E.** 12.0 M

Answer: A

Step 1: Write the balanced equation at equilibrium.

$$N_2(g) + O_2(g) \leftrightarrow 2\,NO(g)$$

Step 2: Write the equilibrium expression.

$$K_{eq} = \frac{[NO]^2}{[N_2][O_2]} = 4.00$$

Step 3: Create a chart that shows the initial concentrations, the final concentrations, and the changes in concentration. Let x represent the concentration (M) of either N_2 or O_2 (their concentrations are in a 1:1 molar ratio) that is transformed through the reaction into NO.

Species	Initial Concentration	Change in Concentration	Final Concentration
N_2	3.00 M	$-x$	$3.00 - x$
O_2	3.00 M	$-x$	$3.00 - x$
NO	0 M	$+2x$	$2x$

Step 4: Take the concentrations at equilibrium and substitute them into the equilibrium expression.

$$K_{eq} = \frac{[NO]^2}{[N_2][O_2]} = \frac{(2x)^2}{(3.00-x)^2} = 4.00$$

Step 5: Solve for x by taking the square root of both sides.

$$\frac{2x}{3.00-x} = 2.00$$
$$2x = 6.0 - 2.00x$$
$$x = 1.50$$

Step 6: Plug the value for *x* into the expression of the equilibrium concentration for NO.

$$[NO] = 2x = 2(1.50) = 3.00 \text{ M}$$

Note: You could also solve *this problem* using equilibrium partial pressures of the gases:

$$K_p = \frac{P_{NO^2}}{(P_{N_2})(P_{O_2})} = 4.00$$

Use $P = CRT$, where C represents the molar concentration of the gas, $\frac{n}{V}$

Try this approach to confirm that $[NO] = 3.00$ M.

---

3. Solid carbon reacts with carbon dioxide gas to produce carbon monoxide. At 1,500°C, the reaction is found to be at equilibrium with a K_p value of 0.50 and a total pressure of 3.5 atm. What is the proper expression for the partial pressure (in atmospheres) of the carbon dioxide?

A. $\dfrac{-0.5 + \sqrt{[(0.50)^2 - 4(1)(-3.5)]}}{2(1)}$

B. $\dfrac{-0.5 + \sqrt{[(0.50)^2 - 4(1)(-1.75)]}}{2(1)}$

C. $\dfrac{-0.5 + \sqrt{[(0.50) - 4(1)(-1.75)]}}{2(1)}$

D. $\dfrac{-0.5 + \sqrt{[(0.50)^2 - 2(1)(3.5)]}}{2(1)}$

E. $\dfrac{0.5 + \sqrt{[(0.50)^2 - 4(1)(-1.75)]}}{2(1)}$

Answer: B

Step 1: Write the balanced equilibrium equation.

$$C(s) + CO_2(g) \leftrightarrow 2\, CO(g)$$

Step 2: Write the equilibrium expression.

$$K_p = \frac{(P_{CO})^2}{P_{CO_2}} = 0.50$$

Step 3: Express the two unknowns, pressure of CO and pressure of CO_2, in terms of a single unknown, pressure of CO.

$$P_{total} = P_{CO} + P_{CO_2} = 3.5 \text{ atm}$$
$$P_{CO_2} = 3.5 \text{ atm} - P_{CO}$$

Step 4: Rewrite the equilibrium expression in terms of the single unknown.

$$K_p = 0.50 = \frac{(P_{CO})^2}{3.5 - P_{CO}}$$

Step 5: Rewrite this relationship in terms of the quadratic equation so that you can solve for the unknown x, the pressure of the CO.

$$x^2 = 1.75 - 0.50x$$

Putting this equation into the standard form, $ax^2 + bx + c = 0$, you get

$$x^2 + 0.50x - 1.75 = 0$$

Step 6: Use the quadratic equation to solve for x.

$$x = \frac{-0.50 \pm \sqrt{[(0.50)^2 - 4(1)(-1.75)]}}{2(1)}$$

For Examples 4 and 5, use the following information:

A student prepared a 1.00 M acetic acid solution ($HC_2H_3O_2$). The student found the pH of the solution to be 2.00.

---

4. What is the K_a value for the solution?

   A 3.00×10^{-7}

   B 2.00×10^{-6}

   C 2.00×10^{-5}

   D 1.00×10^{-4}

   E 1.00×10^{-3}

Answer: D

Step 1: Write the balanced equation in a state of equilibrium.

$$HC_2H_3O_{2(aq)} \rightleftharpoons H^+_{(aq)} + C_2H_3O_2^-{}_{(aq)}$$

Step 2: Write the equilibrium expression.

$$K_a = \frac{[H^+][C_2H_3O_2^-]}{[HC_2H_3O_2]}$$

Step 3: Use the pH of the solution to determine [H^+].

$$pH = -\log[H^+] \quad = 2.00$$
$$H^+ = 10^{-2.00} \quad\quad = 0.0100 \text{ M}$$

Step 4: Determine $\left[C_2H_3O_2^-\right]$

The molar ratio of [H⁺] to $\left[C_2H_3O_2^-\right]$ is 1:1, $\left[C_2H_3O_2^-\right] = 0.0100$ M also.

Step 5: Substitute the concentrations into the equilibrium expression.

$$K_a = \frac{\left[H^+\right]\left[C_2H_3O_2^-\right]}{\left[HC_2H_3O_2\right]} = \frac{(0.0100)^2}{0.99} \approx 1.00 \times 10^{-4}$$

---

5. What is the % dissociation of the acetic acid? (Use the 5% rule.)

  A. 0.05%

  B. 1.00%

  C. 1.50%

  D. 2.00%

  E. 2.50%

Answer: B

Step 1: Write the generic formula for % dissociation.

$$\% \text{ dissociation} = \frac{\text{part}}{\text{whole}} \times 100\% = \frac{M_{HC_2H_3O_2 \text{ dissociated}}}{M_{HC_2H_3O_2 \text{ available}}} \times 100\%$$

Step 2: Substitute the known information into the generic equation and solve.

$$\% \text{ dissociation} = \frac{0.0100}{0.99} \times 100\% \approx 1.00\%$$

Note: The 5% rule states that the approximation $a - x \approx a$ is valid if $x < 0.05a$. The rule depends on the generalization that the value of the constant in the equation in which x appears is seldom known to be better than 5%.

---

6. Given that the first, second, and third dissociation constants for H_3PO_4 are 7.0×10^{-3}, 6.0×10^{-8}, and 5.0×10^{-13}, respectively, calculate K for the overall reaction.

  A. 2.10×10^{-32}

  B. 2.10×10^{-28}

  C. 2.10×10^{-22}

  D. 2.10×10^{-11}

  E. 2.10×10^{22}

Answer: C

This problem involves the concept of multiple equilibria. The dissociation constants given in the example are related to the following reactions:

$$H_3PO_4 \text{ (aq)} \rightleftharpoons H^+\text{(aq)} + H_2PO_4^-\text{(aq)} \qquad K_1 = 7.0 \times 10^{-3}$$
$$H_2PO_4^-\text{(aq)} \rightleftharpoons H^+\text{(aq)} + HPO_4^{2-}\text{(aq)} \qquad K_2 = 6.0 \times 10^{-8}$$
$$HPO_4^{2-}\text{(aq)} \rightleftharpoons H^+\text{(aq)} + PO_4^{3-}\text{(aq)} \qquad K_3 = 5.0 \times 10^{-13}$$

For multiple equilibria dissociation constants (such as polyprotic acids), K for the overall reaction is the product of the equilibrium constants for the individual reactions. Therefore,

$$K = K_1 \times K_2 \times K_3 = \frac{\left(H^+\right)\left(H_2PO_4^-\right)}{\left(H_3PO_4\right)} \times \frac{\left(H^+\right)\left(HPO_4^{2-}\right)}{\left(H_2PO_4^-\right)}$$

$$\times \frac{\left(H^+\right)\left(PO_4^{3-}\right)}{\left(HPO_4^{2-}\right)} = \frac{\left(H^+\right)^3\left(PO_4^{3-}\right)}{\left(H_3PO_4\right)} = 210 \times 10^{-24} = 2.10 \times 10^{-22}$$

which is the equilibrium constant for the sum of three individual reactions:

$$H_3PO_4 \text{ (aq)} \rightleftharpoons 3\, H^+\text{(aq)} + PO_4^{3-}\text{(aq)}$$

---

7. A buffer is found to contain 0.35 M NH_3 ($K_b = 1.8 \cdot 10^{-5}$) and 0.20 M NH_4Cl. What would be the mathematical expression for K_b in terms of $\left[NH_4^+\right]$, $[OH^-]$, and $[NH_3]$?

A. $1.8 \times 10^{-5} = \dfrac{(0.35 + x)(x)}{0.20}$

B. $1.8 \times 10^{-5} = \dfrac{(0.35 - x)(0.20x)}{0.20}$

C. $1.8 \times 10^{-5} = \dfrac{(0.20 + x)(x)}{(0.35 - x)}$

D. $1.8 \times 10^{-5} = \dfrac{(0.20)(x)}{0.35}$

E. $1.8 \times 10^{-5} = \dfrac{(0.20 + x)(0.35 - x)}{0.35}$

Answer: C

Step 1: Get a picture of the solution in equilibrium. Ammonia (NH_3) is a weak base. NH_3 reacts with water in accordance with the following equilibrium equation:

$$NH_{3\,\text{(aq)}} + H_2O_{\text{(}\ell\text{)}} \rightleftharpoons NH_{4\,\text{(aq)}}^+ + OH_{\text{(aq)}}^-$$

Ammonium chloride is soluble in water (see the solubility rules on page 116). Therefore, the concentration NH_4^+ and that of Cl^- are both 0.20 M.

Step 2: Write an equilibrium expression.

$$K_b = \frac{[NH_4^+][OH^-]}{[NH_3]} = 1.8 \times 10^{-5}$$

Step 3: Set up a chart showing initial concentrations of the species and final concentrations at equilibrium (the Cl^- does not contribute to the pH). Let x represent the portion of NH_3 that eventually converts to NH_4^+. x will also represent the amount by which the concentration of NH_4^+ increases.

Species	Initial Concentration	Final Concentration
NH_3	0.35 M	$0.35 - x$
NH_4^+	0.20 M	$0.20 + x$
OH^-	~0 M*	~x

Because NH_3 is a weak base and you are using a relatively weak solution (0.35 M), for calculating purposes you can essentially claim that OH_{init}^- is 0. Further, because [OH^-] and NH_4^+ at equilibrium are in a 1: 1 molar ratio, the concentration of OH^- will increase by the amount x.

Step 4: Substitute these chart values into the equilibrium expression.

$$K_b = \frac{[NH_4^+][OH^-]}{[NH_3]} = 1.8 \times 10^{-5} = \frac{(0.20 + x)(x)}{(0.35 - x)}$$

Since, $K_b <<< 0.20$, $K_b = \frac{(0.20)(x)}{(0.35)} = 1.8 \times 10^{-5}$ could be used to solve for x. Therefore C is the correct expression, but D could be used to solve for the answer.

Note: If this question were in the free-response section (calculators allowed), and if the question asked you to solve for the pH of the buffer, you would need to solve for x using the quadratic equation. x would be 3.2×10^{-5}. Because [OH^-] $\approx 3.2 \times 10^{-5}$ and pOH = -log [OH^-], pOH = 4.5. pH = 14.0 – pOH = 14.0 – 4.5 = 9.5.

8. Copper(II) iodate has a solubility of 3.3×10^{-3} M at 25°C. Calculate its K_{sp} value.

   A. 1.4×10^{-7}

   B. 1.1×10^{-5}

   C. 3.3×10^{-3}

   D. 5.1×10^{-1}

   E. 3.3×10^{3}

Answer: A

Step 1: Write the equation for the dissociation of copper(II) iodate.

$$Cu(IO_3)_{2(s)} \rightleftharpoons Cu_{(aq)}^{2+} + 2IO_{3(aq)}^{-}$$

Step 2: Write down the concentrations during the process of dissociation.

$$3.3 \times 10^{-3} M\, Cu(IO_3)_2 \rightarrow 3.3 \times 10^{-3} M\, Cu^{2+}(aq) + 2\left(3.3 \times 10^{-3} M\right) IO_3^{-}(aq)$$

$$\left[Cu^{2+}\right] = 3.3 \times 10^{-3} M$$
$$\left[IO_3^{-}\right] = 6.6 \times 10^{-3} M$$

Step 3: Write the equilibrium expression.

$$K_{sp} = [Cu^{2+}][IO_3^{-}]^2$$

Step 4: Substitute the equilibrium concentration into the K_{sp} expression.

$$K_{sp} = (3.3 \times 10^{-3})(6.6 \times 10^{-3})^2 = 1.4 \times 10^{-7}$$

---

9. Lead iodide has a K_{sp} value of 1.08×10^{-7} at 20°C. Calculate its solubility at 20°C.

A. 5.00×10^{-8}

B. 3.00×10^{-6}

C. 1.00×10^{-4}

D. 6.00×10^{-3}

E. 3.00×10^{-3}

Answer: E

Step 1: Write the equilibrium equation for the dissociation of lead iodide.

$$PbI_2(s) \leftrightarrow Pb^{2+}(aq) + 2\,I^{-}(aq)$$

Step 2: Write the equilibrium expression.

$$K_{sp} = [Pb^{2+}][I^{-}]^2 = 1.08 \times 10^{-7}$$

Step 3: Set up a chart that expresses initial and final concentrations (at equilibrium) of the $Pb^{2+}(aq)$ and $I^{-}(aq)$.

Species	Initial Concentration	Final Concentration at Equilibrium
Pb^{2+}	0 M	x
I^{-}	0 M	$2x$

Step 4: Substitute the equilibrium concentrations of the ions into the equilibrium expression.

$$K_{sp} = [Pb^{2+}][I^-]^2 = (x)(2x)^2 = 1.08 \times 10^{-7}$$
$$4x^3 = 1.08 \times 10^{-7}$$
$$x^3 = 27 \times 10^{-9}$$
$$x = 3.0 \times 10^{-3}$$

---

10. Will a precipitate form when one mixes 75.0 mL of 0.050 M K_2CrO_4 solution with 75.0 mL of 0.10 M $Sr(NO_3)_2$? K_{sp} for $SrCrO_4 = 3.6 \times 10^{-5}$

   A. Yes, a precipitate will form, $Q > K_{sp}$

   B. Yes, a precipitate will form, $Q < K_{sp}$

   C. Yes, a precipitate will form, $Q = K_{sp}$

   D. No, a precipitate will not form, $Q > K_{sp}$

   E. No, a precipitate will not form, $Q < K_{sp}$

Answer:

Recognize that this problem is one involving the ion product, Q. We calculate Q in the same manner as K_{sp}, except that we use initial concentrations of the species instead of equilibrium concentrations. We then compare the value of Q to that of K_{sp}:

   If $Q < K_{sp}$ — no precipitate.

   If $Q = K_{sp}$ — no precipitate.

   If $Q > K_{sp}$ — a precipitate forms.

At this point you can rule out choices B, C, and D because they do not make sense. If you have forgotten how to do the problem mathematically, you should guess now; you have a 50% chance of getting the answer right.

Step 1: Realize that this problem involves a possible double displacement, the possible precipitate being either KNO_3 or $SrCrO_4$. Rule out the KNO_3 since most nitrates are soluble and that you were provided with the K_{sp} for $SrCrO_4$. To answer these questions, you must know your solubility rules!

Step 2: Write the net ionic equation.

$$Sr^{+2}{}_{(aq)} + CrO_4{}^{2-}{}_{(aq)} \rightarrow SrCrO_{4(s)}$$

Step 3: Write the equilibrium expression.

$$K_{sp} = [Sr^{2+}][CrO_4{}^{2-}]$$

Step 4: Determine the initial concentrations of the ions that may form the precipitate in the mixed solution.

$$Sr^{2+}_{(aq)}: \frac{0.075 \, \text{liter}}{1} \times \frac{0.10 \, \text{mole}}{1 \, \text{liter}} = 7.5 \times 10^{-3} \, \text{mole}$$

$$CrO_4^{2-}_{(aq)}: \frac{0.075 \, \text{liter}}{1} \times \frac{0.05 \, \text{mole}}{1 \, \text{liter}} = 3.8 \times 10^{-3} \, \text{mole}$$

The total liters of solution = 0.075 + 0.075 = 0.15 liter. Therefore,

$$\left[Sr^{2+}\right] = \frac{7.5 \times 10^{-3} \, \text{mole}}{0.15 \, \text{liter}} = 0.050 \, M$$

$$\left[CrO_4^{2-}\right] = \frac{3.8 \times 10^{-3} \, \text{mole}}{0.15 \, \text{liter}} = 0.025 \, M$$

Step 5: Determine Q, the ion product.

$$Q = \left[Sr^{2+}\right]\left[CrO_4^{2-}\right] = (0.05)(0.025) = 1.3 \times 10^{-3}$$

Therefore, since Q (1.3×10^{-3}) > K_{sp} (7.1×10^{-4}), a precipitate will form.

Samples: Free-Response Questions

The equilibrium problem will represent one of three possible types:

> Gaseous equilibrium — K_c or K_p
>
> Acid-base equilibrium — K_a or K_b
>
> Solubility — K_{sp}

---

1. 250.0 grams of solid copper(II) nitrate is placed in an empty 4.0-liter flask. Upon heating the flask to 250°C, some of the solid decomposes into solid copper(II) oxide, gaseous nitrogen(IV) oxide, and oxygen gas. At equilibrium, the pressure is measured and found to be 5.50 atmospheres.

   (a) Write the balanced equation for the reaction.

   (b) Calculate the number of moles of oxygen gas present in the flask at equilibrium.

   (c) Calculate the number of grams of solid copper(II) nitrate that remained in the flask at equilibrium.

   (d) Write the equilibrium expression for K_p and calculate the value of the equilibrium constant.

   (e) If 420.0 grams of the copper(II) nitrate had been placed into the empty flask at 250°C, what would the total pressure have been at equilibrium?

Answer

1. Given: 250.0 g $Cu(NO_3)_2$ in 4.0-liter flask

   Heated to 250°C, reaches equilibrium

   Total pressure at equilibrium = 5.50 atmospheres

   (a) Restatement: Balanced reaction.

   $$2\ Cu(NO_3)_2(s) \leftrightarrow 2\ CuO(s) + 4\ NO_2(g) + O_2(g)$$

(b) Restatement: Moles of $O_2(g)$ at equilibrium.

$$PV = nRT$$

$$n_{gas} = \frac{PV}{RT} = \frac{(5.50\,\text{atm})(4.0\,\text{liters})}{(0.0821\,\text{liter}\cdot\text{atm}/\text{mole}\cdot\text{K})(523\,\text{K})}$$

$$= 0.51\,\text{mole gas}$$

$$\frac{0.51\,\text{mole gas}}{1} \times \frac{1\,\text{mole oxygen gas}}{5\,\text{mole total gas}} = \mathbf{0.10\,mole\ O_2}$$

(c) Restatement: Grams of solid $Cu(NO_3)_2$ in flask at equilibrium.

moles of $Cu(NO_3)_2$ that decomposed:

$$\frac{0.10\,\text{mole O}_2}{1} \times \frac{2\,\text{moles Cu(NO}_3)_2}{1\,\text{mole O}_2} = 0.20\,\text{mole Cu(NO}_3)_2$$

mass of $Cu(NO_3)_2$ that decomposed:

$$\frac{0.20\,\text{mole Cu(NO}_3)_2}{1} \times \frac{187.57\,\text{g Cu(NO}_3)_2}{1\,\text{mole Cu(NO}_3)_2} = 38\,\text{g Cu(NO}_3)_2$$

mass of $Cu(NO_3)_2$ that remains in flask:

250.0 g $Cu(NO_3)_2$ originally − 38 g $Cu(NO_3)_2$ decomposed

$$= \textbf{212 g Cu(NO}_3)_2\ \textbf{remain}$$

(d) Restatement: Equilibrium expression for K_p and value.

$$\mathbf{K_p = (pressure\ NO_2)^4 \times pressure\ O_2}$$

Dalton's law of partial pressures:

$$\frac{4\,\text{moles NO}_2(g)}{5\,\text{total moles gas}} \times \frac{5.50\,\text{atm}}{1} = 4.40\,\text{atm NO}_2$$

5.50 atm$_{tot}$ − 4.40 atm$_{NO_2}$ = 1.10 atm O_2

$$K_p = (4.40\,\text{atm})^4(1.10\,\text{atm}) = \textbf{412 atm}^5$$

(e) Given: 420.0 grams of $Cu(NO_3)_2$ placed in flask.

Restatement: What total pressure at equilibrium?

Because the temperature was kept constant, as was the size of the flask, and because some of the original 250.0 grams of $Cu(NO_3)_2$ was left as solid in the flask at equilibrium, any extra $Cu(NO_3)_2$ introduced into the flask would remain as solid — there would be *no change in the pressure.*

> **2.** Magnesium hydroxide has a solubility of 9.24×10^{-4} grams per 100 mL H_2O when measured at 25°C.
>
> **(a)** Write a balanced equation representing magnesium hydroxide in equilibrium in a water solution.
>
> **(b)** Write an equilibrium expression for magnesium hydroxide in water.
>
> **(c)** Calculate the value of K_{sp} at 25°C for magnesium hydroxide.
>
> **(d)** Calculate the value of pH and pOH for a saturated solution of magnesium hydroxide at 25°C.
>
> **(e)** Show by the use of calculations whether a precipitate would form if one were to add 75.0 mL of a 4.00×10^{-4} M aqueous solution of magnesium chloride to 75.0 mL of a 4.00×10^{-4} M aqueous solution of potassium hydroxide.

Answer

2. Given: $Mg(OH)_2$ solubility = 9.24×10^{-4} g/100 mL H_2O at 25°C.

(a) Restatement: Balanced equation in equilibrium.

$$\mathbf{Mg(OH)_2(s) \leftrightarrow Mg^{2+}(aq) + 2OH^-(aq)}$$

(b) Restatement: Equilibrium expression.

$$\mathbf{K_{sp} = [Mg^{2+}][OH^-]^2}$$

(c) Restatement: Value of K_{sp}.

MW $Mg(OH)_2$ = 58.33 g/mole

$$\frac{9.24 \times 10^{-4} \text{ g Mg(OH)}_2}{100 \text{ mL H}_2O} \times \frac{1 \text{ mole Mg(OH)}_2}{58.33 \text{ g Mg(OH)}_2} \times \frac{1000 \text{ mL H}_2O}{1 \text{ liter H}_2O}$$

$$= 1.58 \times 10^{-4} \text{ M Mg(OH)}_2$$

$$= 1.58 \times 10^{-4} \text{ M Mg}^{2+}$$

$$= 2(1.58 \times 10^{-4}) = 3.16 \times 10^{-4} \text{ M OH}^-$$

$$K_{sp} = [Mg^{2+}][OH^-]^2 = (1.58 \times 10^{-4})(3.16 \times 10^{-4})^2$$

$$\mathbf{= 1.58 \times 10^{-11}}$$

(d) Restatement: pH and pOH.

$$pOH = -\log[OH^-]$$
$$= -\log(3.16 \times 10^{-4}) = 3.5$$
$$pH = 14.0 - pOH = 10.5$$

(e) Given: Add 75.0 mL of 4.00×10^{-4} M $MgCl_2$ to 75.0 mL of 4.00×10^{-4} M KOH.

Restatement: Would a precipitate form?

$$MgCl_2 \rightarrow Mg^{2+}(aq) + 2 \, Cl^-(aq)$$

$$KOH \rightarrow K^+(aq) + OH^-(aq)$$

Total volume of solution = 75.0 mL + 75.0 mL = 150.0 mL

$$M_1V_1 = M_2V_2$$

$(4.00 \times 10^{-4}$ mole/liter$)(0.0750$ liter$) = (x)(0.1500$ liter$)$

$x = [Mg^{2+}] = 2.00 \times 10^{-4}$ M

The same would be true for $[OH^-]$.

$Q = [Mg^{2+}][OH^-]^2 = (2.00 \times 10^{-4})^3 = 8.00 \times 10^{-12}$

$K_{sp} = 1.58 \times 10^{-11}$

A precipitate would *not* form because $Q < K_{sp}$.

---

3. Acetic acid, $HC_2H_3O_2$, which is represented as HA, has an acid ionization constant K_a of 1.74×10^{-5}.

   (a) Calculate the hydrogen ion concentration, $[H^+]$, in a 0.50-molar solution of acetic acid.

   (b) Calculate the pH and pOH of the 0.50-molar solution.

   (c) What percent of the acetic acid molecules do not ionize?

   (d) A buffer solution is designed to have a pH of 6.50. What is the $[HA]:[A^-]$ ratio in this system?

   (e) 0.500 liter of a new buffer is made using sodium acetate. The concentration of sodium acetate in this new buffer is 0.35 M. The acetic acid concentration is 0.50 M. Finally, 1.5 grams of LiOH is added to the solution. Calculate the pH of this new buffer.

Answer

   3. Given: K_a for HA is 1.74×10^{-5}.

   (a) Restatement: $[H^+]$ in 0.50 M HA.

Step 1: Write the balanced equation for the ionization of acetic acid, HA.

$$HA(aq) \leftrightarrow H^+(aq) + A^-(aq)$$

Step 2: Write the equilibrium expression for K_a.

$$K_a = \frac{[H^+][A^-]}{[HA]}$$

Step 3: Substitute into the equilibrium expression known (and unknown) information. Let x equal the amount of H^+ that ionizes from HA. Because the molar ratio of $[H^+]:[A^-]$ is 1:1, $[A^-]$ also equals x, and we can approximate $0.50 - x$ as 0.50 (5% rule).

$$1.74 \times 10^{-5} = \frac{x^2}{0.50}$$

Step 4: Solve for x.

$$x^2 = (1.74 \times 10^{-5})(0.50)$$
$$= 8.7 \times 10^{-6}$$
$$\mathbf{x = 2.9 \times 10^{-3} \, M = [H^+]}$$

(b) Restatement: pH and pOH in 0.50 M HA.

$$pH = -\log[H^+] = -\log(2.9 \times 10^{-3}) = \mathbf{2.47}$$
$$pH + pOH = 14$$
$$pOH = 14.0 - 2.47 = \mathbf{11.53}$$

Note: Significant figures for logarithms is equal to the number of significant figures in the mantissa.

(c) Restatement: % HA that does ionize.

$$\% \, \frac{\text{Part}}{\text{Whole}} \times 100\% = \frac{[H^+]}{[HA]} \times 100\%$$

$$= \frac{2.9 \times 10^{-3}}{0.50} \times 100\% = 0.58\%$$

However, 0.58% represents the percentage of the HA molecules that *do* ionize. Therefore, $100.00 - 0.58 = \mathbf{99.42\%}$ **of the HA molecules do** *not* **ionize.**

(d) Given: Buffer pH = 6.50

Restatement: [HA]/[A⁻]?

Step 1: Recognize the need to use the Henderson-Hasselbalch equation.

$$pH = pK_a + \log_{10} \frac{[\text{base}]}{[\text{acid}]}$$

Step 2: Substitute into the equation the known (and unknown) information.

$$pK_a = -\log K_a = 4.76$$

$$6.5 = 4.76 + \log \frac{[A^-]}{[HA]}$$

$$\log \frac{[A^-]}{[HA]} = 6.50 - 4.76 = 1.74$$

$$\frac{[A^-]}{[HA]} = 5.5 \times 10^1$$

Because the question is asking for [HA] / [A⁻], you need to take the reciprocal: 0.018

(e) Given: 0.500 liter = volume

$$0.35M = [NaC_2H_3O_2] \rightarrow Na^+ + C_2H_3O_2^-$$

[HA] = 0.50 M

1.5 g LiOH added to solution

$$\frac{1.5\,g\,LiOH}{1} \times \frac{1\,mole\,LiOH}{23.95\,g\,LiOH} = 0.063\,mole\,LiOH$$

Restatement: pH = ?

Step 1: Write a balanced equation expressing the reaction betweeen acetic acid and lithium hydroxide.

$$HA + OH^- \rightarrow A^- + H_2O$$

Step 2: Create a chart that expresses initial and final concentrations (at equilibrium) of the species.

Species	Initial Concentration	Final Concentration
HA	0.50 M	$0.50 - \dfrac{0.063}{0.500} = 0.375\,M$
A$^-$	0.35 M	$0.35 + \dfrac{0.063}{0.500} = 0.475\,M$

Step 3: Write an equilibrium expression for the ionization of acetic acid.

$$K_a = \frac{[H^+][A^-]}{[HA]} = 1.74 \times 10^{-5} = \frac{[H^+][0.48]}{[0.374]}$$

$$[H^+] = 1.37 \times 10^{-5}\,M$$

Step 4: Solve for the pH.

$$pH = -\log[H^+] = -\log(1.37 \times 10^{-5}) = \mathbf{4.86}$$

Key Terms

Words that can be used as topics in essays:

acid-base indicator	ion-product constant
acidic oxide	Lewis acid, base
amphoteric substance	neutralization
Arrhenius acid, base	organic acids
autoionization	oxyacids
basic oxide	percent ionization
Brønsted-Lowry acid, base	pH
buffer	pOH
carboxyl group	polyprotic acid
common ion effect	salt
conjugate acid, base	salt hydrolysis
equivalence point	strong acid, base
hydronium ion	weak acid, base

Key Concepts

Equations and relationships that you need to know:

- $pH = -\log[H^+]$

  $pOH = -\log[OH^-]$

  $K_w = [H^+][OH^-] = 10^{-14} = K_a \cdot K_b$

  $pK_w = pH + pOH = 14$ (exact)

  $K_a = \dfrac{[H^+][A^-]}{[HA]}$ for the equation $HA(aq) \rightleftharpoons H(aq)^+ + A(aq)^-$

  $pK_a = -\log K_a$

  $pH = pK_a + \log \dfrac{[\text{conjugate base}]}{[\text{conjugate acid}]} = pK_a + \log \dfrac{[A^-]}{[HA]}$

  $K_b = \dfrac{[BH^+][OH^-]}{[B]}$ for the equation $B(aq) + H_2O(l) \rightleftharpoons BH^+(aq) + OH^-(aq)$

157

$$pK_b = -\log K_b$$

$$pOH = pK_b + \log \frac{[\text{conjugate acid}]}{[\text{conjugate base}]} = pK_b + \log \frac{[BH^+]}{[B]}$$

$$\% \text{ ionization} = \frac{\text{amount ionized (M)}}{\text{initial concentration (M)}} \cdot 100\% = \frac{[H^+]}{[HA]_0} \cdot 100\%$$

$$\text{conjugate acid} \underset{\text{gain of } H^+}{\overset{\text{loss of } H^+}{\rightleftharpoons}} \text{conjugate base}$$

• Acid-Base Theory

	Arrhenius	Brønsted-Lowry	Lewis
acid	H^+ supplied to water	donates H^+	accepts electron pair
base	OH^- supplied to water	accepts H^+	donates electron pair

• Acid-Base Properties of Common Ions in Aqueous Solution

	Anions (−)	Cations (+)
acidic	HSO_4^-, H_2PO^{4-}	NH_4^+, Mg^{2+}, Al^{3+} transition metal ions
basic	$C_2H_3O_2^-$, CN^-, CO_3^{2-}, F^-, HCO_3^{-*}, HPO_4^{2-} HS^-, NO_2, PO_4^{3-}, S^{2-}	none
neutral	Cl^-, Br^-, I^-, ClO_4^-, NO_3^-, SO_4^{2-}	Li^+, Na^+, K^+, Ca^{2+}, Ba^{2+}

Close to neutral

Samples: Multiple-Choice Questions

1. What is the OH⁻ concentration (M) of a solution that contains 5.00×10^{-3} mole of H^+ per liter? $K_w = 1.00 \times 10^{-14}$.

   A. 7.00×10^{-14} M

   B. 1.00×10^{-12} M

   C. 2.00×10^{-12} M

   D. 1.00×10^{-11} M

   E. 2.00×10^{-11} M

Answer: C

$[H^+][OH^-] = 10^{-14}$. Substituting gives $(5.00 \times 10^{-3})(x) = 10^{-14}$. Solving for x yields $x = 2.00 \times 10^{-12}$.

2. Given the following equation, identify the conjugate acid found in the products.

   $$NH_3(g) + H_2O(\ell) \rightarrow NH_4^+(aq) + OH^-(aq)$$

   A. NH_3

   B. H_2O

   C. NH_4^+

   D. OH^-

   E. H^+

Answer: C

The conjugate acid of a base is formed when the base acquires a proton from the acid. In this reaction, water acts as an acid because it donates a proton to the ammonia molecule. The ammonium ion (NH_4^+) is the conjugate acid of ammonia (NH_3), a base, which receives a proton from water. The hydroxide ion (OH^-) is the conjugate base.

3. Arrange the following oxyacids in order of decreasing acid strength.

$HClO$, HIO, $HBrO$, $HClO_3$, $HClO_2$

A. $HClO > HIO > HBrO > HClO_3 > HClO_2$

B. $HClO > HClO_2 > HClO_3 > HBrO > HIO$

C. $HIO > HBrO > HClO > HClO_2 > HClO_3$

D. $HBrO > HClO > HClO_3 > HClO_2 > HIO$

E. $HClO_3 > HClO_2 > HClO > HBrO > HIO$

Answer: E

For a series of oxyacids of the same structure that differ only in the halogen, the acid strength increases with the electronegativity of the halogen. Because the electronegativity of the halogens increases as we move up the column, the order at this point would be $HClO > HBrO > HIO$. For a series of oxyacids containing the same halogen, the H—O bond polarity, and hence the acid strength, increases with the oxidation state of the halogen. Therefore, in the series $HClO$, $HClO_2$, and $HClO_3$, the oxidation states of the chlorine are +1, +3, and +5, respectively. And thus the correct order for decreasing acid strength is $HClO_3 > HClO_2 > HClO > HBrO > HIO$.

4. Given the following reversible equation, determine which species is or are Brønsted acids.

$$CO_3^{2-}(aq) + H_2O(\ell) \rightleftharpoons HCO_3^{-}(aq) + OH^{-}{}_{(aq)}$$

A. $CO_3^{2-}(aq)$

B. $H_2O(\ell)$ and $OH^{-}(aq)$

C. $H_2O(\ell)$ and $HCO_3^{-}(aq)$

D. $CO_3^{2-}(aq)$ and $OH^{-}(aq)$

E. $H_2O(\ell)$

Answer: C

Brønsted acids donate protons (H^+). In the equation, both H_2O and HCO_3^{-} donate H^+.

5. Which of the following salts contains a basic anion?

A. $NaCl$

B. $Ba(HSO_4)_2$

C. KI

D. Li_2CO_3

E. NH_4ClO_4

Answer: D

See the chart on page 158. Any anion derived from a weak acid acts as a base in a water solution. The carbonate polyatomic anion, $CO_3^{2-}{}_{(aq)}$, is derived from the weak acid carbonic acid, H_2CO_3. There are no common basic cations.

---

6. Identify the net ionic product(s) produced when solutions of potassium bicarbonate ($KHCO_3$) and hydrobromic acid (HBr) are mixed.

    A. KBr and H_2CO_3

    B. H_2CO_3, K^+, and Br^-

    C. KBr, H_2O, and CO_2

    D. K^+, Br^-, H_2O, and CO_2

    E. H_2CO_3

Answer: E

To write a net ionic equation for an acid-base reaction between two solutions, use the following three steps:

1. Determine the nature of the principal species in both solutions. $KHCO_3$ would ionize to produce K^+ and HCO_3^-. HBr would ionize to produce H^+ and Br^-.

2. Determine which species take part in the acid-base reaction. The bicarbonate anion (HCO_3^-) is basic, and the H^+ from the hydrobromic acid is acidic.

3. Write a balanced net ionic equation. Because H_2CO_3 is a weak acid, it would tend to remain as carbonic acid, especially in the presence of a strong acid like HBr. KBr is an ionic solid, soluble in water; therefore, it would exist as separate ions known as spectator ions, which are *not* written in the net ionic equation.

---

7. All of the following choices are strong bases EXCEPT

    A. CsOH

    B. RbOH

    C. $Ca(OH)_2$

    D. $Ba(OH)_2$

    E. $Mg(OH)_2$

Answer: E

All hydroxides of the Group I metals are strong bases. The hydroxides of the heavier Group II metals (Ca, Sr, and Ba) are also strong bases. $Mg(OH)_2$ is not very soluble in water, yielding relatively little $OH^-(aq)$.

8. A solution is prepared by adding 0.600 liter of 1.0×10^{-3} M HCl to 0.400 liter of 1.0×10^{-3} M HNO_3. What is the pH of the final solution?

   A. 1.00

   B. 2.00

   C. 3.00

   D. 4.00

   E. 5.00

Answer: C

First, determine the volume of the mixture.

$$0.600 \text{ liter} + 0.400 \text{ liter} = 1.000 \text{ liter}$$

Next, determine the concentration of each acid.

HCl: $\dfrac{0.600 \text{ liter} \times 1.0 \times 10^{-3} \text{M}}{1.00 \text{ liter}} = 0.000600 \text{ M}$

HNO_3: $\dfrac{0.400 \text{ liter} \times 1.0 \times 10^{-3} \text{M}}{1.00 \text{ liter}} = 0.000400 \text{ M}$

Because both acids are strong (and monoprotic), the H^+ concentration is equal to the concentration of the acid. Therefore, $[H^+] = 6.00 \times 10^{-4}$ M $+ 4.00 \times 10^{-4}$ M $= 1.00 \times 10^{-3}$ M, and pH $= -\log[H^+] = 3.00$.

9. Suppose that 0.500 liter of 0.0200 M HCl is mixed with 0.100 liter of 0.100 M $Ba(OH)_2$. What is the pH in the final solution after neutralization has occurred?

   A. 3.00

   B. 5.00

   C. 7.00

   D. 9.00

   E. 12.00

Answer: E

Step 1: Write a balanced reaction.

$$2 \text{ HCl} + Ba(OH)_2 \rightarrow BaCl_2 + 2 \text{ H}_2O$$

Step 2: Calculate the number of moles of H^+.

$$\frac{0.500 \text{ liter}}{1} \times \frac{0.0200 \text{ mole}}{1 \text{ liter}} = 1.00 \times 10^{-2} \text{ mole } H^+$$

Step 3: Calculate the number of moles of OH⁻.

There should be twice as many moles of OH^- as moles of $Ba(OH)_2$.

$$\frac{2 \text{ moles OH}^-}{1 \text{ mole Ba (OH)}_2} \times \frac{0.100 \text{ liter}}{1} \times \frac{0.100 \text{ mole Ba (OH)}_2}{1 \text{ liter}} = 0.0200 \text{ mole OH}^-$$

Step 4: Write the net ionic equation.

$$H^+ + OH^- \rightarrow H_2O$$

Step 5: Because every mole of H^+ uses 1 mole of OH^-, calculate the number of moles of excess H^+ or OH^-.

$$2.00 \times 10^{-2} \text{ mole OH}^- - 1.00 \times 10^{-2} \text{ mole H}^+$$
$$= 1.00 \times 10^{-2} \text{ mole OH}^- \text{ excess}$$

Step 6: What is the approximate pH in the final solution?

$$pOH = -\log[OH^-] = -\log[1.00 \times 10^{-2}] = 2$$
$$pH = 14.00 - pOH = 14.00 - 2.00 = 12.00$$

Another way to do this step, if you could use a calculator, would be

$$[OH^-] = \frac{1.00 \times 10^{-2} \text{ mole}}{0.600 \text{ liter}} = 0.0167M$$

$$pOH = 1.778$$
$$pH = 12.222$$

---

10. A student wants to make up 250 mL of an HNO_3 solution that has a pH of 2.00. How many milliliters of the 2.00 M HNO_3 should the student use? (The remainder of the solution is pure water.)

    A. 0.50 mL

    B. 0.75 mL

    C. 1.0 mL

    D. 1.3 mL

    E. This can't be done. The 2.00M acid is weaker than the solution required.

Answer: D

Step 1: Calculate the number of moles of H^+ in 250 mL of a HNO_3 solution which has a pH of 2.00. HNO_3 is a monoprotic acid. $pH = -\log[H^+]$, so $2.00 = -\log[H^+]$ and $[H^+] = 1.00 \times 10^{-2}$ M.

$$\text{mole H}^+ = \frac{1.00 \times 10^{-2} \text{ mole H}^+}{1 \text{ liter}} \times \frac{0.25 \text{ liter}}{1} = 2.5 \times 10^{-3} \text{ mole H}^+$$

Step 2: Determine the number of milliliters of concentrated HNO_3 solution that is needed.

$$\frac{2.5 \times 10^{-3} \, \text{mole H}^+}{1} \times \frac{1000 \, \text{mL sol'n}}{2.00 \, \text{mole H}^+} = 1.3 \, \text{mL sol'n}$$

---

11. Calculate the mass of 1 equivalent of $Sr(OH)_2$. Assume complete ionization:

$$Sr(OH)_{2\,(aq)} \rightarrow Sr^{2+}_{(aq)} + 2OH^-_{(aq)}$$

   A.   15.21 g

   B.   30.41 g

   C.   60.82 g

   D.   121.64 g

   E.   243.28 g

Answer: C

The gram-equivalent weight (GEW) of a base is the mass of the base (in grams) that will provide 1 mole of hydroxide ions in a reaction or that will react with 1 mole of H^+ ions. This problem can be done by using the factor-label method.

$$\frac{121.64 \, \text{g Sr(OH)}_2}{1 \, \text{mole Sr(OH)}_2} \times \frac{1 \, \text{mole Sr(OH)}_2}{2 \, \text{equiv. Sr(OH)}_2} = \frac{60.82 \, \text{g Sr(OH)}_2}{1 \, \text{equiv. Sr(OH)}_2}$$

Note: Had the reaction been

$$Sr(OH)_{2\,(aq)} + H^+_{(aq)} \rightarrow Sr(OH)^+_{(aq)} + H_2O_{(l)}$$

there would have been only 1 equivalent per mole. In other words, you must know the specific reaction in order to use the concept of equivalents.

Samples: Free-Response Questions

1. In an experiment to determine the equivalent mass of an unknown acid, a student measured out a 0.250-gram sample of an unknown solid acid and then used 45.77 mL of 0.150 M NaOH solution for neutralization to a phenolphthalein end point. Phenolphthalein is colorless in acid solutions but becomes pink when the pH of the solution reaches 9 or higher. During the course of the experiment, a back-titration was further required using 1.50 mL of 0.010 M HCl.

   (a) How many moles of OH^- were used in the titration?

   (b) How many moles of H^+ were used in the back-titration?

   (c) How many moles of H^+ are there in the solid acid?

   (d) What is the equivalent mass of the unknown acid?

Answer

1. Given: 0.250 g solid acid

   45.77 mL of 0.150 M NaOH required for phenolphthalein end point

   Back-titration: 1.50 mL of 0.010 M HCl

   (a) Restatement: Moles of OH^- used in the titration.

   moles of OH^- = moles of NaOH

   $$= \frac{0.04577 \text{ liter}}{1} \times \frac{0.150 \text{ mole NaOH}}{1 \text{ liter}}$$

   $$= 6.87 \times 10^{-3} \text{ mole } OH^-$$

   (b) Restatement: Moles of H^+ used in back-titration.

   moles of H^+ = moles of HCl

   $$= \frac{0.00150 \text{ liter}}{1} \times \frac{0.010 \text{ mole HCl}}{1 \text{ liter}}$$

   $$= 1.5 \times 10^{-5} \text{ mole } H^+$$

   (c) Restatement: Moles of H^+ in solid acid.

   moles H^+ in solid acid = moles OH^- − moles H^+

   $$= (M_{NaOH} \times V_{NaOH}) \times (M_{HCl} \times V_{HCl})$$

   $$= 6.87 \times 10^{-3} \text{ mole } OH^- \times 1.5 \times 10^{-5} \text{ mole } H^+$$

   $$= 6.86 \times 10^{-3} \text{ mole } H^+$$

(d) Restatement: Equivalent mass of unknown acid.

$$GEM = \frac{\text{grams of acid}}{\text{moles of } H^+ \text{ furnished}}$$

$$= \frac{0.250 \text{ g acid}}{6.86 \times 10^{-3} \text{ mole } H^+}$$

$$= \textbf{36.4 g/mole acid}$$

---

2. A student wanted to determine the molecular weight of a monoprotic, solid acid, symbolized as HA. The student carefully measured out 25.000 grams of HA and dissolved it in distilled H_2O to bring the volume of the solution to exactly 500.00 mL. The student next measured out several fifty-mL aliquots of the acid solution and then titrated it against standardized 0.100 M NaOH solution. The results of the three titrations are given in the table.

Trial	Milliliters of HA Solution	Milliliters of NaOH Solution
1	49.12	87.45
2	49.00	84.68
3	48.84	91.23

(a) Calculate the number of moles of HA in the fifty-mL aliquots.

(b) Calculate the molecular weight of the acid, HA.

(c) Calculate the pH of the fifty-mL aliquot solution (assume complete ionization).

(d) Calculate the pOH of the fifty-mL aliquot solution (assume complete ionization).

(e) Discuss how each of the following errors would affect the determination of the molecular weight of the acid, HA.

   (1) The balance that the student used in measuring out the 25.000 grams of HA was reading 0.010 gram too low.

   (2) There was an impurity in the acid, HA.

   (3) The NaOH solution used in titration was actually 0.150 M instead of 0.100 M.

Answer:

2. Given: 25.000 grams of HA.

Dissolved in H_2O to make 500.00 mL of solution.

Fifty-mL aliquots (samples) of acidic solution.

0.100 M NaOH used for titration.

See results of titrations in given table.

(a) Restatement: Moles of HA in fifty-mL aliquots.

At the end of titration, moles of HA = moles of NaOH.

$$\text{average volume of NaOH} = \frac{87.45 + 84.68 + 91.23}{3}$$

$$= 87.79 \text{ mL}$$

$$\text{moles HA} = \text{moles NaOH} = V_{NaOH} \times M_{NaOH}$$

$$= \frac{0.08779 \text{ liter}}{1} \times \frac{0.100 \text{ mole}}{1 \text{ liter}} = \mathbf{8.78 \times 10^{-3} \text{ mole}}$$

(b) Restatement: Molecular weight of HA.

$$MW = \frac{48.99 \text{ mL HA sol'n}^*}{8.78 \times 10^{-3} \text{ mole HA}} \times \frac{25.00 \text{ g HA}}{500.00 \text{ mL HA sol'n}}$$

$$= \mathbf{279 \text{ g/mole}}$$

 * = average

(c) Restatement: pH of fifty-mL aliquots (assume 100% ionization).

Average volume of fifty-mL aliquots = 48.99 mL

$$\left[H^+\right] = \frac{\text{moles } H^+}{\text{liters solution}} = \frac{8.78 \times 10^{-3} \text{ mole } H^+}{0.04899 \text{ liter HA sol'n}} = 0.179 \text{ M}$$

$$pH = -\log[0.179] = \mathbf{0.747}$$

(d) Restatement: pOH of fifty-mL aliquot.

$$pOH = 14.000 - pH = 14.000 - 0.747 = \mathbf{13.253}$$

(e) Restatement: Effects of following errors.

(1) Balance reading 0.010 gram too low.

- Student would think she or he had 25.000 grams when there were actually 25.010 grams.

- In the calculation of molecular weight, grams/mole, grams would be too low, so the effect would be a **lower MW** than expected.

(2) An impurity in the sample of HA.

- Student would have less HA than expected.

- In the calculation of molecular weight, g/mole, there would be less HA available than expected. Therefore, in the titration against NaOH, it would take less NaOH than expected to reach the equivalence point. This error would cause a **larger MW** than expected, because the denominator (moles) would be smaller.

- These results assume that the impurity does not have more H^+/mass of impurity than the HA.

(3) NaOH was 0.150 M instead of 0.100 M.

- It would take less NaOH to reach the equivalence point because the NaOH is stronger.

- Because it would take less NaOH, the number of moles of NaOH would be less than expected, causing the denominator (moles) to be smaller than expected, making the calculated **MW larger** than expected.

Note: Using volume averages in the design of this particular experiment can lead to inaccuracy. A better design would be to calculate three values for molecular weight from three separate runs and average the results.

Energy and Spontaneity

Key Terms

Words that can be used as topics in essays:

chemical thermodynamics

endothermic

enthalpy change, ΔH

entropy change, ΔS

exothermic

first law of thermodynamics

free-energy change, ΔG

free energy of formation, ΔG_f

Gibbs-Helmholtz equation

second law of thermodynamics

surroundings

system

third law of thermodynamics

work

Key Concepts

Equations and relationships that you need to know:

- $\Delta H° = \Sigma \Delta H_f°{}_{\text{products}} - \Sigma \Delta H_f°{}_{\text{reactants}}$

  $\Delta S° = \Sigma \Delta S°_{\text{products}} - \Sigma \Delta S°_{\text{reactants}}$

  $\Delta S_{\text{univ}} = \Delta S_{\text{sys}} + \Delta S_{\text{surr}}$

  $\Delta S_{\text{surr}} = \dfrac{-\Delta H_{\text{sys}}}{\text{T}}$ if system and surroundings are at same T

  $\Delta G° = \Sigma \Delta G°_{\text{products}} - \Sigma \Delta G°_{\text{reactants}}$

  $\Delta G° < 0$: spontaneous in forward direction

  $\Delta G° > 0$: nonspontaneous in forward direction

  $\Delta G° = 0$: equilibrium

  $\Delta G° = \Delta H° - T\Delta S°$

  $\Delta E = q + w$

  $\Delta G° = -RT \ln K$ where $R = 8.314 \; J/(K \cdot \text{mole})$

  if $K > 1$, then $\Delta G°$ is negative; products favored at equilibrium

  if $K = 1$; then $\Delta G° = 0$; reactants and products equally favored at equilibrium

  if $K < 1$, then $\Delta G°$ is positive; reactants are favored at equilibrium

$\Delta G = \Delta G° + RT \cdot \ln Q$ where Q = reaction quotient

$\Delta G = \Delta G° + 2.303 \, RT \log K$

$\Delta G° = -n \, \mathscr{F} E°$

where $\mathscr{F}$ = Faradays

$1 \, \mathscr{F} = 96,500 \, J \cdot mole^{-1} \cdot V^{-1}$

$E°$ = standard cell potential, volts (V)

n = number of electrons in the half-reaction

$\Delta S = 2.303 C_p \log \dfrac{T_2}{T_1}$

$\dfrac{\Delta H° (T_2 - T_1)}{2.303 \, R \cdot T_1 \cdot T_2} = \log \dfrac{K_{T_2}}{K_{T_1}}$

where $R = 8.314 \, J \cdot K^{-1}$

• Temperature's Effect on Spontaneity: $\Delta G° = \Delta H° - T\Delta S°$

Case	$\Delta H°$	$\Delta S°$	$\Delta G°$
I	−	+	− spontaneous at all temperatures ex: $2H_2O_2(l) \rightarrow 2H_2O(l) + O_2(g)$
II	+	−	+ nonspontaneous at all temperatures ex: $3O_2(g) \rightarrow 2O_3(g)$
III	+	+	+ nonspontaneous at low temperatures − spontaneous at high temperatures ex: $H_2(g) + I_2(g) \rightarrow 2HI(g)$
IV	−	−	− spontaneous at low temperatures + nonspontaneous at high temperatures ex: $NH_3(g) + HCl(g) \rightarrow NH_4Cl(s)$

Samples: Multiple-Choice Questions

---

1. Given the following standard molar entropies measured at 25°C and 1 atm pressure, calculate $\Delta S°$ in (J/K) for the reaction

$$2 \, Al(s) + 3 \, MgO(s) \rightarrow 3 \, Mg(s) + Al_2O_3(s)$$

$Al(s) = 28.0$ J/K

$MgO(s) = 27.0$ J/K

$Mg(s) = 33.0$ J/K

$Al_2O_3(s) = 51.0$ J/K

   A.   −29.0 J/K

   B.   −13.0 J/K

   C.   13.0 J/K

   D.   69.0 J/K

   E.   139 J/K

Answer: C

$\Delta S° = \Sigma \Delta S°_{products} - \Sigma \Delta S°_{reactants}$

$\Delta S° = [3(33.0) + 51.0] - [2(28.0) + 3(27.0)] = 13.0$ J/K

---

2. For the given reaction and the following information, calculate $\Delta G°$ at 25°C.

$$2 \, PbO(s) + 2 \, SO_2(g) \rightarrow 2 \, PbS(s) + 3 \, O_2(g)$$

Species	ΔH° (kJ/mole) at 25°C and 1 atm	ΔS° (J/mole · K) at 25°C and 1 atm
PbO(s)	−218.0	70.0
SO₂(g)	−297.0	248.0
PbS(s)	−100.0	91.0
O₂(g)	−	205.0

   A.   273.0 kJ

   B.   438.0 kJ

   C.   634.0 kJ

   D.   782.0 kJ

   E.   830.0 kJ

Answer: D

This problem requires us to use the Gibbs-Helmholtz equation:

$$\Delta G° = \Delta H° - T\Delta S°$$

Step 1: Using the given information, calculate $\Delta H°$.

$$\Delta H° = \Sigma\Delta H°_{products} - \Sigma\Delta H°_{reactants}$$
$$= [2(-100.0)] - [2(-218.0) + 2(-297.0)] = 830.0 \text{ kJ/mole}$$

Step 2: Calculate $\Delta S°$.

$$\Delta S° = \Sigma\Delta S°_{products} - \Sigma\Delta S°_{reactants}$$
$$= [2(91.0) + 3(205.0)] - [2(70.0) + 2(248.0)]$$
$$= 797.0 - 636.0 = 161.0 \text{ J/mole} \cdot \text{K} = 0.161 \text{ kJ/mole} \cdot \text{K}$$

Step 3: Substitute into the Gibbs-Helmholtz equation.

$$\Delta G° = \Delta H° - T\Delta S°$$
$$\frac{830.0 \text{ kJ}}{1 \text{ mole}} - \frac{298 \cancel{\text{K}} \cdot (0.161 \text{ kJ})}{\text{mole} \cdot \cancel{\text{K}}} = 782.0 \text{ kJ/mole}$$

---

3. Given the information that follows, calculate the standard free energy change, $\Delta G°$, for the reaction

$$CH_4(g) + 2\,O_2(g) \rightarrow 2\,H_2O(\ell) + CO_2(g)$$

Species	$\Delta H°$ (kJ/mole) at 25°C and 1 atm	$\Delta G°$ (kJ/mole) at 25°C and 1 atm
$CH_4(g)$	−75.00	−51.00
$O_2(g)$	0	0
$H_2O(\ell)$	−286.00	−237.00
$CO_2(g)$	−394.00	−394.00

A. −919.00 kJ/mole

B. −817.00 kJ/mole

C. −408.50 kJ/mole

D. 459.50 kJ/mole

E. 919.00 kJ/mole

Answer: B

$$\Delta G^\circ = \Sigma \Delta G^\circ_{products} - \Sigma \Delta G^\circ_{reactants}$$
$$= [2(-237.00) + (-394.00)] - (-51.00)$$
$$= -817.00 \text{ kJ/mole}$$

---

4. Calculate the approximate standard free energy change for the ionization of hydrofluoric acid, HF ($K_a = 1.0 \times 10^{-3}$), at 25°C.

A. -9.0 kJ

B. -4.0 kJ

C. 0.050 kJ

D. 4.0 kJ

E. 17 kJ

Answer: E

At equilibrium, $\Delta G = 0 = \Delta G^\circ + 2.303 \, RT \log K$ (at equilibrium $Q = K$).

$$\Delta G^\circ = -2.303(8.314 \text{ J} \cdot \text{K}^{-1})(298 \text{ K})(\log 1.0 \times 10^{-3})$$

Rounding,

$$\sim -2.3(8.3)(300)(-3.0) = 17,181 \text{ J} \approx 17 \text{ kJ}$$

---

5. Arrange the following reactions according to increasing ΔS°_{rxn} values.

1. $H_2O(g) \rightarrow H_2O(\ell)$
2. $2 \, HCl(g) \rightarrow H_2(g) + Cl_2(g)$
3. $SiO_2(s) \rightarrow Si(s) + O_2(g)$

lowest $\rightarrow$ highest

A. $\Delta S^\circ(1) < \Delta S^\circ(2) < \Delta S^\circ(3)$

B. $\Delta S^\circ(2) < \Delta S^\circ(3) < \Delta S^\circ(1)$

C. $\Delta S^\circ(3) < \Delta S^\circ(1) < \Delta S^\circ(2)$

D. $\Delta S^\circ(1) < \Delta S^\circ(3) < \Delta S^\circ(2)$

E. $\Delta S^\circ(3) < \Delta S^\circ(2) < \Delta S^\circ(1)$

Answer: A

Entropy is a measure of the randomness or disorder of a system. The greater the disorder of a system, the greater its entropy.

In $H_2O(g) \rightarrow H_2O(\ell)$, the reaction is going from a disordered state (g) to a more ordered state (ℓ); low entropy, $\Delta S < 0$.

In $2\,HCl(g) \rightarrow H_2(g) + Cl_2(g)$, the change in energy will be very small since there are two moles of gas molecules on each side of the equation.

In $SiO_2(s) \rightarrow Si(s) + O_2(g)$, the system is becoming more disordered, apparent from the presence of gas molecules on the product side; high entropy, $\Delta S > 0$.

---

6. Given for the reaction $Hg(\ell) \rightarrow Hg(g)$ that $\Delta H° = 63.0\,kJ \cdot mole^{-1}$ and $\Delta S° = 100.\,J \cdot K^{-1} \cdot mole^{-1}$, calculate the normal boiling point of Hg.

A. 6.30 K

B. 63.0 K

C. 630 K

D. 6.30×10^3 K

E. cannot be determined from the information provided

Answer: C

At equilibrium $Hg(\ell) \leftrightarrow Hg(g)$, which represents the condition of boiling, and at equilibrium $\Delta G° = 0$. The word *normal* in the question refers to conditions at 1 atm of pressure, which is reflected in the notation for standardized conditions for $\Delta S°$ and $\Delta H°$. Therefore, using the Gibbs-Helmholtz equation, $\Delta G° = \Delta H° - T\Delta S°$, we can substitute 0 for $\Delta G°$ and solve for T.

$$T = \frac{\Delta H°}{\Delta S°} = \frac{63,000\,J \cdot \cancel{mole^{-1}}}{100\,J \cdot K^{-1} \cdot \cancel{mole^{-1}}} = 630\,K$$

7. Given the following data:

$$Fe_2O_3(s) + 3\,CO(g) \rightarrow 2\,Fe(s) + 3\,CO_2(g) \qquad \Delta H° = -27\ kJ/mole$$
$$3\,Fe_2O_3(s) + CO(g) \rightarrow 2\,Fe_3O_4(s) + CO_2(g) \qquad \Delta H° = -61\ kJ/mole$$
$$Fe_3O_4(s) + CO(g) \rightarrow 3\,FeO(s) + CO_2(g) \qquad \Delta H° = 38\ kJ/mole$$

Species	$\Delta S°\ (J \cdot K^{-1} \cdot mole^{-1})$
$Fe_2O_3(s)$	87.0
$CO(g)$	190.0
$Fe(s)$	27.0
$CO_2(g)$	214.0
$Fe_3O_4(s)$	146.0
$FeO(s)$	61.0

Calculate the approximate $\Delta G°$ (at 25°C) for the reaction

$$FeO(s) + CO(g) \rightarrow Fe(s) + CO_2(g)$$

A.   −26 kJ/mole

B.   −13 kJ/mole

C.   13 kJ/mole

D.   26 kJ/mole

E.   39 kJ/mole

Answer: B

To solve this problem, use the Gibbs-Helmholtz equation:

$$\Delta G° = \Delta H° - T\Delta S°$$

Step 1: Solve for $\Delta H°$. Realize that you will have to use Hess's law to determine $\Delta H°$. Be sure to multiply through the stepwise equations to achieve the lowest common denominator (6), and reverse equations where necessary.

$$3Fe_2O_3(s) + 9CO(g) \rightarrow 6Fe(s) + 9CO_2(g) \qquad -81\,kJ/mole$$
$$2Fe_3O_4(s) + CO_2(g) \rightarrow 3Fe_2O_3(s) + CO(g) \qquad 61\,kJ/mole$$
$$6FeO(s) + 2CO_2(g) \rightarrow 2Fe_3O_4(s) + 2CO(g) \qquad -76\,kJ/mole$$
$$6FeO(s) + 6CO(g) \rightarrow 6Fe(s) + 6CO_2(g) \qquad -96\,kJ/mole$$

$$\Delta H° = \frac{-96\,kJ/mole}{6} = -16\,kJ/mole$$

Step 2: Solve for $\Delta S°$.

$$FeO(s) + CO(g) \rightarrow Fe(s) + CO_2(g)$$

$\Delta S° = \Sigma\Delta S°_{products} - \Sigma\Delta S°_{reactants}$
$= (27.0 + 214.0) - (61.0 + 190.0) = -10.0 \text{ J} \cdot \text{K}^{-1} \cdot \text{mole}^{-1}$

Step 3: Substitute $\Delta S°$ and $\Delta H°$ into the Gibbs-Helmholtz equation.

$\Delta G° = \Delta H° - T\Delta S°$
$= -16 \text{ kJ/mole} - 298 \text{ K}(-0.0100 \text{ kJ} \cdot \text{K}^{-1} \cdot \text{mole}^{-1})$
$\approx -13 \text{ kJ/mole}$

---

8. Given the balanced equation

$$H_2(g) + F_2(g) \leftrightarrow 2\,HF(g) \quad \Delta G° = -546 \text{ kJ/mole}$$

Calculate ΔG if the pressures were changed from the standard 1 atm to the following and the temperature was changed to 500°C.

$H_2(g) = 0.50$ atm $F_2(g) = 2.00$ atm $HF(g) = 1.00$ atm

A. -1090 kJ/mole

B. -546 kJ/mole

C. -273 kJ/mole

D. 546 kJ/mole

E. 1090 kJ/mole

Answer: B

Realize that you will need to use the equation

$$\Delta G = \Delta G° + RT \ln Q$$

Step 1: Solve for the reaction quotient, Q.

$$Q = \frac{(P_{HF})^2}{(P_{H_2})(P_{F_2})} = \frac{(1.00)^2}{(0.50)(2.00)} = 1.00$$

$\ln 1.00 = 0$

Step 2: Substitute into the equation.

$\Delta G = \Delta G° + RT \ln Q$
$= -546,000 \text{ J} + (8.3148 \text{ J} \cdot \text{K}^{-1} \cdot \text{mole}^{-1}) \cdot 773 \text{ K}(0)$
$= -546 \text{ kJ/mole}$

---

9. If $\Delta H°$ and $\Delta S°$ are both negative, then $\Delta G°$ is

   A.   always negative

   B.   always positive

   C.   positive at low temperatures and negative at high temperatures

   D.   negative at low temperatures and positive at high temperatures

   E.   zero

Answer: D

Examine the Gibbs-Helmholtz equation, $\Delta G° = \Delta H° - T\Delta S°$, to see the mathematical relationships of negative $\Delta H°$'s and $\Delta S°$'s. (Refer to page 170.)

---

10. Determine the entropy change that takes place when 50.0 grams of compound *x* are heated from 50.°C to 2,957°C. It is found that 290.7 kilojoules of heat are absorbed.

   A.   −461 J/K

   B.   0.00 J/K

   C.   230 J/K

   D.   461 J/K

   E.   921 J/K

Answer: C

Use the equation

$$\Delta S = 2.303 C_p \log \frac{T_2}{T_1}$$

where C_p represents the heat capacity

$(C_p = \Delta H / \Delta T = 290{,}700 \text{ J} / 2907 \text{ K} = 100.0 \text{ J/K}).$

Substituting into the equation yields

$$\Delta S = 2.303 C_p \log \frac{T_2}{T_1}$$

$$= 2.303 \left(100.0 \frac{\text{J}}{\text{K}}\right) \log \frac{3230.\cancel{\text{K}}}{323\cancel{\text{K}}} = 230 \text{ J/K}$$

Samples: Free-Response Questions

1. Given the equation $N_2O_4(g) \rightarrow 2\,NO_2(g)$ and the following data:

Species	$\Delta H°_f$ (kJ · mole^{-1})	$G°_f$ (kJ · mole^{-1})
$N_2O_4(g)$	9.16	97.82
$NO_2(g)$	33.2	51.30

(a) Calculate $\Delta G°$.

(b) Calculate $\Delta H°$.

(c) Calculate the equilibrium constant K_p at 298 K and 1 atm.

(d) Calculate K at 500°C and 1 atm.

(e) Calculate $\Delta S°$ at 298 K and 1 atm.

(f) Calculate the temperature at which $\Delta G°$ is equal to zero at 1 atm, assuming that $\Delta H°$ and $\Delta S°$ do not change significantly as the temperature increases.

Answer

1. Given: $\Delta H_f°$ and $\Delta G_f°$ information for the equation

$$N_2O_4(g) \rightarrow 2\,NO_2(g)$$

(a) Restatement: Calculate $\Delta G°$.

$\Delta G° = \Sigma \Delta G_f°$ products $- \Sigma \Delta G_f°$ reactants
$\quad = 2(51.30) - (97.82) = 4.78$ kJ · mole^{-1}

(b) Restatement: Calculate $\Delta H°$.

$\Delta H° = \Sigma \Delta H_f°$ products $- \Sigma \Delta H_f°$ reactants
$\quad = 2(33.2) - 9.16 = 57.2$ kJ · mole^{-1}

(c) Restatement: Calculate the equilibrium constant K_p at 298 K and 1 atm.

$K_p = \dfrac{P_{NO_2}^2}{P_{N_2O_4}}$ (where P represents the partial pressure of a gas in atmospheres)

$\Delta G° = -2.303\,RT \log K \ (R = 8.314$ J · K$^{-1})$

$\log K_p = \dfrac{\Delta G°}{-2.303\,RT} = \dfrac{4.78\,\cancel{kJ} \cdot \text{mole}^{-1}}{-2.303\left(0.008314\,\cancel{kJ} \cdot \cancel{K}^{-1}\right)\left(298\,\cancel{K}\right)}$

$\quad = -0.838$

$K_p = 0.145$ (at standard temperature of 298 K)

(d) Restatement: Calculate K at 500°C and 1 atm.

$$\frac{\Delta H°(T_2 - T_1)}{2.303 R T_1 T_2} = \log \frac{K_{T_2}}{K_{T_1}}$$

$$\frac{57,200\,J\left(773\,K - 298\,K\right)}{(2.303)(8.314\,J \cdot K^{-1})(773\,K)(298\,K)} = \log \frac{K_{773}}{K_{298}}$$

$$6.16 = \log \frac{K_{773}}{K_{298}}$$

$\log K_{773} - \log K_{298} = 6.16$

$\log K_{298} = -0.838$ from part (c)

$\log K_{773} = 6.16 + (-0.838) = 5.32$

$K = 2.09 \times 10^5$

(e) Restatement: Calculate $\Delta S°$ at 298 K and 1 atm.

$\Delta G° = \Delta H° - T\Delta S°$

$$\Delta S° = \frac{\Delta H° - \Delta G°}{T} = \frac{57,200\,J - 4,780\,J}{298\,K} = 176\,J \cdot K^{-1}$$

(f) Restatement: Calculate the temperature at which $\Delta G°$ is equal to zero at 1 atm, assuming that $\Delta H°$ and $\Delta S°$ do not change significantly as the temperature increases.

$\Delta G° = \Delta H° - T\Delta S°$

$0 = 57,200\,J - T(176\,J \cdot K^{-1})$

$$T = \frac{57,200\,J}{176\,J \cdot K^{-1}} = 325\,K$$

---

2. (a) Define the concept of entropy.

(b) From each of the pairs of substances listed, and assuming 1 mole of each substance, choose the one that would be expected to have the lower absolute entropy. Explain your choice in each case.

(1) $H_2O(s)$ or $SiC(s)$ at the same temperature and pressure

(2) $O_2(g)$ at 3.0 atm or $O_2(g)$ at 1.0 atm, both at the same temperature

(3) $NH_3(\ell)$ or $C_6H_6(\ell)$ at the same temperature and pressure

(4) $Na(s)$ or $SiO_2(s)$

Answer

2. (a) Restatement: Define entropy.

Entropy, which has the symbol S, is a thermodynamic function that is a measure of the disorder of a system. Entropy, like enthalpy, is a state function. State functions are those quantities whose changed values are determined by their initial and final values. The quantity of entropy of a system depends on the temperature and pressure of the system. The units of entropy are commonly $J \cdot K^{-1} \cdot mole^{-1}$. If S has a ° ($S°$), then it is referred to as standard molar entropy and represents the entropy at 298 K and 1 atm of pressure; for solutions, it would be at a concentration of 1 molar. The larger the value of the entropy, the greater the disorder of the system.

(b) Restatement: In each set, choose which would have the lower entropy (greatest order) and explain.

(1) **SiC(s)**

- $H_2O(s)$ is a polar covalent molecule. Between the individual molecules would be hydrogen bonds.

- SiC(s) exists as a structured and ordered covalent network.

- Melting point of SiC(s) is much higher than that of $H_2O_{(s)}$, so it would take more energy to vaporize the more ordered SiC(s) than to vaporize $H_2O(s)$.

(2) **O_2(g) at 3.0 atm**

- At higher pressures, the oxygen molecules have less space to move within and are thus more ordered.

(3) **$NH_3(\ell)$**

- $NH_3(\ell)$ has hydrogen bonds (favors order).

- $C_6H_6(\ell)$ has more atoms and so more vibrations — thus greater disorder.

(4) **SiO_2(s)**

- Na(s) has high entropy. It exhibits metallic bonding, forming soft crystals with high amplitudes of vibration.

- SiO_2(s) forms an ordered, structured covalent network.

- SiO_2(s) has a very high melting point, so much more energy is necessary to break the ordered system.

Reduction and Oxidation

Key Terms

Words that can be used as topics in essays:

anode

cathode

disproportionation

electrolysis

electrolytic cell

electromotive force (emf)

Faraday, $\mathscr{F}$

galvanic cell

half-equation

Nernst equation

nonspontaneous reaction

oxidation number

oxidizing agent

redox reaction

redox titration

reducing agent

spontaneous reaction

standard oxidation voltage, E_{ox}°

standard reduction voltage, E_{red}°

standard voltage, E°

voltage

voltaic cell

Key Concepts

Equations and relationships that you need to know:

- $E^{\circ} = E_{ox}^{\circ} + E_{red}^{\circ}$

- For the reaction: $aA + bB \rightarrow cC + dD$

  Nernst equation:

  $$E = E^{\circ} - \frac{0.0592}{n} \log \frac{(C)^c (D)^d}{(A)^a (B)^b} = E^{\circ} - \frac{0.0592}{n} \log Q \text{ at } 25°C$$

  $E^{\circ} - (RT/n\mathscr{F}) \ln Q = E^{\circ} - (2.303 \ RT/n\mathscr{F}) \log Q$

  total charge $= n\mathscr{F}$

  $1\mathscr{F} = 96{,}487$ coulombs/mole $= 96{,}487$ J/V $\cdot$ mole

- $\text{potential}(V) = \frac{\text{work}(J)}{\text{charge}(C)}$

  1 ampere = rate of flow of electrons, measured in coulombs/sec or amps

  electrical energy (work) = volts $\times$ coulombs = joules

- $E = \frac{-w}{q}$ where w = work, q = change

- $\Delta G = w_{max} = -n\mathscr{F}E$

- At equilibrium: $Q = K$; $E_{cell} = 0$

$$\ln K = \frac{n \cdot E_{cell}^{\circ}}{0.0257}$$

- OIL RIG: Oxidation Is Losing; Reduction Is Gaining (electrons)

- AN OX (ANode is where OXidation occurs)

  RED CAT (REDuction occurs at the CAThode)

- Electrolytic cell — electrical energy is used to bring about a nonspontaneous electrical change. Anode is (+) electrode; cathode is (−) electrode.

- Voltaic (chemical) cell — electrical energy is produced by a spontaneous redox reaction. Anode is (−) terminal; cathode is (+) terminal.

- **Relation Between ΔG, K, and E_{cell}°**

ΔG	K	E_{cell}°	Reaction Under Standard State Conditions
−	>1	+	Spontaneous
0	=1	0	Equilibrium
+	<1	−	Nonspontaneous; spontaneous in reverse direction

Samples: Multiple-Choice Questions

1. What mass of copper would be produced by the reduction of the $Cu^{2+}(aq)$ ion by passing 96.487 amperes of current through a solution of copper(II) chloride for 100.00 minutes? (1 Faraday = 96,487 coulombs)

   A. 95.325 g

   B. 190.65 g

   C. 285.975 g

   D. 381.30 g

   E. cannot be determined from the information provided

Answer: B

Step 1: Write the reaction that would occur at the cathode.

$$Cu^{2+}(aq) + 2e^- \rightarrow Cu(s)$$

Step 2: This problem can be solved by using the factor-label method:

(Note all the conversion factors that you should be comfortable with.)

$$\frac{96.487 \cancel{\text{amperes}}}{1} \times \frac{100.0 \cancel{\text{minutes}}}{1} \times \frac{60 \cancel{\text{seconds}}}{1 \cancel{\text{minutes}}}$$

$$\times \frac{1 \cancel{\text{coulomb}}}{1 \cancel{\text{ampere} \cdot \text{second}}} \times \frac{1 \cancel{\text{Faraday}}}{96,487 \cancel{\text{coulombs}}} \times \frac{1 \text{ mole } \cancel{e^-}}{1 \cancel{\text{Faraday}}}$$

$$\times \frac{1 \cancel{\text{mole Cu}}}{2 \text{ moles } \cancel{e^-}} \times \frac{63.55 \text{ g Cu}}{1 \cancel{\text{mole Cu}}} = 190.65 \text{ g Cu}$$

2. Find $E°$ for a cell composed of silver and gold electrodes in 1 molar solutions of their respective ions: $E°_{red}$ Ag = +0.7991 volts; $E°_{red}$ Au = +1.68 volts.

   A. −0.44 volt

   B. 0 volt

   C. 0.44 volt

   D. 0.88 volt

   E. 2.48 volts

Answer: D

Notice that $E°_{red}$ for silver is lower than $E°_{red}$ for gold. This means that because silver is higher in the activity series, silver metal will reduce the gold ion.

Step 1: Write the net cell reaction.

$$Ag(s) + Au^+(aq) \rightleftharpoons Ag^+(aq) + Au(s)$$

Step 2: Write the two half-reactions and include the $E°_{red}$ and $E°_{ox}$ values.

ox: $Ag(s) \rightleftharpoons Ag^+(aq) + e^-$ $E°_{ox} = -0.7991$ volt

red: $Au^+(aq) + e^- \rightleftharpoons Au(s)$ $E°_{red} = 1.68$ volts

$$\overline{Ag(s) + Au^+(aq) \rightleftharpoons Ag^+(aq) + Au(s)} \qquad \overline{E° = +0.88 \text{ volt}}$$

Because the sign of $E°$ is positive, the reaction will proceed spontaneously.

---

3. ... $FeCl_2$ + ... $KMnO_4$ + ... $HCl \rightarrow$ ... $FeCl_3$ + ... KCl + ... $MnCl_2$ + ? H_2O

When the equation for this reaction is balanced with the lowest whole-number coefficients, the coefficient for H_2O is

A. 1

B. 2

C. 3

D. 4

E. 5

Answer: D

Step 1: Decide what elements are undergoing oxidation and what elements are undergoing reduction.

$$\text{ox: } Fe^{2+} \rightarrow Fe^{3+}$$
$$\text{red: } MnO_4^- \rightarrow Mn^{2+}$$

Step 2: Balance each half-reaction with respect to atoms and then charges.

ox: $Fe^{2+} \rightarrow Fe^{3+} + e^-$

Balance the reduction half-reaction, using water to balance the O's.

red: $MnO_4^- \rightarrow Mn^{2+} + 4H_2O$

Balance the H atoms with H^+ ions.

red: $MnO_4^- + 8H^+ \rightarrow Mn^{2+} + 4H_2O$

Balance charges with electrons.

red: $MnO_4^- + 8H^+ + 5e^- \rightarrow Mn^{2+} + 4H_2O$

Step 3: Equalize the number of electrons lost and gained. There were 5 e^- gained in the reduction half-reaction, so there must be 5 e^- lost in the oxidation half-reaction

ox: $5 Fe^{2+} \rightarrow 5 Fe^{3+} + 5 e^-$

Step 4: Add the two half-reactions (cancel the electrons).

ox: $5 Fe^{2+} \rightarrow 5 Fe^{3+} + \cancel{5e^-}$

red: $MnO_4^- + 8H^+ + \cancel{5e^-} \rightarrow Mn^{2+} + 4H_2O$

$5Fe^{2+} + MnO_4^- + 8H^+ \rightarrow 5Fe^{3+} + Mn^{2+} + 4H_2O$

---

4. Given the following notation for an electrochemical cell:

$Pt(s) \mid H_2(g) \mid H^+(aq) \parallel Ag^+(aq) \mid Ag(s)$

Which of the following represents the overall balanced (net) cell reaction?

A. $H_2(g) + Ag^+(aq) \rightarrow 2 H^+(aq) + Ag(s)$

B. $H_2(g) + Ag(s) \rightarrow H^+(aq) + Ag^+(aq)$

C. $Ag(s) + H^+(aq) \rightarrow Ag^+(aq) + H_2(g)$

D. $2 H^+(aq) + Ag(s) \rightarrow H_2(g) + Ag^+(aq)$

E. none of the above

Answer: E

The vertical lines represent phase boundaries. By convention, the anode is written first, at the left of the double vertical lines, followed by the other components of the cell as they would appear in order from the anode to the cathode. The platinum is present to represent the presence of an inert anode. The two half-reactions that occur are

anode: $H_2(g) \rightarrow 2 H^+(aq) + 2 e^-$ oxidation
OIL (Oxidation Is Losing electrons)
AN OX (ANode is where OXidation occurs)

cathode: $Ag^+(aq) + e^- \rightarrow Ag(s)$ reduction
RIG (Reduction Is Gaining electrons)
RED CAT (REDuction occurs at the CAThode)

In adding the two half-reactions, multiply the reduction half-reaction by 2 so the electrons are in balance, giving the overall reaction

$H_2(g) + 2 Ag^+(aq) \rightarrow 2 H^+(aq) + 2 Ag(s)$

5. Given the following information, which of the statements is true?

$$Cu^{2+}(aq) + e^- \rightarrow Cu^+(aq) \qquad\qquad E^\circ_{red} = 0.34 \text{ V}$$
$$2 H^+(aq) + 2 e^- \rightarrow H_2(g) \qquad\qquad E^\circ_{red} = 0.00 \text{ V}$$
$$Fe^{2+}(aq) + 2 e^- \rightarrow Fe(s) \qquad\qquad E^\circ_{red} = -0.44 \text{ V}$$
$$Ni(s) \rightarrow Ni^{2+}(aq) + 2 e^- \qquad\qquad E^\circ_{ox} = 0.25 \text{ V}$$

A. $Cu^{2+}(aq)$ is the strongest oxidizing agent.

B. $Cu^{2+}(aq)$ is the weakest oxidizing agent.

C. $Ni(s)$ is the strongest oxidizing agent.

D. $Fe(s)$ would be the weakest reducing agent.

E. $H^+(aq)$ would be the strongest oxidizing agent.

Answer: A

The more positive the E°_{red} value, the greater the tendency for the substance to be reduced, and conversely, the less likely it is to be oxidized. It would help in this example to reverse the last equation so it can be easily compared to the other E°_{red} values. The last equation becomes $Ni^{2+}(aq) + 2e^-(aq) \rightarrow Ni(s)$; $E^\circ_{red} = -0.25$ V.

The equation with the largest E°_{red} is $Cu^{2+}(aq) + e^- \rightarrow Cu^+(aq)$; $E^\circ_{red} = 0.34$ V. Thus, $Cu^{2+}(aq)$ is the strongest oxidizing agent of those listed because it has the greatest tendency to be reduced. Conversely, $Cu^+(aq)$ would be the weakest reducing agent. $Fe^{2+}(aq)$ would be the weakest oxidizing agent because it would be the most difficult species to reduce. Conversely, $Fe(s)$ would be the strongest reducing agent.

6. A cell has been set up as shown in the following diagram, and E° has been measured as 1.00 V at 25°C. Calculate $\Delta G°$ for the reaction.

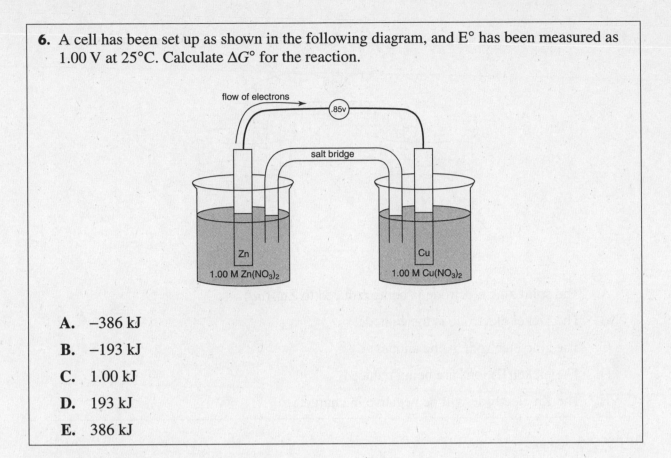

A. −386 kJ

B. −193 kJ

C. 1.00 kJ

D. 193 kJ

E. 386 kJ

Answer: B

The formula you need for this problem is $\Delta G° = -n\mathscr{F}E°$. The Faraday constant, $\mathscr{F}$, is equal to 9.65×10^4 joules · volt^{-1} · mole^{-1}. n is the number of electrons transferred between oxidizing and reducing agents in a balanced redox equation.

Step 1: Write the balanced redox equation.

$$Zn(s) + Cu^{2+}(aq) \rightarrow Zn^{2+}(aq) + Cu(s)$$

Step 2: Identify the variables needed for the equation.

$$\Delta G° = ? \qquad \mathscr{F} = 9.65 \times 10^4 \text{ joules · volt}^{-1} \text{ · mole}^{-1}$$
$$n = 2 \qquad E° = 1.00 \text{ volt}$$

Step 3: Substitute into the equation and solve.

$$\Delta G° = -\frac{\left(2 \text{ moles e}^-\right)}{1} \times \frac{9.65 \times 10^4 \text{ joules}}{\text{volt · mole e}^-} \times \frac{1.00 \text{ volt}}{1} = -1.93 \times 10^5 \text{ joules} = -193 \text{ kJ}$$

7. Given the following diagram, determine which of the following statements is FALSE.

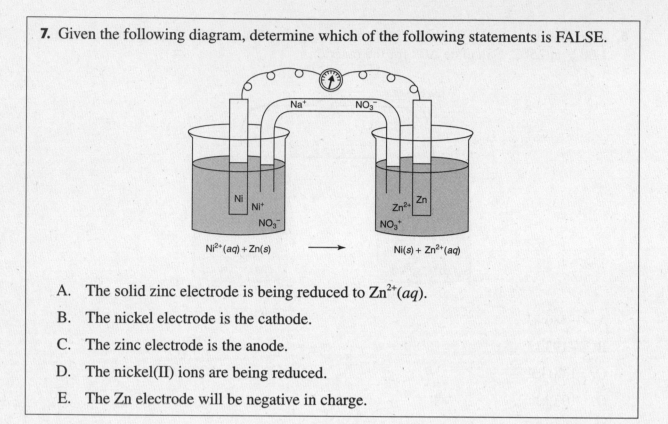

A. The solid zinc electrode is being reduced to $Zn^{2+}(aq)$.

B. The nickel electrode is the cathode.

C. The zinc electrode is the anode.

D. The nickel(II) ions are being reduced.

E. The Zn electrode will be negative in charge.

Answer: A

There are two types of cells: electrolytic (which requires a battery or external power source) and voltaic (which requires no battery or external power source). The reaction in the diagram is voltaic and therefore spontaneous. In a voltaic cell, the anode is the negative terminal, and oxidation occurs at the anode. Remember the OIL portion of OIL RIG (Oxidation Is Losing electrons) and AN OX (ANode is where OXidation occurs).

$$Zn(s) \rightarrow Zn^{2+}(aq) + 2\ e^-$$

The cathode is the positive terminal in a voltaic cell, and reduction occurs at the cathode. Remember the RIG portion of OIL RIG (Reduction Is Gaining electrons) and RED CAT (REDuction occurs at the CAThode).

$$Ni^{2+}(aq) + 2\ e^- \rightarrow Ni(s)$$

8. Given that

$$Zn^{2+}(aq) + 2e^- \rightarrow Zn(s) \quad E^\circ_{red} = -0.76 \text{ V}$$
$$Cr^{3+}(aq) + 3e^- \rightarrow Cr(s) \quad E^\circ_{red} = -0.74 \text{ V}$$

calculate the equilibrium constant K for the following balanced reaction:

$$3 \text{ Zn}(s) + 2 \text{ Cr}^{3+}(aq) \rightarrow 3 \text{ Zn}^{2+}(aq) + 2 \text{ Cr}(s)$$

A. $K = e^{-0.02}$

B. $K = e^{0.02}$

C. $K = e^{4.7}$

D. $K = e^{8.0}$

E. cannot be determined from the information provided

Answer: C

Step 1: Determine the oxidation and reduction half-reactions and E°_{cell}.

$$\text{ox: } 3\left[Zn(s) \rightarrow Zn^{2+}(aq) + 2e^-\right] \qquad E^\circ_{ox} = +0.76\text{V}$$

$$\underline{\text{red: } 2\left[Cr^{3+}(aq) + 3e^- \rightarrow Cr(s)\right]} \qquad \underline{E^\circ_{red} = -0.74\text{V}}$$

$$3Zn(s) + 2Cr^{3+}(aq) \rightarrow 3Zn^{2+}(aq) + 2Cr(s) \quad E^\circ_{cell} = 0.02\text{V}$$

Step 2: Use the equation.

$$\ln K = \frac{n \cdot E^\circ_{cell}}{0.0257} = \frac{6 \cdot 0.02 \text{ V}}{0.0257} = 4.7$$
$$K = e^{4.7}$$

9. An electric current is applied to an aqueous solution of $FeCl_2$ and $ZnCl_2$. Which of the following reactions occurs at the cathode?

A. $Fe^{2+}(aq) + 2 e^-(aq) \rightarrow Fe(s)$ $\qquad\qquad E^\circ_{red} = -0.44$ V

B. $Fe(s) \rightarrow Fe^{2+}(aq) + 2 e^-$ $\qquad\qquad\quad E^\circ_{ox} = 0.44$ V

C. $Zn^{2+}(aq) + 2 e^-(aq) \rightarrow Zn(s)$ $\qquad\quad E^\circ_{red} = -0.76$ V

D. $Zn(s) \rightarrow Zn^{2+}(aq) + 2 e^-$ $\qquad\qquad\quad E^\circ_{ox} = 0.76$ V

E. $2 H_2O(\ell) \rightarrow O_2(g) + 4 H^+(aq) + 4 e^-$ $\quad E^\circ_{ox} = -1.23$ V

Answer: A

Reduction occurs at the cathode. You can eliminate choices (B), (D), and (E) because these reactions are oxidations. E°_{red} for $Fe^{2+}(aq)$ is -0.44 V, and E°_{red} for $Zn^{2+}(aq)$ is -0.76 V. Because $Fe^{2+}(aq)$ has the more positive E°_{red} of the two choices. $Fe^{2+}(aq)$ is the more easily reduced and therefore plates out on the cathode.

189

10. For the reaction

$$Pb(s) + PbO_2(s) + 4H^+(aq) + 2SO_4^{2-}(aq) \rightarrow 2PbSO_4(s) + 2H_2O(\ell)$$

which is the overall reaction in a lead storage battery, $\Delta H° = -315.9$ kJ/mole and $\Delta S° = 263.5$ J/K · mole. What is the proper setup to find $E°$ at 75°C?

A. $\dfrac{-315.9 - 348(0.2635)}{-2(96.487)}$

B. $\dfrac{-348 + 315.9(0.2635)}{2(96.487)}$

C. $\dfrac{-348 + 315.9(0.2635)}{96.487}$

D. $\dfrac{-2(-348) + 263.5}{96.487 + 315.9}$

E. $\dfrac{2(315.9) - 263.5}{(96.487)(348)}$

Answer: A

Use the relationships

$$\Delta G° = -n\mathscr{F}E° = \Delta H° - T\Delta S°$$

to derive the formula

$$E° = \frac{\Delta H° - T\Delta S°}{-n\mathscr{F}}$$

Next, take the given equation and break it down into the oxidation and reduction half-reactions so that you can discover the value for *n*, the number of moles of electrons either lost or gained.

Anode reaction (oxidation)

$$Pb(s) + SO_4^{2-} \rightarrow PbSO_4(s) + 2 e^-$$

Cathode reaction (reduction)

$$\frac{PbO_2(s) + SO_4^{2-}(aq) + 4H^+(aq) + 2e^- \rightarrow PbSO_4(s) + 2H_2O(\ell)}{Pb(s) + PbO_2(s) + 4H^+(aq) + 2SO_4^{2-}(aq) \rightarrow 2PbSO_4(s) + 2H_2O(\ell)}$$

Now substitute all the known information into the derived equation.

$$E° = H° - T\Delta S°/-n\mathscr{F} = (-315.9 \text{ kJ/mole} - 348 \text{ K} \cdot 0.2635 \text{ kJ/K} \cdot \text{mole} / -2(96.487 \text{ kJ/V} \cdot \text{mole})$$

Samples: Free-Response Questions

1. A student places a copper electrode in a 1 M solution of $CuSO_4$ and in another beaker places a silver electrode in a 1 M solution of $AgNO_3$. A salt bridge composed of Na_2SO_4 connects the two beakers. The voltage measured across the electrodes is found to be +0.42 volt.

   (a) Draw a diagram of this cell.

   (b) Describe what is happening at the cathode. (Include any equations that may be useful.)

   (c) Describe what is happening at the anode. (Include any equations that may be useful.)

   (d) Write the balanced overall cell equation.

   (e) Write the standard cell notation.

   (f) The student adds 4 M ammonia to the copper sulfate solution, producing the complex ion $Cu(NH_3)_4^{2+}(aq)$. The student remeasures the cell potential and discovers the voltage to be 0.88 volt at 25°C. What is the $Cu^{2+}(aq)$ concentration in the cell after the ammonia has been added?

Answer

1. Given: Cu electrode in 1 M $CuSO_4$.

   Ag electrode in 1 M $AgNO_3$.

   Voltage = 0.42 volt.

   (a) Restatement: Diagram the cell.

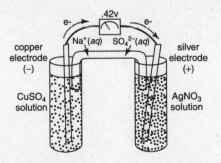

   (b) Restatement: What happens at the cathode?

   Reduction always occurs at the cathode. Note that $E°_{red}$ for silver is +0.7991 volt, according to the Table of Standard Reduction Potentials. $E°_{red}$ for copper is +0.337. This means that the copper metal is higher in the activity series than the silver metal, so copper metal will reduce the silver ion. The equation that describes reduction (or the cathode reaction) is therefore

   red: $Ag^+(aq) + e^- \rightarrow Ag(s)$

(c) Restatement: What happens at the anode?

Oxidation occurs at the anode. Silver is lower in the activity series than copper. Therefore, the oxidation half-reaction is

ox: $Cu(s) \rightarrow Cu^{2+}(aq) + 2\ e^{-}$

(d) Restatement: Overall cell equation.

Combining these two half-equations into one cell equation produces

ox: $Cu(s) \rightarrow Cu^{2+}(aq) + \cancel{2e^{-}}$ $\qquad\qquad$ $E^{\circ}_{cell} = -0.337$

red: $2Ag^{+}(aq) + \cancel{2e^{-}} \rightarrow 2Ag(s)$ $\qquad\qquad$ $E^{\circ}_{red} = +0.7991$

$\overline{\qquad\qquad Cu(s) + 2Ag^{+}(aq) \rightarrow Cu^{2+}(aq) + 2Ag(s) \qquad\qquad E^{\circ}_{cell} = 0.462}$

The overall theoretical E°_{cell} is $0.7991 - 0.337 = +0.462$ volt. Because this reaction is spontaneous, as was discovered when the cell was measured, producing the experimentally determined 0.42 volt, this is in agreement with our cell designations.

(e) Restatement: Cell notation.

$Cu(s) \mid Cu^{2+}(aq) \parallel Ag^{+}(aq) \mid Ag(s)$

(f) Given:

$\quad$ 4 M $NH_3(aq)$ added to $CuSO_4$ solution $\rightarrow Cu(NH_3)_4{}^{2+}(aq)$

$\quad$ Voltage = 0.88 volt

$\quad$ Restatement: Calculate $[Cu^{2+}(aq)]$.

Because this cell is not operating under standard conditions, you will need to use the Nernst equation:

$$E_{cell} = E^{\circ}_{cell} - \frac{0.0592}{n} \log Q \text{ at } 25^{\circ}C$$

The variables take on the following values:

E_{cell} = 0.88 volt

E°_{cell} = 0.46 volt

$$Q = \frac{\left[Cu^{2+}(aq) \right]}{\left[Ag^{+}(aq) \right]^{2}} = \frac{x}{1^2} = x$$

Substituting into the Nernst equation yields

$0.88 = 0.46 - \dfrac{0.0592}{2} \log x$

$0.0296 \log x = -0.42$

$\log x = \dfrac{0.42}{0.0296} = -14$

So $[Cu^{2+}(aq)] = 1.0 \times 10^{-14}$ M

2. The ferrous ion, $Fe^{2+}(aq)$, reacts with the permanganate ion, $MnO_4^-(aq)$, in an acidic solution to produce the ferric ion, $Fe^{3+}(aq)$. A 6.893-gram sample of ore was mechanically crushed and then treated with concentrated hydrochloric acid, which oxidized all of the iron in the ore to the ferrous ion, $Fe^{2+}(aq)$. Next, the acid solution containing all of the ferrous ions was titrated with 0.100 M $KMnO_4$ solution. The end point was reached when 13.889 mL of the potassium permanganate solution was used.

(a) Write the oxidation half-reaction.

(b) Write the reduction half-reaction.

(c) Write the balanced final redox reaction.

(d) Identify the oxidizing agent, the reducing agent, the species oxidized, and the species reduced.

(e) Calculate the number of moles of iron in the sample of ore.

(f) Calculate the mass percent of iron in the ore.

Answer

2. Given: $Fe^{2+}(aq) + MnO_4^-(aq) \rightarrow Fe^{3+}(aq)$

6.893 grams of ore.

All of the iron in the ore was converted to $Fe^{2+}(aq)$.

$Fe^{2+}(aq)$ treated with 0.001 M $KMnO_4$ solution $\rightarrow Fe^{3+}(aq)$.

13.889 mL of $KMnO_4$ required to reach end point.

(a) Restatement: Oxidation half-reaction.

$Fe^{2+}(aq) \rightarrow Fe^{3+}(aq) + e^-$ OIL (Oxidation Is Losing)

(b) Restatement: Reduction half-reaction.

$MnO_4^-(aq) \rightarrow Mn^{2+}(aq) + 4H_2O(1)$ (balance O's)

$MnO_4^-(aq) + 8H^+(aq) \rightarrow Mn^{2+}(aq) + 4H_2O(1)$ (balance H's)

$MnO_4^-(aq) + 8H^+(aq) + 5e^- \rightarrow Mn^{2+}(aq) + 4H_2O(1)$ (balance charge)

(c) Restatement: Balanced redox reaction.

ox: $5Fe^{2+} \rightarrow 5Fe^{3+} + 5e^-$

red: $MnO_4^- + 8H^+ + 5e^- \rightarrow Mn^{2+} + 4H_2O$

$\overline{MnO_4^- + 8H^+ + 5Fe^{2+} \rightarrow Mn^{2+} + 4H_2O + 5Fe^{3+}}$

(d) Restatement: Identify:

oxidizing agent: $MnO_4^-(aq)$ a species that accepts electrons from another

reducing agent: $Fe^{2+}(aq)$ a species that furnishes electrons to another

species oxidized: $Fe^{2+}(aq)$ Oxidation Is Losing (electrons)

species reduced: $MnO_4^-(aq)$ Reduction Is Gaining (electrons)

(e) Restatement: Moles of iron in the sample of ore.

This problem can be done by using the factor-label method.

$$\frac{13.889 \cancel{\text{mL KMnO}_4}}{1} \times \frac{1 \cancel{\text{liter KMnO}_4 \text{ sol'n}}}{1000 \cancel{\text{mL KMnO}_4 \text{ sol'n}}}$$

$$\times \frac{0.100 \cancel{\text{mole KMnO}_4}}{1 \cancel{\text{liter KMnO}_4 \text{ sol'n}}} \times \frac{5 \text{ moles Fe}^{2+}}{1 \cancel{\text{mole KMnO}_4}}$$

$= 6.94 \times 10^{-3}$ mole $Fe^{2+} = 6.94 \times 10^{-3}$ mole Fe
(because all of the Fe was converted to Fe^{2+})

(f) Restatement: Mass percent of iron in the ore.

$$\% = \frac{\text{part}}{\text{whole}} \times 100\% = \frac{0.00694 \cancel{\text{mole Fe}}}{6.893 \text{ g ore}} \times \frac{55.85 \text{ g Fe}}{1 \cancel{\text{mole Fe}}} \times 100\% = 5.62\%$$

Organic Chemistry

Note to the student: You will find *very little* organic chemistry in either Section I or Section II of the AP chemistry exam. According to the Educational Testing Service, "Physical and chemical properties of simple organic compounds should be included as exemplary material for the study of other areas such as bonding, equilibria involving weak acids, kinetics, colligative properties, and stoichiometric determinations of empirical and molecular formulas. *(Reactions specific to organic chemistry are NOT tested.)*"

What this means is that you will *not* be tested on advanced organic chemistry concepts such as organic synthesis, nucleophilic substitutions, electrophilic additions, or molecular rearrangements. Rather, if you were given an acid-dissociation constant (K_a) problem, the AP chemistry exam might include an organic acid such as acetic acid ($HC_2H_3O_2$) instead of an inorganic acid such as boric acid (H_3BO_3). However, *the mechanisms of the problem would remain unchanged.*

In a thermochemistry problem, you might be given the equation

$$C_2H_4(g) + H_2(g) \rightarrow C_2H_6(g) \ \Delta H° = -137 \text{ kJ}$$

to work with, but the mechanics of the operations are the same regardless of whether you work with inorganic or organic species.

Every once in a while you may run into an organic chemistry problem in Section II, on writing equations. An example is "Write an equation that describes burning methanol in air." You would need to know the chemical formula and structural formula of methanol in order to do this problem. Writing organic reactions is covered in more detail in the chapter entitled "Writing and Predicting Chemical Reactions."

In short, then, in the area of organic chemistry you should know

- simple nomenclature
- functional groups
- various types of reactions (addition, substitution, elimination, condensation, and polymerization)
- how to draw various types of isomers: geometric, positional, functional, structural, stereo, and optical.

Key Terms

Words that can be used as topics in essays:

addition reaction

alcohol

aldehyde

aliphatic hydrocarbon

alkane

alkene

alkyne

amine

aromatic hydrocarbon

aryl group

branched-chain hydrocarbon

carboxylic acid

chiral center

condensation reaction

cycloalkane

dimer

elimination reaction

enantiomer

ester

ether

free radical

functional group

hydrocarbon

isomerization

isomers (functional, geometric, positional, structural, optical)

ketone

saturated hydrocarbon

substitution reaction

thiols

unsaturated hydrocarbon

Key Concepts

Equations and relationships that you need to know:

- **Functional Groups**

	Functional Group	*General Formula*	*Example*
Acids	O ‖ −C−O−H	O ‖ R−C−O−H	acetic acid H O | ‖ H−C−C−OH | H
Alcohol	−O−H	R−O−H	methanol H | H−C−OH | H
Aldehyde	O ‖ −C−H	O ‖ R−C−H	formaldehyde O ‖ H−C−H

	Functional Group	General Formula	Example
Amide	$-\overset{\overset{\displaystyle O}{\|\|}}{C}-\overset{H}{\underset{H}{N}}$	$R-\overset{\overset{\displaystyle O}{\|\|}}{C}-\overset{H}{\underset{H}{N}}$	acetamide $H-\overset{\overset{\displaystyle H}{\|}}{\underset{H}{C}}-\overset{\overset{\displaystyle O}{\|\|}}{C}-\overset{H}{\underset{H}{N}}$
Amine	$-\overset{\|}{N}-$	$R-N\overset{R'}{\underset{R''}{}}$	methylamine $H-\overset{\overset{\displaystyle H}{\|}}{\underset{H}{C}}-\overset{H}{\underset{H}{N}}$
Disulfide	$-S-S-$	$R-S-S-R'$	dimethyl disulfide $H-\overset{\overset{\displaystyle H}{\|}}{\underset{H}{C}}-S-S-\overset{\overset{\displaystyle H}{\|}}{\underset{H}{C}}-H$
Ester	$-\overset{\overset{\displaystyle O}{\|\|}}{C}-O-$	$R-\overset{\overset{\displaystyle O}{\|\|}}{C}-O-R'$	ethyl acetate $H-\overset{\overset{\displaystyle H}{\|}}{\underset{H}{C}}-\overset{\overset{\displaystyle O}{\|\|}}{C}-O-\overset{\overset{\displaystyle H}{\|}}{\underset{H}{C}}-\overset{\overset{\displaystyle H}{\|}}{\underset{H}{C}}-H$
Ether	$-O-$	$R-O-R'$	ethyl methyl ether $H-\overset{\overset{\displaystyle H}{\|}}{\underset{H}{C}}-O-\overset{\overset{\displaystyle H}{\|}}{\underset{H}{C}}-\overset{\overset{\displaystyle H}{\|}}{\underset{H}{C}}-H$
Ketone	$-\overset{\overset{\displaystyle O}{\|\|}}{C}-$	$R-\overset{\overset{\displaystyle O}{\|\|}}{C}-R'$	methyl ethyl ketone $H-\overset{\overset{\displaystyle H}{\|}}{\underset{H}{C}}-\overset{\overset{\displaystyle O}{\|\|}}{C}-\overset{\overset{\displaystyle H}{\|}}{\underset{H}{C}}-\overset{\overset{\displaystyle H}{\|}}{\underset{H}{C}}-H$
Salt (Carboxylate)	$-\overset{\overset{\displaystyle O}{\|\|}}{C}-O^-\cdots M^+$	$R-\overset{\overset{\displaystyle O}{\|\|}}{C}-O^-\cdots M^+$	sodium acetate $H-\overset{\overset{\displaystyle H}{\|}}{\underset{H}{C}}-\overset{\overset{\displaystyle O}{\|\|}}{C}-O^-\cdots Na^+$
Thiol	$-S-H$	$R-S-H$	ethanethiol $H-\overset{\overset{\displaystyle H}{\|}}{\underset{H}{C}}-\overset{\overset{\displaystyle H}{\|}}{\underset{H}{C}}-S-H$

197

- ## Types of Isomers

Structural

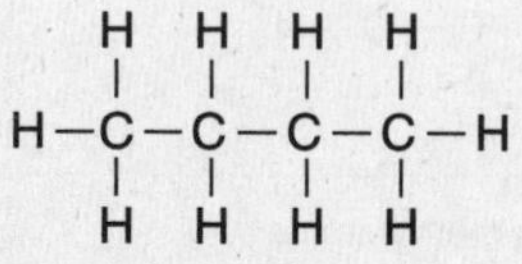

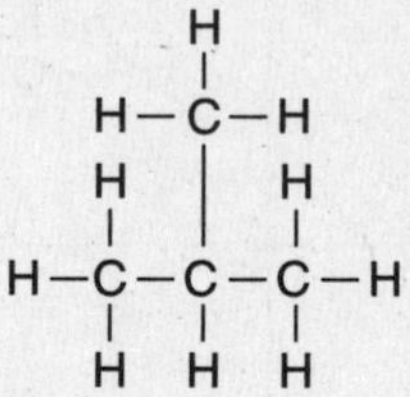

H H H H
| | | |
H—C—C—C—C—H
| | | |
H H H H

```
        H
        |
     H—C—H
     H  |  H
     |  |  |
  H—C——C——C—H
     |  |  |
     H  H  H
```

Geometric

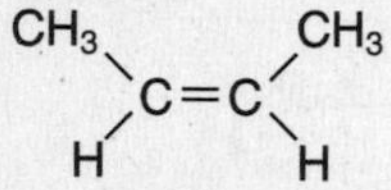

CH₃   CH₃
    \   /
     C = C
    /   \
   H   H
        cis

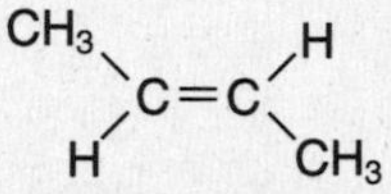

CH₃   H
    \   /
     C = C
    /   \
   H   CH₃
        trans

Positional

H H H
| | |
H—C—C—C—OH
| | |
H H H

H OH H
| | |
H—C—C—C—H
| | |
H H H

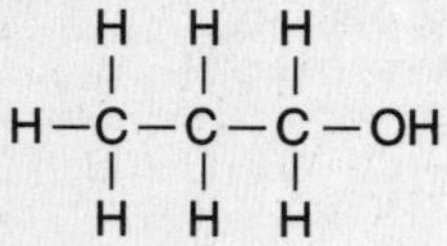

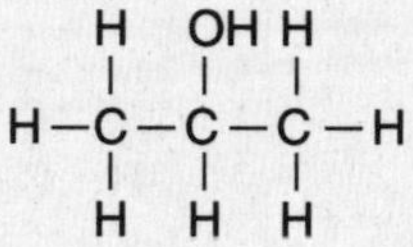

Functional

H H
| |
H—C—C—OH
| |
H H

H   H
| |
H—C—O—C—H
| |
H   H

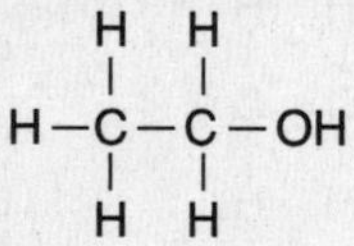

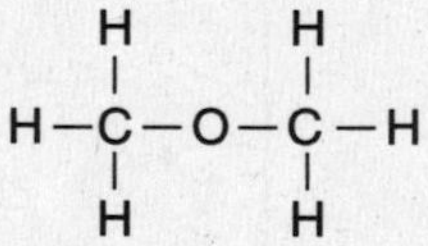

Samples: Multiple-Choice Questions

Directions: The first lettered list consists of the answer choices for questions 1–5, the second list gives the answer choices for questions 6–10, and the third list gives the answer choices for questions 11–14. Select the letter of the one choice that properly identifies the given compound (questions 1–5), describes the given reaction (questions 6–10), or identifies the product of the reaction (questions 11–14). A choice may be used once, more than once, or not at all in each set.

Questions 1–5

   A. Alcohol

   B. Aldehyde

   C. Carboxylic acid

   D. Ester

   E. Ether

1.
$$HO-\overset{\overset{\textstyle O}{\|}}{C}-\overset{\overset{\textstyle O}{\|}}{C}-OH$$

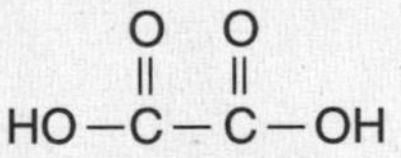

Answer: C

The functional group of a carboxylic acid is

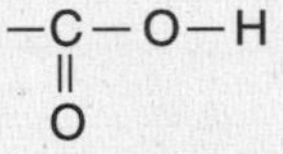

$$-\overset{}{\underset{\underset{\textstyle O}{\|}}{C}}-O-H$$

The name of this compound is oxalic acid, which is an aliphatic dicarboxylic acid.

2.
$$H-\overset{\overset{\textstyle H}{|}}{\underset{\underset{\textstyle H}{|}}{C}}-\overset{\overset{\textstyle H}{|}}{\underset{\underset{\textstyle H}{|}}{C}}-O-H$$

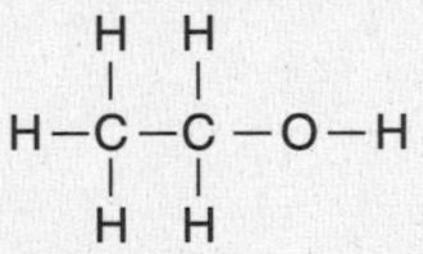

Answer: A

The functional group of an alcohol is

$$-O-H$$

The name of this alcohol is ethyl alcohol.

3.
$$H-\overset{\overset{\displaystyle H}{|}}{\underset{\underset{\displaystyle H}{|}}{C}}-\overset{\overset{\displaystyle H}{|}}{\underset{\underset{\displaystyle H}{|}}{C}}-O-\overset{\overset{\displaystyle H}{|}}{\underset{\underset{\displaystyle H}{|}}{C}}-\overset{\overset{\displaystyle H}{|}}{\underset{\underset{\displaystyle H}{|}}{C}}-H$$

Answer: E

The functional group of an ether is

$$-O-$$

The name of this ether is diethyl ether.

4.
$$H-\overset{\overset{\displaystyle H}{|}}{\underset{\underset{\displaystyle H}{|}}{C}}-\overset{\overset{\displaystyle }{}}{\underset{\underset{\displaystyle O}{\|}}{C}}-H$$

Answer: B

The functional group of an aldehyde is

$$\underset{\underset{\displaystyle O}{\|}}{-C}-H$$

The name of this aldehyde is acetaldehyde.

5.
$$H-\overset{\overset{\displaystyle H}{|}}{\underset{\underset{\displaystyle H}{|}}{C}}-\underset{\underset{\displaystyle O}{\|}}{C}-O-C_8H_{17}$$

Answer: D

The functional group of an ester is

$$\underset{\underset{\displaystyle O}{\|}}{-C}-O-$$

The name of this ester is n-octyl acetate, which is the odor of oranges. Esters are formed from the reaction of a carboxylic acid with an alcohol.

Questions 6–10

    A. Addition reaction

    B. Elimination reaction

    C. Condensation reaction

    D. Substitution reaction

    E. Polymerization reaction

---

6. $C_2H_5OH(aq) + HCl(aq) \rightarrow C_2H_5Cl(l) + H_2O(l)$

Answer: D

A substitution reaction occurs when an atom or group of atoms in a molecule is replaced by a different atom or group.

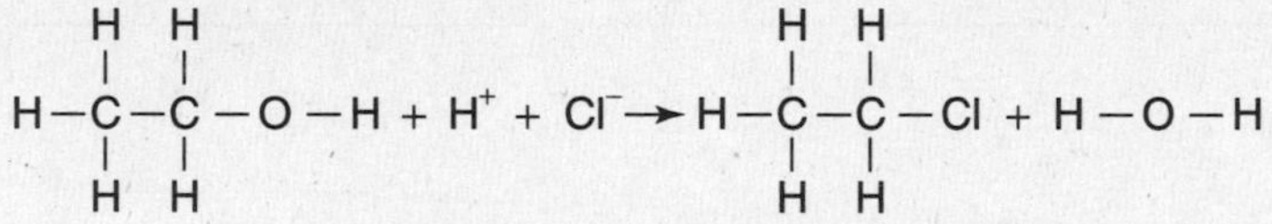

---

7. $CH_2{=}CH_2 + Br_2 \rightarrow CH_2{-}CH_2$ (with Br, Br substituents)

Answer: A

In addition reactions, a small molecule adds across a double or triple bond.

---

8. $C_2H_5Cl(g) \underset{\text{base}}{\rightleftharpoons} C_2H_4(g) + HCl(g)$

Answer: B

An elimination reaction involves the elimination of two groups from adjacent carbon atoms, converting a saturated molecule into an unsaturated molecule. It is essentially the reverse of an addition reaction.

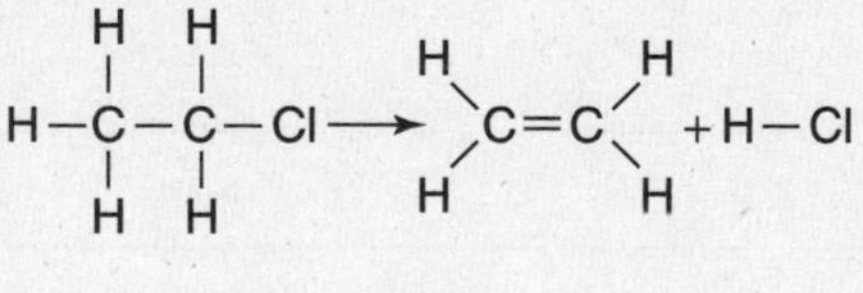

---

9. $CH_3OH(\ell) + CH_3OH(\ell) \underset{\text{acid}}{\rightleftharpoons} CH_3OCH_3(\ell) + H_2O(\ell)$

Answer: C

Condensation reactions occur when two molecules combine by splitting out a small molecule such as water.

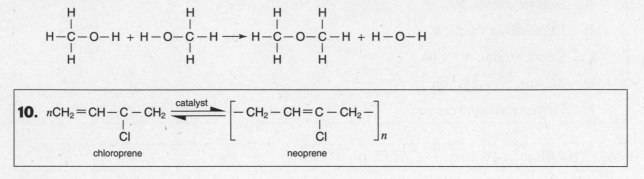

10. $nCH_2{=}CH{-}C{-}CH_2 \xrightleftharpoons{\text{catalyst}} \left[-CH_2{-}CH{=}C{-}CH_2{-}\right]_n$

chloroprene   neoprene

Answer: E

Polymers are built up of a large number of simple molecules known as monomers, which have reacted with one another.

> *Questions 11–14*
>
> **A.** Alcohol
>
> **B.** Carboxylic acid
>
> **C.** Ester
>
> **D.** Ether
>
> **E.** Ketone

11. The product of the reaction of an alcohol and a carboxylic acid.

Answer: C

An example of esterification is the production of ethyl acetate by the reaction of ethanol with acetic acid.

12. The product of the reaction of an alkene and water.

Answer: A

An example is the production of ethanol by the addition of water to ethylene.

$$\underset{\substack{}}{\overset{H}{\underset{H}{C}}}{=}\underset{\substack{}}{\overset{H}{\underset{H}{C}}} + H{-}O{-}H \xrightarrow{\substack{\text{acid}\\\text{catalyst}}} H{-}\overset{H}{\underset{H}{C}}{-}\overset{H}{\underset{H}{C}}{-}O{-}H$$

13. The product formed by the oxidation of a secondary alcohol.

Answer: E

A secondary alcohol has the general structure

$$\begin{array}{c} R \\ \diagdown \\ R' \diagup \end{array} \overset{\textstyle H}{\underset{\textstyle}{C}} - O - H$$

where the R and R' (which may be the same or different) represent hydrocarbon fragments. An example is the oxidation of isopropyl alcohol to acetone.

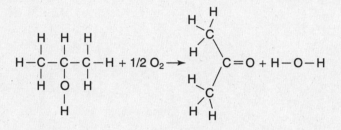

14. The product formed by the condensation reaction of alcohols.

Answer: D

A condensation reaction is characterized by the joining of two molecules and the elimination of a water molecule. In the example below, two methyl alcohol molecules react to form dimethyl ether.

$$\underset{\displaystyle \underset{H}{|}}{\overset{\displaystyle \overset{H}{|}}{H-C}} - O - H \; + \; H - O - \underset{\displaystyle \underset{H}{|}}{\overset{\displaystyle \overset{H}{|}}{C}} - H \; \xrightarrow[\text{catalyst}]{H_2SO_4} \; \underset{\displaystyle \underset{H}{|}}{\overset{\displaystyle \overset{H}{|}}{H-C}} - O - \underset{\displaystyle \underset{H}{|}}{\overset{\displaystyle \overset{H}{|}}{C}} - H \; + \; H - O - H$$

Nuclear Chemistry

Key Terms

Words that can be used as topics in essays:

alpha particle, α

beta particle, β

binding energy

chain reaction

decay series

Einstein equation

electron capture

first-order rate constant

fission

fusion

gamma radiation, γ

half-life

isotopes

mass defect

nuclear transformation

nucleon

nuclide

positron

rate of decay

transmutation

transuranium elements

zone of stability

Key Concepts

- $t_{1/2} = \dfrac{0.693}{k}$

- $\ln\left(\dfrac{x_0}{x}\right) = kt$

    where

    x_0 = original number of nuclides at time 0

    x = number of nuclides at time t

- $\Delta E = \Delta mc^2$

    where

    $c = 3.00 \times 10^8 \ \text{m} \cdot \text{sec}^{-1}$

    or

    $\Delta E = 9.00 \times 10^{10} \ \text{kJ} \cdot \text{g}^{-1} \cdot \Delta m$

- **Symbols**

proton	^1_1p or ^1_1H
neutron	^1_0n
electron (β particle)	$^{\ 0}_{-1}e$ or $^{\ 0}_{-1}\beta$ or e^-
positron	^0_1e or $^0_1\beta$
α particle	^4_2He or $^4_2\alpha$
gamma rays	γ
deuteron	^2_1H
triton	^3_1H

Samples: Multiple-Choice Questions

> Questions 1–3
>
> A. γ
>
> B. ^1_1H
>
> C. ^1_0n
>
> D. ^4_2He
>
> E. $2\,^1_0\text{n}$

> **1.** $^{54}_{26}\text{Fe} + ? \rightarrow\, ^{56}_{26}\text{Fe} + 2\,^1_1\text{H}$

Answer: D

Think of the yield sign (arrow) as an equal sign. The superscript represents the mass number. The sum of the mass numbers on both sides of the arrow must be equal. The subscript represents the atomic number, and as with mass numbers, the sum of the numbers on both sides of the arrow must be equal.

mass: $54 + \underline{4} = 56 + 2(1)$

atomic number: $26 + \underline{2} = 26 + 2(1)$

> **2.** $^{65}_{29}\text{Cu} + ^1_0\text{n} \rightarrow\, ^{64}_{29}\text{Cu} + ?$

Answer: E

mass: $65 + 1 = 64 + \underline{2(1)}$

atomic number: $29 + 0 = 29 + \underline{2(0)}$

> **3.** $^{14}_7\text{N} + ^1_1\text{H} \rightarrow\, ^{15}_8\text{O} + ?$

Answer. A

mass: $14 + 1 = 15 + \underline{0}$

atomic number: $7 + 1 = 8 + \underline{0}$

---

4. Choose the one FALSE statement.

    A. Nuclei with an even number of protons and an even number of neutrons tend to be stable.

    B. "γ"-rays are high-energy photons.

    C. Nuclei with too few neutrons per proton tend to undergo positron ($_1^0\beta$) emission.

    D. Light nuclides are stable when the atomic number (Z) equals the mass number minus the atomic number (A–Z).

    E. Nuclei with too many neutrons per proton tend to undergo β-particle ($_{-1}^0$e) emission.

Answer: E

Positron production occurs for nuclides that are below the zone of stability (those nuclides whose neutron/proton ratios are too small). The net effect of positron emission is to change a proton to a neutron. An example of positron emission would be

$$_{11}^{22}\text{Na} \rightarrow {}_1^0\text{e} + {}_{10}^{22}\text{Ne}$$

Nuclides with too many neutrons per proton tend to undergo β-particle production. The net effect of β-particle production is to change a neutron to a proton. Examples of β-particle production are

$$_{90}^{234}\text{Th} \rightarrow {}_{91}^{234}\text{Pa} + {}_{-1}^0\text{e}$$
$$_{53}^{131}\text{I} \rightarrow {}_{54}^{131}\text{Xe} + {}_{-1}^0\text{e}$$

Questions 5–9

    A. $_{80}^{201}\text{Hg} + {}_{-1}^0\text{e} \rightarrow {}_{79}^{201}\text{Au} + {}_0^0\gamma$

    B. $_6^{11}\text{C} \rightarrow {}_1^0\text{e} + {}_5^{11}\text{B}$

    C. $_{93}^{237}\text{Np} \rightarrow {}_2^4\text{He} + {}_{91}^{233}\text{Pa}$

    D. $_{92}^{235}\text{U} + {}_0^1\text{n} \rightarrow {}_{56}^{141}\text{Ba} + {}_{36}^{92}\text{Kr} + 3\,{}_0^1\text{n}$

    E. $_{83}^{214}\text{Bi} \rightarrow {}_{84}^{214}\text{Po} + {}_{-1}^0\text{e}$

---

5. Alpha (α)-particle production.

Answer: C

An alpha particle is a helium nucleus.

---

6. Beta (β)-particle production.

Answer: E

A β-particle is an electron. An unstable nuclide in β-particle production creates an electron as it releases energy in the decay process. This electron is created from the decay process, rather than being present before the decay occurs.

7. Electron capture.

Answer: A

Electron capture is a process by which one of the inner-orbital electrons is captured by the nucleus.

8. Fission.

Answer: D

Fission is the process whereby a heavy nucleus splits into two nuclei (not including He nuclei) with smaller mass numbers.

9. Positron production.

Answer: B

A positron is a particle with the same mass as an electron but with the opposite charge. The net effect of positron production is to change a proton to a neutron.

10. The half-life of C is 5770 years. What percent of the original radioactivity would be present after 28,850 years?

    A. 1.56%

    B. 3.12%

    C. 6.26%

    D. 12.5%

    E. 25.0%

Answer: B

This problem can be solved using the factor-label method.

$$\frac{28,850 \text{ years}}{1} \times \frac{1 \text{ half-life}}{5770 \text{ years}} = 5.00 \text{ half-lives}$$

In 5.00 half-lives, the radioactivity is reduced by

$$\left(\frac{1}{2}\right)^{5.00} = 0.0312 \times 100 = 3.12\%$$

---

11. Which of the following choices correctly describes the *decreasing* ability of the radiation to penetrate a sheet of lead that is 3 inches thick?

    A.   alpha particles > beta particles > gamma rays

    B.   gamma rays > alpha particles > beta particles

    C.   alpha particles > gamma rays > beta particles

    D.   beta particles > alpha particles > gamma rays

    E.   gamma rays > beta particles > alpha particles

Answer: E

Gamma rays (γ) have high penetrating power and are not deflected by electric or magnetic fields. Beta particles (β) have a lower ionizing power and greater penetrating power than alpha particles (α).

Samples: Free-Response Questions

---

1. $^{208}_{81}\text{Tl}$ undergoes β-decay. The half-life is 3.1 min.

  (a) Write the nuclear equation.

  (b) How long will it take for 1.00 gram of $^{208}_{81}\text{Tl}$ to be reduced to 0.20 gram by decay?

  (c) Given 11.00 grams of pure $^{208}_{81}\text{Tl}$ initially, how many grams of pure $^{208}_{81}\text{Tl}$ will remain after 23.0 min?

Answer

1. Given: $^{208}_{81}\text{Tl}\left(\beta\text{-decay}\right)$ $t_{1/2} = 3.1\,\text{min}$

  (a) Restatement: Nuclear equation.

$$^{208}_{81}\text{Tl} \rightarrow\, ^{208}_{82}\text{Pb} +\, ^{0}_{-1}\text{e}$$

  (b) Restatement: Time for 1.00 g to reduce to 0.20 g.

$$\ln\frac{[A]_0}{[A]} = \ln\frac{1.00}{0.20} = kt$$

$$0.693 = kt_{1/2}$$

$$k = \frac{0.693}{3.1\,\text{min}} = 0.22\,\text{min}^{-1}$$

$$t = \ln\frac{1.00}{0.20} \times \frac{1}{0.22\,\text{min}^{-1}} = 7.3\,\text{min}$$

  (c) Given: 11.00 g of $^{208}_{81}\text{Tl}$ initially.

Restatement: How many grams remain after 23 min?

$$\log\frac{x_0}{x} = \frac{kt}{2.303} \qquad\qquad k = \frac{0.693}{3.1\,\text{min}} = 0.22\,\text{min}^{-1}$$

$$\log\frac{11.00}{x} = \frac{0.22\,\text{min}^{-1} \times 23\,\text{min}}{2.303} = 2.2$$

$$\log\frac{11}{x_0} = 2.2 \rightarrow \frac{11}{x_0} = 10^{2.2}$$

$$x_0 = 0.070\,\text{g}$$

2. Radon-222 can be produced from the α-decay of radium-226.

   (a) Write the nuclear reaction.

   (b) Calculate ΔE (in kJ) when 7.00 g of $^{226}_{88}$Ra decays.

   $^{4}_{2}$He = 4.0015 g/mole

   $^{222}_{86}$Rn = 221.9703 g/mole

   $^{226}_{88}$Ra = 225.9771 g/mole

   (c) Calculate the mass defect of $^{226}_{88}$Ra.

   1 mole protons = 1.00728 g

   1 mole neutrons = 1.00867 g

   atomic mass $^{226}_{88}$Ra = 225.9771 g/mole

   (d) Calculate the binding energy (in kJ/mole) of $^{226}_{88}$Ra.

   (e) $^{226}_{88}$Ra has a half-life of 1.62×10^{3} yr. Calculate the first-order rate constant.

   (f) Calculate the fraction of $^{226}_{88}$Ra that will remain after 100.0 yr.

Answer

2. Given: $^{222}_{86}$Rn produced by α-decay of $^{226}_{88}$Ra.

   (a) Restatement: Nuclear reaction.

   $$^{226}_{88}\text{Ra} \rightarrow {}^{222}_{86}\text{Rn} + {}^{4}_{2}\text{He}$$

   (b) Given: 7.00 g of $^{226}_{88}$Ra decays.

   Restatement: Calculate ΔE (in kJ).

   $$\Delta E = \Delta mc^2$$

   $$= 9.00 \times 10^{10} \frac{\text{kJ}}{\text{g}} \times \Delta m$$

   Δm = mass products − mass reactants

   $$= 4.0015 \text{ g} + 221.9703 \text{ g} - 225.9771 \text{ g}$$

   $$= -0.0053 \text{ g}$$

   $$\Delta E = \frac{7.00 \text{ g } ^{226}_{88}\text{Ra}}{1} \times \frac{1 \text{ mole } ^{226}_{88}\text{Ra}}{225.9771 \text{ g } ^{226}_{88}\text{Ra}} \times \frac{-0.0053 \text{ g}}{1 \text{ mole } ^{226}_{88}\text{Ra}} \times \frac{9.00 \times 10^{10} \text{ kJ}}{\text{g}} = -1.5 \times 10^{7} \text{ kJ}$$

   (c) Restatement: Mass defect of $^{226}_{88}$Ra.

   88 moles protons = 88 × 1.00728 g =   88.6406 g

   138 moles neutrons = 138 × 1.00867 g = + 139.196 g

   227.837 g

   mass defect = 227.837 g − 225.9771 g = 1.860 g

(d) Restatement: Calculate binding energy (kJ/mole) of $^{226}_{88}Ra$.

$$\Delta E = \frac{9.00 \times 10^{10}\,kJ}{g} \times \frac{1.860\,g}{1\,mole} = 1.67 \times 10^{11}\,kJ/mole$$

(e) Given $^{226}_{88}Ra$ has $t_{1/2} = 1.62 \times 10^3$ yr.

Restatement: Rate constant for $^{226}_{88}Ra$.

$$k = \frac{0.693}{1.62 \times 10^3\,yr} = 4.28 \times 10^{-4}\,yr$$

(f) Restatement: Fraction of $^{226}_{88}Ra$ that remains after 100.0 yr.

$$\ln \frac{x_0}{x} = kt = 4.28 \times 10^{-4}\,yr^{-1} \times 100.0\,yr = 4.28 \times 10^{-2}$$

$$e^{0.0428} = 1.04 = \frac{x_0}{x}$$

fraction remaining: $\frac{x}{x_0} = \frac{1}{1.04} = 0.962 = 96.2\%$

Writing and Predicting Chemical Reactions

Question 1 of Section II (also known as Part A) of the AP chemistry exam is *always* a question on writing reactions. The directions for this question follow.

1. You may NOT use a calculator for this section of the exam. You will be given 10 minutes to answer FIVE of the eight options in this part. (Answers to more than five options will not be scored.)

Give the formulas to show the reactants and the products for FIVE of the following chemical reactions. Each of the reactions occurs in aqueous solution unless otherwise indicated. Represent substances in solution as ions if the substance is extensively ionized. Omit formulas for any ions or molecules that are unchanged by the reaction. In all cases a reaction occurs. You need not balance.

Example: A strip of magnesium is added to a solution of silver nitrate.

$$Mg + Ag^+ \rightarrow Mg^{2+} + Ag$$

The question is scored according to the following guide:

- Each of the five reactions you choose is worth a maximum of 3 points — 15 points in all.
- One point is given if the reactants are correct.
- Two points are given if the products are correct.
- If the equation is correct but the charge on the ions is incorrect, 1 point is deducted.
- Part A represents 15% of your score for Section II.

In order to do well on this question, *you must know the solubility rules!* See page 116 of this book. Also, in the sections that follow, I have tried to give you *general* rules for predicting the products when the reactants are given. These guidelines will not work for 100% of the questions on the AP exam, but they should get you through at least four or five of the eight you are presented with — and that's all you need.

I. Metals Combining with Nonmetals

Metal + Nonmetal → Salt* (Metal Ion$^+$ + Nonmetal Ion$^-$)

Reactions of Alkali Metals and Alkaline Earth Metals

Group IA Metal (M)	Combining Substance	Reaction
All	Hydrogen	$2\,M(s) + H_2(g) \rightarrow 2\,MH(s)$
All	Halogen	$2\,M(s) + X_2 \rightarrow 2\,MX(s)$
Li	Nitrogen	$6\,Li(s) + N_2(g) \rightarrow 2\,Li_3N(s)$
All	Sulfur	$2\,M(s) + S(s) \rightarrow M_2S(s)$
Li	Oxygen	$4\,Li(s) + O_2(g) \rightarrow 2\,Li_2O(s)$
Na	Oxygen	$2\,Na(s) + O_2(g) \rightarrow Na_2O_2(s)$**
Na, K, Rb, Cs	Oxygen	$M(s) + O_2(g) \rightarrow MO_2(s)$
All	Water	$2\,M(s) + H_2O \rightarrow 2\,M^+(aq) + 2\,OH^-(aq) + H_2(g)$

Group IIA Metal (M)	Combining Substance	Reaction
Ca, Sr, Ba	Hydrogen	$M(s) + H_2(g) \rightarrow MH_2(s)$
All	Halogens	$M(s) + X_2 \rightarrow MX_2(s)$
Mg, Ca, Sr, Ba	Nitrogen	$3\,M(s) + N_2(g) \rightarrow M_3N_2(s)$
Mg, Ca, Sr, Ba	Sulfur	$M(s) + S(s) \rightarrow MS(s)$
Be, Mg, Ca, Sr, Ba	Oxygen	$2\,M(s) + O_2(g) \rightarrow 2\,MO(s)$
Ba	Oxygen	$Ba(s) + O_2(g) \rightarrow BaO_2(s)$
Ca, Sr, Ba	Water	$M(s) + 2H_2O \rightarrow M^{2+}(aq) + 2OH^-(aq) + H_2(g)$
Mg	Water	$Mg(s) + H_2O(g) \rightarrow MgO(s) + H_2(g)$

M = symbol for the metal, *X* = any halogen

By definition, a salt is an ionic compound made up of a cation other than H^+ and an anion other than OH^- or O^{2-}. However, in these examples a "salt" represents any ionic (metal-nonmetal) compound.

**Superoxides need special conditions.*

1. Calcium metal is heated in the presence of nitrogen gas:

   $Ca + N_2 \rightarrow Ca_3N_2$

2. Solid potassium is added to a flask of oxygen gas:

   $K + O_2 \rightarrow KO_2$

   This reaction is also true for rubidium and cesium. (Note: These are superoxides.)

3. A piece of solid magnesium, which is ignited, is added to water:

   $Mg + H_2O \rightarrow MgO + H_2$

II. Combustion

Substance + Oxygen Gas → Oxide of Element

The usual products are the oxides of the elements present in the original substance in their *higher* valence states. When N, Cl, Br, and I are present in the original compound, they are usually released as free elements, *not* as the oxides.

1. Solid copper(II) sulfide is heated strongly in oxygen gas:

   $CuS + O_2 \rightarrow CuO + SO_2$

   or

   $\rightarrow Cu_2O + SO_2$

2. Carbon disulfide gas is burned in excess oxygen gas:

   $CS_2 + O_2 \rightarrow CO_2 + SO_2$

3. Methanol is burned completely in air:

   $CH_3OH + O_2 \rightarrow CO_2 + H_2O$

   All alcohols (as well as hydrocarbons and carbohydrates) burn in oxygen gas to produce CO_2 and H_2O.

4. Silane is combusted in a stream of oxygen gas:

   $SiH_4 + O_2 \rightarrow SiO_2 + H_2O$

III. Metallic Oxide + H₂O → Base (Metallic Hydroxide)

1. Lithium oxide is added to water:

   $Li_2O + H_2O \rightarrow Li^+ + OH^-$

2. A solid piece of potassium oxide is dropped into cold water:

   $K_2O + H_2O \rightarrow K^+ + OH^-$

IV. Nonmetallic Oxide + H₂O → Acid

1. Dinitrogen pentoxide is added to water:

   $N_2O_5 + H_2O \rightarrow H^+ + NO_3^-$

2. Carbon dioxide gas is bubbled through water:

$CO_2 + H_2O \rightarrow H_2CO_3$ (carbonic acid)

or

$\rightarrow H^+ + HCO_3^-$ (weak acid)

3. Sulfur trioxide gas is bubbled through water:

$SO_3 + H_2O \rightarrow H^+ + SO_4^{2-}$

4. Phosphorus(V) oxytrichloride is added to water:

$POCl_3 + H_2O \rightarrow H^+ + HPO_4^{2-} + Cl^-$

V. Metallic Oxide + Acid → Salt + H₂O

1. Ferric oxide is added to hydrochloric acid:

$Fe_2O_3 + H^+ \rightarrow Fe^{3+} + H_2O$

2. Copper(II) oxide is added to nitric acid:

$CuO + H^+ \rightarrow Cu^{2+} + H_2O$

VI. Nonmetallic Oxide + Base → Salt + H₂O

1. Carbon dioxide gas is bubbled through a solution of sodium hydroxide:

$CO_2 + OH^- \rightarrow CO_3^{2-} + H_2O$

2. Sulfur dioxide gas is bubbled through a solution of lithium hydroxide:

$SO_2 + OH^- \rightarrow SO_3^{2-} + H_2O$

VII. Metallic Oxide + Nonmetallic Oxide → Salt

1. Magnesium oxide is heated strongly in carbon dioxide gas:

$MgO + CO_2 \rightarrow MgCO_3$

2. Calcium oxide is heated in an environment of sulfur trioxide gas:

$CaO + SO_3 \rightarrow CaSO_4$

VIII. Acid + Base → Salt + Water (Neutralization)

1. Hydrochloric acid is added to potassium hydroxide:

$H^+ + OH^- \rightarrow H_2O$

2. Equal volumes of 0.5 M sulfuric acid and 0.5 M sodium hydroxide are mixed:

$HSO_4^- + OH^- \rightarrow SO_4^{2-} + H_2O$

or

$H^+ + OH^- \rightarrow H_2O$

3. Hot nitric acid is added to solid sodium hydroxide:

$H^+ + NaOH \rightarrow HOH + Na^+$

IX. Acid + Metal → Salt + Hydrogen

1. Sulfuric acid is added to a solid strip of zinc:

$$H^+ + Zn \rightarrow H_2 + Zn^{2+}$$

2. A piece of magnesium is dropped into a beaker of 6 M hydrochloric acid:

$$Mg + H^+ \rightarrow Mg^{2+} + H_2$$

3. Calcium metal is added to a solution of 4 M HCl:

$$Ca + H^+ \rightarrow Ca^{2+} + H_2$$

Some Common Departures

4. Lead shot is dropped into hot, concentrated sulfuric acid:

$$Pb + H^+ \, SO_4^{2-} \rightarrow PbSO_4 + SO_2 + H_2O$$

Note the departure from the general guideline.

5. Solid copper shavings are added to a concentrated nitric acid solution. (This reaction is well known and is covered quite extensively in textbooks. Note how it departs from the guideline.)

$$Cu + H^+ + NO_3^- \rightarrow Cu^{2+} + H_2O + NO_2$$

X. Base + Amphoteric Metal → Salt + Hydrogen Gas

Amphoteric metals (such as Al, Zn, Pb, and Hg) have properties that may be intermediate between those of metals and those of nonmetals. They will react with a base to form a complex ion with oxygen. *This is a rare problem.*

1. A piece of solid aluminum is added to a 6 M solution of sodium hydroxide:

$$Al + OH^- \rightarrow AlO_3^{3-} + H_2$$

2. A solid piece of zinc is added to a 6 M solution of potassium hydroxide:

$$Zn + OH^- \rightarrow ZnO_2^{2-} + H_2$$

XI. Strong Acid + Salt of Weak Acid → Salt of Strong Acid + Weak Acid

1. Hydrochloric acid is added to potassium acetate:

$$H^+ + Ac^- \rightarrow HAc \text{ (Do not dissociate a weak acid.)}$$

2. A 9 M nitric acid solution is added to a solution of potassium carbonate:

$$H^+ + CO_3^{2-} \rightarrow H_2CO_3$$

XII. Weak Acid + Weak Base → Conjugate Base + Conjugate Acid

1. A solution of ammonia is mixed with an equimolar solution of hydrofluoric acid:

$$NH_3 + HF \rightarrow NH_4^+ + F^-$$

2. Acetic acid is added to a solution of ammonia:

$$HC_2H_3O_2 + NH_3 \rightarrow C_2H_3O_2^- + NH_4^+$$

XIII.   Weak Acid + Strong Base → Water + Conjugate Base

1. Hydrofluoric acid is added to a solution of sodium hydroxide:

$$HF + OH^- \rightarrow H_2O + F^-$$

2. A solution of vinegar (acetic acid) is titrated with lye (sodium hydroxide):

$$HC_2H_3O_2 + OH^- \rightarrow H_2O + C_2H_3O_2^-$$

3. Nitrous acid is added to sodium hydroxide:

$$HNO_2 + OH^- \rightarrow H_2O + NO_2^-$$

XIV.   Strong Acid + Weak Base → Conjugate Acid (Weak Acid)

1. Hydrochloric acid is added to a solution of ammonia:

$$H^+ + NH_3 \rightarrow NH_4^+$$

2. Sulfuric acid is added to a solution of sodium fluoride:

$$H^+ + F^- \rightarrow HF$$

3. Hydrochloric acid is added to sodium carbonate:

$$H^+ + CO_3^{2-} \rightarrow HCO_3^-$$

or

$$\rightarrow H_2CO_3$$

or

$$\rightarrow CO_2 + H_2O$$

4. Sodium acetate is added to a weak solution of nitric acid:

$$H^+ + C_2H_3O_2^- \rightarrow HC_2H_3O_2$$

XV.   Acid + Carbonate → Salt + CO₂ + Water

1. Hydrochloric acid is added to a sodium carbonate solution.

$$H^+ + CO_3^{2-} \rightarrow CO_2 + H_2O$$

2. Hydroiodic acid is mixed with solid calcium carbonate.

$$H^+ + CaCO_3 \rightarrow Ca^{2+} + CO_2 + H_2O$$

XVI.   Acidic Anhydride + Base → Salt + Water

1. Dinitrogen pentoxide gas is bubbled through a solution of concentrated ammonia.

$$N_2O_5 + NH_3 \rightarrow NH_4^+ + NO_3^- + H_2O$$

2. Carbon dioxide gas is bubbled through a solution of lye.

$$CO_2 + OH^- \rightarrow CO_3^{2-} + H_2O$$

XVII. Acidic Anhydride + Basic Anhydride → Salt

1. Sulfur trioxide gas is heated in the presence of sodium oxide.

   $SO_3 \text{ (g)} + Na_2O \text{ (s)} \rightarrow Na_2SO_4 \text{ (s)}$

2. Carbon dioxide gas is heated in the presence of solid magnesium oxide.

   $CO_2 \text{ (g)} + MgO \text{ (s)} \rightarrow MgCO_3 \text{ (s)}$

XVIII. Basic Anhydride + Acid → Salt + Water

1. Solid copper (II) oxide is dropped into sulfuric acid.

   $CuO + H^+ \rightarrow Cu^{2+} + H_2O$

2. Chunks of magnesium oxide are dropped into hydrochloric acid.

   $MgO + H^+ \rightarrow Mg^{2+} + H_2O$

XIX. Precipitation Reactions

These problems involve mixing two solutions. Each solution is a water solution of an ionic compound. From the mixture of the two solutions, at least one insoluble precipitate will form. The other ions present are probably soluble and are called spectator ions; they are not included in the net ionic equation. *You must know your solubility rules to do these problems.*

1. A solution of silver nitrate is added to a solution of hydrochloric acid:

   $Ag^+ + Cl^- \rightarrow AgCl$ (All nitrates are soluble.)

2. Silver nitrate is added to a solution of potassium chromate:

   $Ag^+ + CrO_4^{2-} \rightarrow Ag_2CrO_4$

3. Lead nitrate is added to a solution of sodium chloride:

   $Pb^{2+} + Cl^- \rightarrow PbCl_2$

4. Iron(III) nitrate is added to a strong sodium hydroxide solution:

   $Fe^{3+} + OH^- \rightarrow Fe(OH)_3$

5. Strontium chloride is added to a solution of sodium sulfate:

   $Sr^{2+} + SO_4^{2-} \rightarrow SrSO_4$

XX. Redox Reactions

1. A strip of zinc metal is added to a solution of copper(II) nitrate:

   $Zn + Cu^{2+} \rightarrow Zn^{2+} + Cu$

2. A piece of aluminum is dropped into a solution of lead chloride:

   $Al + Pb^{2+} \rightarrow Pb + Al^{3+}$

3. Chlorine gas is bubbled through a solution of sodium iodide:

   $Cl_2 + I^- \rightarrow I_2 + Cl^-$

(4–20: Common Oxidation State Changes)

4. $Cl_2 \xrightarrow{\text{dilute basic sol'n}} ClO^-$

5. $Cl_2 \xrightarrow{\text{conc. basic sol'n}} ClO_2^-$

6. $Cl^- \rightarrow Cl_2$ (or other halide)

7. $Na \rightarrow Na^+$ (free metal)

8. $NO_2^- \rightarrow NO_3^-$

9. $HNO_3 \rightarrow NO_2$
   (conc.)

10. $HNO_3 \rightarrow NO$
    (dilute)

11. $Sn^{2+} \rightarrow Sn^{4+}$

12. $MnO_4^- \rightarrow Mn^{2+}$
    (in acid sol'n)

13. $MnO_2 \rightarrow Mn^{2+}$
    (in acid sol'n)

14. $MnO_4^- \rightarrow MnO_2$
    (in neutral or
    basic sol'n)

15. $H_2SO_4 \rightarrow SO_2$
    (hot, conc.)

16. SO_3^{2-} or $SO_2(aq) \rightarrow SO_4^{2-}$

17. $Na_2O_2 \rightarrow NaOH$

18. $H_2O_2 \rightarrow H_2O + O_2$

19. $HClO_4 \rightarrow Cl^-$

20. $Cr_2O_7^{2-} \rightarrow Cr^{3+}$
    (in acid)

XXI.   Synthesis Reactions (A + B → AB)

1. Solid calcium oxide is added to silicon dioxide and the mixture is heated strongly:

   $$CaO + SiO_2 \rightarrow CaSiO_3$$

2. Hydrogen gas is mixed with oxygen gas and the mixture is sparked:

   $$H_2 + O_2 \rightarrow H_2O$$

3. Boron trichloride gas and ammonia gas are mixed:

   $$BCl_3 + NH_3 \rightarrow BCl_3NH_3$$

   If this compound is not familiar, try drawing a Lewis diagram to see how it is put together.

XXII.   Decomposition (AB → A + B)

1. A sample of calcium carbonate is heated:

$$CaCO_3 \rightarrow CaO + CO_2$$

2. Hydrogen peroxide is gently warmed:

$$H_2O_2 \rightarrow H_2O + O_2$$

3. Manganese dioxide (acting as a catalyst) is added to a solid sample of potassium chlorate and the mixture is then heated:

$$KClO_3 \xrightarrow{MnO_2} KCl + O_2$$

4. Solid aluminum hydroxide is heated:

$$Al(OH)_3 \rightarrow Al_2O_3 + H_2O$$

XXIII.   Single Displacement (A + BC → AC + B)

1. Chlorine gas is bubbled through a strong solution of potassium bromide:

$$Cl_2 + Br^- \rightarrow Cl^- + Br_2$$

Note: This is also a redox reaction.

2. Powdered lead is added to a warm solution of copper(II) sulfate:

$$Pb + Cu^{2+} + SO_4^{2-} \rightarrow PbSO_4 + Cu$$

3. Strontium turnings are added to a 4 M sulfuric acid solution:

$$Sr + H^+ + SO_4^{2-} \rightarrow SrSO_4 + H_2$$

4. Silver is added to a solution of hot hydrochloric acid:

$$Ag + H^+ + Cl^- \rightarrow AgCl + H_2$$

XXIV.   Complex Ions*

1. A concentrated solution of ammonia is added to a solution of zinc nitrate:

$$Zn^2 + NH_3 \rightarrow Zn(NH_3)_4^{2+}$$

or

$$Zn^{2+} + NH_3 + H_2O \rightarrow Zn(OH)_2 + NH_4^+$$

The concentration of NH_3 determines the product.

2. A solution of iron(III) iodide is added to a solution of ammonium thiocyanate:

$$Fe^{3+} + SCN^- \rightarrow Fe(SCN)_6^{3-}$$

3. A solution of copper(II) nitrate is added to a strong solution of ammonia:

$$Cu^{2+} + NH_3 \rightarrow Cu(NH_3)_4^{2+}$$

4. A concentrated potassium hydroxide solution is added to solid aluminum hydroxide:

$$OH^- + Al(OH)_3 \rightarrow Al(OH)_4^-$$

Ligands are generally electron pair donors (Lewis bases). Important ligands are NH_3, CN^-, and OH^-. Ligands bond to a central atom that is usually the positive ion of a transition metal, forming complex ions and coordination compounds. On the AP exam, the number of ligands attached to a central metal ion is often twice the oxidation number of the central metal ion.

XXV. Organic Substitution

1. Chlorine gas is added to a flask of methane gas and heated:

$$CH_4 + Cl_2 \rightarrow CH_3Cl + HCl$$

or

$$\rightarrow CH_2Cl_2$$

or

$$\rightarrow CHCl_3$$

or

$$\rightarrow CCl_4$$

2. Chloromethane is bubbled through a solution of warm ammonia:

$$CH_3Cl + NH_3 \rightarrow H^+ + Cl^-$$

or

$$[CH_3NH_3]^+Cl^-$$

or

$$\rightarrow [CH_3NH_3]^+ + Cl^-$$

XXVI. Organic Addition

1. Hot steam is mixed with propene gas:

$$C_3H_6 + H_2O \rightarrow C_3H_7OH$$

The major isomeric product is:

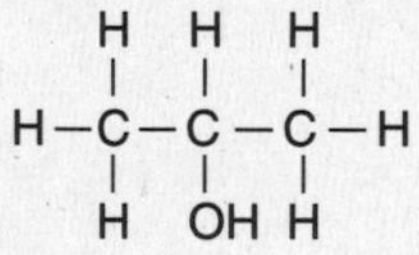

$$
\begin{array}{ccc}
\text{H} & \text{H} & \text{H} \\
| & | & | \\
\text{H}-\text{C}-\text{C}-\text{C}-\text{H} \\
| & | & | \\
\text{H} & \text{OH} & \text{H}
\end{array}
$$

2. Ethene gas is mixed with chlorine gas:

$$C_2H_4 + Cl_2 \rightarrow C_2H_4Cl_2$$

XXVII. Organic Elimination

1. Ethanol is heated in the presence of sulfuric acid:

$$C_2H_5OH \xrightarrow{H_2SO_4} C_2H_5-O-C_2H_5$$

2. Chloroethane is heated:

$$C_2H_5Cl \xrightarrow{base} C_2H_4 + HCl$$

XXVIII. Organic Condensation

1. Methanol is mixed with acetic acid and then gently warmed:

$$CH_3OH + CH_3COOH \xrightarrow[\text{esterification}]{H^+} CH_3COOCH_3 + H_2O$$

2. Methyl alcohol is mixed with a small amount of sulfuric acid and then warmed gently:

$$CH_3OH + HOCH_3 \rightarrow CH_3OCH_3 + H_2O$$

XXIX. Important Organic Reactions

Addition of Br_2 or Cl_2 to alkenes or alkynes	$-\overset{\mid}{C}=\overset{\mid}{C}- + X_2 \longrightarrow -\overset{\mid}{\underset{X}{C}}-\overset{\mid}{\underset{X}{C}}-$ X = Br or Cl
Addition of H_2 to alkenes or alkynes	$-\overset{\mid}{C}=\overset{\mid}{C}- + H_2 \xrightarrow{catalyst} -\overset{\mid}{\underset{H}{C}}-\overset{\mid}{\underset{H}{C}}-$
Addition of H_2O to alkenes	$-\overset{\mid}{C}=\overset{\mid}{C}- + H_2O \xrightarrow{H^+} -\overset{\mid}{\underset{H}{C}}-\overset{\mid}{\underset{OH}{C}}-$
Addition of HX to alkenes or alkynes	$-\overset{\mid}{C}=\overset{\mid}{C}- + HX \longrightarrow -\overset{\mid}{\underset{X}{C}}-\overset{\mid}{\underset{X}{C}}-$ X = F, Cl, Br, or I
Dehydration of alcohols to give alkenes	$-\overset{\mid}{\underset{H}{C}}=\overset{\mid}{\underset{OH}{C}}- \xrightarrow{H_2SO_4} -\overset{\mid}{C}=\overset{\mid}{C}- + H_2O$
Dehydration of alcohols to give ethers	$2ROH \xrightarrow{H_2SO_4} R-O-R + H_2O$
Formation of acetals	$R-\overset{\mid\mid}{\underset{O}{C}}-H + R'OH \longrightarrow R-\overset{\overset{OR'}{\mid}}{\underset{\underset{OH}{\mid}}{C}}-H \xrightarrow[R'OH]{H^+} R-\overset{\overset{OR'}{\mid}}{\underset{\underset{OR'}{\mid}}{C}}-H$
Formation of ketals	$R-\overset{\mid\mid}{\underset{O}{C}}-R'' + R'OH \longrightarrow R-\overset{\overset{OR'}{\mid}}{\underset{\underset{OH}{\mid}}{C}}-R'' \xrightarrow[R'OH]{H^+} R-\overset{\overset{OR'}{\mid}}{\underset{\underset{OR'}{\mid}}{C}}-R''$ Ketone Hemiketal Ketal
Halogenation of aromatic rings	(benzene ring) $+ X_2 \xrightarrow{Fe}$ (benzene ring)—X X = Br or Cl
Hydrolysis of acetals	$R-\overset{\overset{OR'}{\mid}}{\underset{\underset{OR'}{\mid}}{C}}-H + H_2O \xrightarrow{H^+} R-\overset{\mid\mid}{\underset{O}{C}}-H + 2R'OH$ Acetal Aldehyde Alcohol

225

Hydrolysis of amides-acid catalyzed	$R-\underset{\underset{O}{\|}}{C}-NH_2 + H_2O \xrightarrow{H^+} R-\underset{\underset{O}{\|}}{C}-OH + NH_4^+$
Hydrolysis of amides-base catalyzed	$R-\underset{\underset{O}{\|}}{C}-NHR' + H_2O \xrightarrow{OH^-} R-\underset{\underset{O}{\|}}{C}-O^- + R'NH_2$
Hydrolysis of carboxylic esters — acid catalyzed	$R-\underset{\underset{O}{\|}}{C}-OR' + H_2O \underset{}{\overset{H^+}{\rightleftharpoons}} R-\underset{\underset{O}{\|}}{C}-OH + R'OH$
Hydrolysis of carboxylic esters — base catalyzed (saponifications)	$R-\underset{\underset{O}{\|}}{C}-OR' + OH^- \longrightarrow R-\underset{\underset{O}{\|}}{C}-O^- + R'OH$
Neutralization of carboxylate salts	$R-\underset{\underset{O}{\|}}{C}-O^- + HCl \longrightarrow R-\underset{\underset{O}{\|}}{C}-OH + Cl^-$
Nitration of aromatic rings	⬡ $+ HNO_3 \xrightarrow{H_2SO_4}$ ⬡$-NO_2$
Oxidation of alcohols	**Primary** $RCH_2OH \xrightarrow{[O]} R-\underset{\underset{O}{\|}}{C}-H \xrightarrow{[O]} R-\underset{\underset{O}{\|}}{C}-OH$ **Secondary** $R-\underset{\underset{OH}{\|}}{CH}-R' \xrightarrow{[O]} R-\underset{\underset{O}{\|}}{C}-R'$ **Tertiary** $R-\underset{\underset{R''}{\|}}{\overset{\overset{R'}{\|}}{C}}-OH \xrightarrow{[O]}$ no reaction
Oxidation of aldehydes	$R-\underset{\underset{O}{\|}}{C}-H \xrightarrow{[O]} R-\underset{\underset{O}{\|}}{C}-OH$ Aldehyde Carboxylic acid
Oxidation of ketones	$R-\underset{\underset{O}{\|}}{C}-R' \xrightarrow[\text{Mild conditions}]{[O]}$ no reaction Ketone
Oxidation of thiols	$2RSH \xrightarrow{[O]} R-S-S-R$

Preparation of amides from acyl chlorides and ammonia or amines	$R-\underset{O}{\overset{\parallel}{C}}-Cl + R'_2NH \longrightarrow R-\underset{O}{\overset{\parallel}{C}}-NR'_2 + HCl$
Preparation of amides from anhydrides and ammonia or amines	$R-\underset{O}{\overset{\parallel}{C}}-O-\underset{O}{\overset{\parallel}{C}}-R + NH_3 \longrightarrow R-\underset{O}{\overset{\parallel}{C}}-NH_2 + R-\underset{O}{\overset{\parallel}{C}}-OH$
Preparation of amides from carboxylic acids and ammonia or amines (with DCC-dicyclohexylcarbodiimide)	$R-\underset{O}{\overset{\parallel}{C}}-OH + R'NH_2 + DCC \longrightarrow R-\underset{O}{\overset{\parallel}{C}}-NH-R' + DHU$
Preparation of carboxylic esters from acylchlorides and alcohols	$R-\underset{O}{\overset{\parallel}{C}}-Cl + R'OH \longrightarrow R-\underset{O}{\overset{\parallel}{C}}-OR' + HCl$
Preparation of carboxylic esters from anhydrides and alcohols	$R-\underset{O}{\overset{\parallel}{C}}-O-\underset{O}{\overset{\parallel}{C}}-R + R'OH \longrightarrow R-\underset{O}{\overset{\parallel}{C}}-OR' + R-\underset{O}{\overset{\parallel}{C}}-OH$
Preparation of carboxylic esters from carboxylic acids and alcohols	$R-\underset{O}{\overset{\parallel}{C}}-OH + R'OH \underset{}{\overset{H^+}{\rightleftharpoons}} R-\underset{O}{\overset{\parallel}{C}}-OR' + H_2O$

Reaction of amines with acids
  Primary
  Secondary
  Tertiary
	Primary $RNH_2 + HCL \rightarrow RNH_3 + Cl^-$
	Secondary $R_2NH + R'COOH \rightarrow R_2NH_2 + R'COO^-$
	Tertiary $R_3N + HNO_3 \rightarrow R_3NH + NO_3^-$

Reaction of amines with nitrous acids
  Primary
  Secondary
	Primary $RNH_2 + HONO \rightarrow N_2 +$ other products
	Secondary $R_2NH + HONO \rightarrow R_2N-NO$

Reaction of carboxylic acids with bases	$R-\underset{O}{\overset{\parallel}{C}}-OH + OH^- \longrightarrow R-\underset{O}{\overset{\parallel}{C}}-O^- + H_2O$ $R-\underset{O}{\overset{\parallel}{C}}-OH + NH_3 \longrightarrow R-\underset{O}{\overset{\parallel}{C}}-O^- + NH_4^+$
Reaction of tertiary amines with alkyl halides	$R_3N + R'Cl \longrightarrow R-\underset{R}{\overset{R}{\underset{\vert}{\overset{\vert}{N}}}}-R' + Cl^-$
Reduction of aldehydes	$R-\underset{O}{\overset{\parallel}{C}}-H \xrightarrow{[H]} R-CH_2-OH$ Aldehyde Primary alcohol
Reduction of disulfides	$R-S-S-R \xrightarrow{[H]} 2RSH$

| Reduction of ketones | 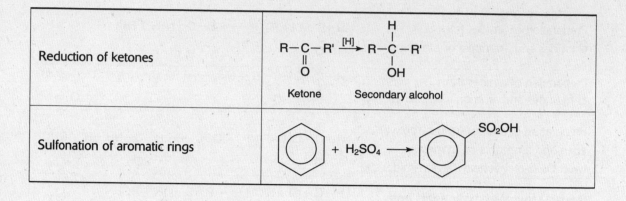 |
| Sulfonation of aromatic rings | |

PART III

AP CHEMISTRY LABORATORY EXPERIMENTS

Laboratory Experiments

Approximately 5–10% of the AP Chemistry Exam is devoted to questions involving laboratory experiments. Understanding basic laboratory concepts and being able to analyze sample data will likely make the difference between a 4 and a 5 on the exam. The questions involving laboratory experiments can be categorized into four main groups:

1. making observations of chemical reactions and substances

2. recording data

3. calculating and interpreting results based on the quantitative data obtained

4. communicating effectively the results of experimental work

Some colleges report that some students, while doing well on the written exam, have been at serious disadvantage when they arrive at college because of inadequate or nonexistent laboratory experience in AP Chemistry at the high school. Completion of all recommended 22 laboratory experiments is *essential* to doing the best possible on the AP Chemistry Exam. Meaningful laboratory work is important in fulfilling the requirements of a college-level course of a laboratory science and in preparing you for sophomore-level chemistry courses in college. Issues of college accreditation are also factors that must be considered, since colleges are giving you college laboratory credit for work that was *supposed* to be completed in the high school setting. "Dry labs," that is substituting real laboratory situations with worksheets or slide shows, are unacceptable in preparing you for college-level work. Because chemistry professors at some institutions will ask to see a record of the laboratory work that you did in AP Chemistry before making a decision about granting credit, placement, or both, you must keep a neat, well-organized laboratory notebook in such a fashion that the reports can be readily reviewed. See page 233, on "The Laboratory Notebook" and page 256–282 for "A Sample Lab Report".

> *"To play a violin, one needs to know how to handle it properly. To do a meaningful experiment, one must mix and measure just as properly."*

> Sienko, Plane and Marcus, 1984

Students involved in laboratory experiments in AP Chemistry should be familiar with basic laboratory equipment, which will include but not be limited to:

- beakers, flasks, test tubes, crucibles, evaporating dishes, watch glasses, burners, plastic and glass tubing, stoppers, valves, spot plates, funnels, reagent bottles, wash bottles and droppers.

Students should also be familiar with measuring equipment, which will include, but not be limited to:

- balances (to the nearest 0.001 g), thermometers, barometers, graduated cylinders, burets, volumetric pipets, graduated pipets, volumetric flasks, ammeters and voltmeters, pH meters, and spectrophotometers.

Students should also have familiarity, involving more than a single day's experience, in such general types of chemical laboratory work as:

- synthesis of compounds (solid and gas)

- separations (precipitations and filtration, dehydration, centrifugation, distillation, chromatography)

- observing and recording phase changes (solid-liquid-gas)

- titration using indicators and meters

- spectophotometry/colorimetry

- devising and utilizing a scheme for qualitative analysis of ions in solution

- gravimetric analysis

Note: The review of laboratory experiments in this book are not meant to replace detailed laboratory procedures, disposal and safety concerns. Under NO circumstances are the scenarios for the labs to be used as directions for actually performing the laboratory work. See the bibliography at the end of this book for a list of resources that provide detailed directions.

The Laboratory Notebook

As was mentioned earlier in this book, laboratory experiments are an indispensable component of AP Chemistry and the most effective way of learning complex concepts. You may be required to submit your AP Chemistry laboratory notebook to a university in order to receive credit for college chemistry. How that notebook is organized and the presentation of the laboratory work is critical for receiving full college credit. Following are some guidelines that should be considered in presenting your work for the year. You will also find an actual lab write-up from a student to be used as a guide in creating your own notebook. This sample lab write-up is found on pages 256–282 of this book.

Obtain a bound notebook with quadrille paper (¼" squares). The "graph paper" will help you line up tables, charts, and so on. Never remove sheets from the notebook.

Consult your instructor as to whether you are to use pen or pencil. Pen is generally preferred (and "cross-outs" are permitted).

Always use a ruler to make straight lines. A "French Curve" is often useful for other types of graphical representations; also, see VI (below).

On the outside of the notebook, and in ink, print your name, class title, instructor's name and course title as neatly as you can in block letters.

The first page of the notebook is the title page, where you, as neatly as possible, place your name, class title, instructor's name, semester and year, institution, and (if you wish) your address, e-mail address, and phone number. Consult your instructor for placement of these items.

Leave the next three pages blank as they will be used for the Table of Contents. In block letters, print "Table of Contents" on the first of these pages. Update the Table of Contents after each lab write-up.

In the upper right hand corner of the fourth sheet in the notebook, place a "1". Continue to number the remaining pages in the notebook. Do not write on the back of the paper. The back is used by the teaching assistant or instructor for comments to you about the facing page.

As you examine the actual lab write-up included in this book, pay careful attention to how neatly the lab was written. Neatness is fundamental for obtaining college credit.

Do not erase errors or mistakes. Instead, draw a single straight line, with a ruler, through any error(s).

The sections of the lab write-up include, but are not limited to:

I. Title

II. Date:

III. Purpose: A brief paragraph describing what is to be accomplished.

IV. Theory: Necessary equations and background for the lab. It "brings up to speed" the reader on concepts that will be investigated. It is also helpful to you, when you review the labs in preparation for the AP Chemistry Exam, to have a short background to the material in your own words.

V. Procedure: It is not necessary to copy the procedure from the lab manual. Such work is "busy work" and accomplishes nothing but taking up room in the notebook. Instead, you may simply say, "Refer to . . . " and list the lab in standard bibliographic style.

VI. Results: Before you write the results down, think how you will organize them. Tables of information are the best way to present data. Label the tables and indicate parts or section numbers that correspond to the procedures in the lab manual. You may include graphs in this section. In graphing, try to use a commercial graphing program to construct your graphs. Excel™ and Cricket Graph™ are examples of programs that make nice presentations of data. Also, try to use more advanced features of these programs such as "best fit", correlation, regression, slope, and so on.

VII. Calculations: Present the calculations necessary to answer the questions in the lab manual. Be sure to rewrite the question so that the reader understands what is being done. If possible, begin with the generic equation followed by substitution of the actual data. Including any or all equations, if necessary for the question, is essential. Do not leave it up to the reader to figure out the equation(s).

VIII. Conclusions and/or Post-lab: Summarize in a paragraph what the laboratory data showed. Include percent error(s) if possible. This is also a good area to include any statistical work comparing your work with class averages or expected/theoretical results. Some laboratory manuals include Post-Lab questions, which would be included here.

Experiment 1: Determination of the Empirical Formula of a Compound

Background: Many elements combine with oxygen or other nonmetals in various ratios; i.e. FeO, Fe_2O_3, Fe_3O_4. This phenomenon demonstrates the Law of Multiple Proportions. In this experiment, you will analyze the ratios in which lead and chlorine can combine and from the data provided, be able to determine the empirical formulas of the compounds produced.

Scenario: A student was given 2.982 grams of a sample of a pure anhydrous lead chloride, Compound A. The student added a small amount of water to the test tube and heated it in a fume hood liberating chlorine gas and creating Compound B. A flask filled with 0.5 M NaOH and gently swirled helped to trap the chlorine gas (see Figure 1).

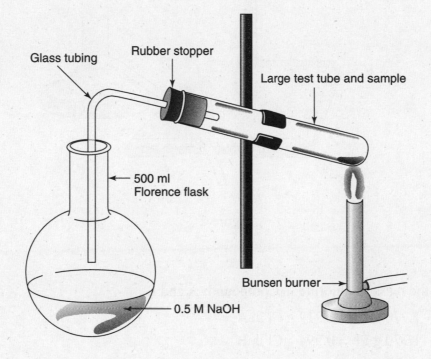

Figure 1

The student then heated Compound B, driving off the remaining water, and determined the mass of Compound B as 2.364 g. The student then reduced Compound B with hydrogen gas to form elemental lead (see Figure 2). The chlorine in Compound B was driven off as $HCl_{(g)}$. This lead was massed and found to weigh 1.770 g.

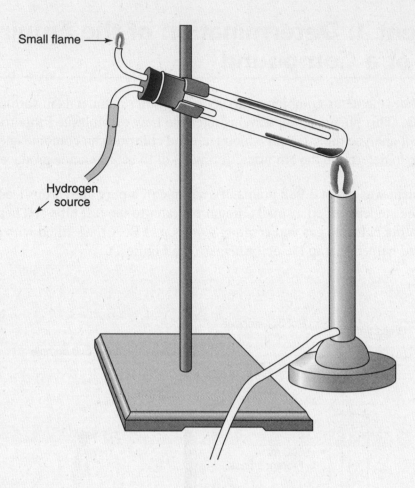

Figure 2

Analysis:

1. Determine the mass of chlorine in Compounds A and B.

   2.982 g A − 1.770 g Pb = 1.212 g Cl in A
   2.364 g B − 1.770 g Pb = 0.594 g Cl in B

2. Using the masses of lead and chlorine from the experiment, what relationship exists for Compounds A and B?

   A: 1.770 g Pb : 1.212 g Cl
   B: 1.770 g Pb : 0.594 g Cl

3. Using information from Question 2, calculate the number of grams of chlorine that would combine with 1 mole of lead in Compounds A and B.

   A: $\dfrac{1.212 \text{ g Cl}}{1.770 \text{ g Pb}} \times \dfrac{207.2 \text{ g Pb}}{1 \text{ mol Pb}} = 141.9 \text{ g Cl} \cdot \text{mol Pb}^{-1}$

   B: $\dfrac{0.594 \text{ g Cl}}{1.770 \text{ g Pb}} \times \dfrac{207.2 \text{ g Pb}}{1 \text{ mol Pb}} = 69.5 \text{ g Cl} \cdot \text{mol Pb}^{-1}$

4. Calculate the number of moles of chlorine that combined with one mole of lead in Compounds A and B.

A: $\dfrac{141.9 \ \text{g Cl}}{35.45 \ \text{g Cl/mol}} = 4.003 \ \text{mol Cl} : 1 \ \text{mol Pb}$

B: $\dfrac{69.5 \ \text{g Cl}}{35.45 \ \text{g Cl/mol}} = 1.96 \ \text{mol Cl} : 1 \ \text{mol Pb}$

5. What is the empirical formula of Compound A?

$PbCl_4$

6. What is the empirical formula of Compound B?

$PbCl_2$

7. Write the equations for the experiment and then identify what is being oxidized and what is being reduced.

$PbCl_4 \rightarrow PbCl_2 + Cl_2$

Oxidation: Chlorine (from −1 to 0)

Reduction: Lead (from +4 to +2)

$PbCl_2 + H_2 \rightarrow Pb + 2HCl$

Oxidation: Hydrogen (from 0 to +1)

Reduction: Lead (from +2 to 0)

Experiment 2: Determination of the Percentage of Water in a Hydrate

Background: Many solid chemical compounds will absorb some water from the air over time. In many cases, this amount is small and is only found on the surface of the crystals. Other chemical compounds, however, absorb large amounts of water from the air and chemically bind the water into the crystal structure. The majority of these compounds are ionic salts (metallic cation other than H^+ and nonmetallic anion other than OH^- or O^{2-}). To remove the water from many of these hydrates, one only needs to gently heat the compound to slightly above the temperature of boiling water. When one heats the hydrate, the crystalline structure will change and often a color change occurs. For example, $CuCl_2 \cdot 2\ H_2O$, copper (II) chloride dihydrate is green as the hydrate; as the anhydride, $CuCl_2$, it is brownish-yellow. When hydrates lose water spontaneously, they are said to effloresce. The degree of efflorescence is a function of the relative humidity. Some compounds will spontaneously absorb water from the air. These compounds are known as desiccants and are said to be hygroscopic (hydrophilic). When desiccants absorb so much water that they dissolve, they are said to be deliquescent. Do not confuse production of water vapor as an absolute indicator of dehydration since many organic compounds such as carbohydrates will produce water vapor upon decomposition. This decomposition is not reversible but is usually reversible in the case of an anhydride (that is, adding water to the anhydride will convert it back to the hydrate). In general, the proportion of water in the hydrate is a mole ratio (whole integer or a multiple of ½) to that of the salt.

Scenario: A student placed some blue $CoCl_2$ on a watch glass and observed that over time it changed to violet and then to red. Upon heating the red compound gently, she observed that it changed back to violet and then to blue. She then took a sample of sodium sulfate decahydrate whose large crystals appeared colorless and transparent and observed over time that they changed to a fine white powder. She then added water to the fine white powder, dissolving it. Upon gently heating this mixture, the large, colorless transparent crystals reappeared. Then, she took 1.700 g of a green nickel sulfate hydrate and heated the hydrate completely in a crucible (see Figure 1).

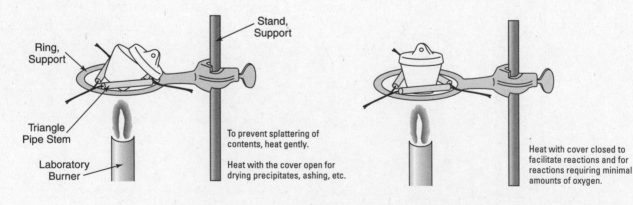

Figure 1

She allowed it to cool in a desiccator (see Figure 2),

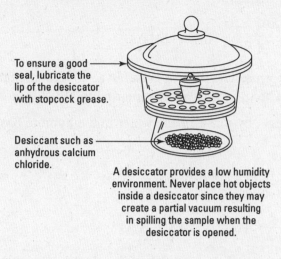

To ensure a good seal, lubricate the lip of the desiccator with stopcock grease.

Desiccant such as anhydrous calcium chloride.

A desiccator provides a low humidity environment. Never place hot objects inside a desiccator since they may create a partial vacuum resulting in spilling the sample when the desiccator is opened.

Figure 2

and then reweighed it and discovered it weighed 1.006 grams. The anhydride produced was yellow.

Analysis:

1. Describe the processes occurring with the cobalt chloride.

   Three color changes were observed: blue, violet and red. From the information provided, the blue form appeared to be the anhydride, $CoCl_2$. When exposed briefly to air containing water vapor, the anhydrous cobalt chloride turned violet, $CoCl_2 \cdot 2H_2O$. Upon further standing in air, the dihydrate was converted to the fully hydrated form, $CoCl_2 \cdot 6 H_2O$. The color of the anhydride-hydrate depended upon the relative humidity of the air to which it was exposed.

2. Describe the processes occurring with the sodium sulfate decahydrate.

   It appeared that the large, well-defined crystals of sodium sulfate decahydrate, $Na_2SO_4 \cdot 10 H_2O$, lost water when exposed to air and converted to the anhydrous form, Na_2SO_4. This process is known as efflorescence. It would appear that the relative humidity of the air was rather low for this to occur. When rehydrated and gently dried, the hydrate reappeared indicating a reversible reaction:

   $Na_2SO_4 \cdot 10 H_2O + heat \leftrightarrow Na_2SO_4 + 10 H_2O$

3. What was the formula of the green nickel sulfate hydrate?

   $$\begin{array}{l} 1.700 \text{ g } NiSO_4 \cdot ?H_2O \\ \underline{-1.006 \text{ g } NiSO_4} \\ 0.694 \text{ g } H_2O \end{array}$$

   $$\frac{1.006 \text{ g } NiSO_4}{1} \times \frac{1 \text{ mol } NiSO_4}{154.76 \text{ g } NiSO_4} = 0.006500 \text{ mol } NiSO_4$$

$$\frac{0.694 \text{ g H}_2\text{O}}{1} \times \frac{1 \text{ mol H}_2\text{O}}{18.02 \text{ g H}_2\text{O}} = 0.0385 \text{ mol H}_2\text{O}$$

$$\frac{0.0385 \text{ mol H}_2\text{O}}{0.006500 \text{ mol NiSO}_4} = \frac{5.93}{1.00} = \text{NiSO}_4 \cdot 6\text{H}_2\text{O} *$$

*NiSO$_4$ exists in two forms. The β-form is stable at 40°C and changes to the blue α-form at room temperature. The α-form is an aquamarine to bluish-green color.

4. What was the percentage, by mass, of the water in the hydrate based upon this experiment?

$$\% \text{ water} = \frac{\text{mass of H}_2\text{O in sample}}{\text{mass of hydrate}} \times 100\%$$

$$\frac{0.694 \text{ g H}_2\text{O}}{1.700 \text{ g NiSO}_4 \cdot 6 \text{ H}_2\text{O}} \times 100\% = 40.8\% \text{ H}_2\text{O}$$

5. What would be the theoretical percent of water in the hydrate and what was the percent error?

$$\% \text{ H}_2\text{O} = \frac{\text{part water}}{\text{whole mass}} \times 100\% = \frac{6 \times 18.02}{262.87} \times 100\% = 41.1\%$$

$$\% \text{ error} = \frac{\text{obs} - \text{exp}}{\text{exp}} \times 100\% = \frac{40.8 - 41.1}{41.1} \times 100\% = -0.7\%$$

6. What was the formula for the yellow anhydride produced?

NiSO$_4$

Experiment 3: Determination of Molar Mass by Vapor Density

Background: The Ideal Gas Law can be used to determine the approximate* molecular mass of a gas through the equation $MM = \dfrac{g \cdot R \cdot T}{P \cdot V}$. In this experiment, a small amount of a volatile liquid is placed into a flask (of which you know the volume) and allowed to vaporize. A small pinhole in the system allows the pressure of the gas to equalize with the air pressure. The temperature of the vapor will be assumed to be the temperature of the boiling water in which the flask is immersed. At this point, you know g, R, T, V, and P. You will then be able to calculate the apparent MM.

Scenario: A student obtained a sample of an unknown volatile liquid and placed it into a 0.264 L flask. The student covered the flask with a piece of aluminum foil which contained a very small pin hole. The student then immersed the flask into a boiling water bath, which she measured as 100.5 °C** and allowed the liquid to vaporize, forcing out the air in the flask. (See Figure 1 for set-up.)

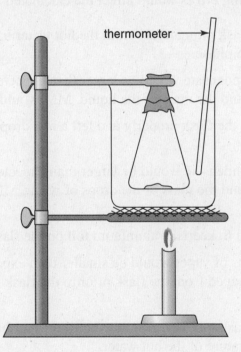

thermometer →

Figure 1

*The gas or vapor may deviate significantly from ideality since it is not that far above the temperature of its boiling point.

**The temperature of boiling water at 100.5°C at 750. torr may indicate that the water that was used in the boiling water bath was not distilled water but rather tap water and approached 1m in ionic concentration.

When there was no more liquid visible in the flask, the student then placed the flask into cold water so that the vapor could condense. She dried the flask thoroughly and then reweighed the flask and determined the mass of the condensed vapor to be 0.430 g. The student then checked the barometric pressure of the room and found it to be 750. torr.

Analysis:

1. Calculate the molecular mass of the unknown. What possible sources of error were not accounted for in the calculations?

$$MM = \frac{g \cdot R \cdot T}{P \cdot V} = \frac{0 \cdot 0.430 \text{ g} \cdot 0.0821 \text{ L} \cdot \text{atm} \cdot 373.5 \text{ K}}{\text{mol} \cdot \text{K} \cdot (750./760 \text{ atm}) \cdot 0.264 \text{ L}} = 50.6 \text{ g} \cdot \text{mol}^{-1}$$

Possible sources of error might include:

(a) mass of air

(b) thermometer immersion point

(c) diffusion effects (air/vapor)

(d) the B.P. of the water at 100.5°C at 750. torr would indicate either a faulty thermometer or that she was using tap water with an ionic concentration of approximately 1m. Calibrations factored in the analysis for the thermometer or correctional factors for the type of water could have been made.

2. Explain how the following errors would affect the calculated molecular mass:

(a) She removed the flask prematurely from the hot water bath leaving some of the unknown in the liquid phase.

The mass of the condensate would be larger than expected, since it would include the mass of the vapor and the mass of the liquid. MM would be higher than expected.

(b) She did not dry off the flask properly and left a few drops of water on the outside of the flask.

The mass of the condensate would be larger than expected, since it would include the mass of the vapor and the mass of the drops of water. MM would be higher than expected.

(c) She was not careful to keep the aluminum foil on the flask while it was cooling.

The calculated mass of vapor would be smaller than expected, since some of the vapor would have escaped from the flask prior to the flask being weighed. MM would be too small.

(d) She removed the flask too soon from the boiling water bath, not allowing it a chance to reach the temperature of the hot water.

Since the temperature of the water bath did not equal the temperature of the vapor, the temperature value used would be too large since the sample was collected at a lower temperature than the boiling water. MM would be too large.

Experiment 4: Determination of Molecular Mass by Freezing-Point Depression

Background: If a nonvolatile solid is dissolved in a liquid, the vapor pressure of the liquid solvent is lowered and can be determined through the use of Raoult's Law, $P_1 = X_1 P_1°$. Raoult's Law is valid for ideal solutions wherein $\Delta H = 0$ and in which there is no chemical interaction among the components of the dilute solution (see Figure 1).

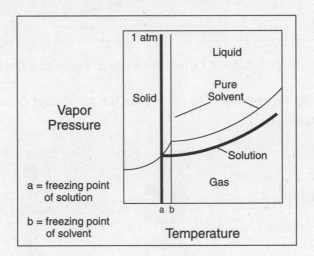

Figure 1

This phenomenon results in higher boiling points and lower freezing points for the solution as compared to the pure solvent. Vapor pressure, freezing-point depressions and boiling-point elevations are known as colligative properties. Colligative properties depend only on the number of particles present, not on what type of particles they are. Each solvent has its own unique freezing-point depression and boiling-point elevation constants — values that must be factored into an equation to solve for molecular mass. For water, the constant $k_b = 0.52\ °C \cdot m^{-1}$ and $k_f = 1.86\ °C \cdot m^{-1}$. To determine the molecular mass of a solute from a freezing-point depression you use the equation

$$\Delta T_f = i \cdot k_f \cdot m = i \cdot k_f \cdot \frac{\text{moles of solute}}{\text{kg of solvent}} = i \cdot k_f \cdot \frac{\dfrac{\text{g solute}}{\text{MM}_{solute}}}{\text{kg solvent}}$$

which can be rearranged to produce

$$\text{MM}_{solute} = \frac{i \cdot k_f \cdot \text{g solute}}{\text{kg of solvent} \cdot \Delta T_f}$$

where i, known as the van't Hoff or dissociation factor, represents the degree to which the solute ionizes. For non-ionic compounds, $i = 1$; for NaCl, $i = 2$ since for every one mole of NaCl there results two moles of ions, Na^+ and Cl^-; $i = 3$ for $NiBr_2$, and so on. The van't Hoff factor is only approximate except in infinitely dilute solutions. Otherwise, one must use activity coefficients for the ions at their concentrations.

Figure 2 shows the cooling curve for a pure solvent and for a solution. Supercooling may result. Should this occur, as the crystals begin to form, the temperature will increase slightly and then remain relatively constant as the pure solvent freezes.

243

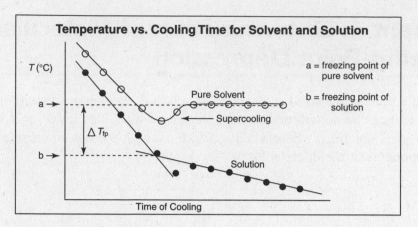

Figure 2: Freezing Point Graph for Pure Solvent and for Solution

Scenario: A student set up an apparatus to determine the molecular mass through freezing point depression (see Figure 3)

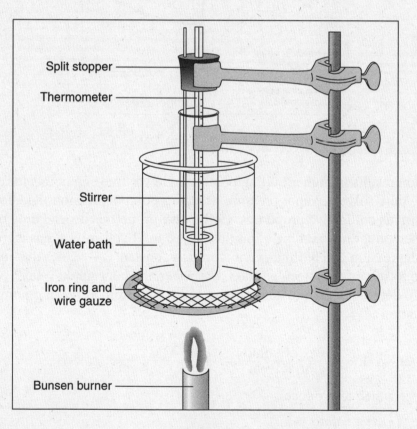

Figure 3

of naphthalene (see Figure 4). The student measured out 52.0 grams of paradichlorobenzene and 4.0 grams of naphthalene, placed them in a test tube in a hot-water bath and allowed the mixture to completely melt. The mixture was stirred well. The tube was then removed from the hot-water bath, dried and allowed to cool while it was gently stirred to minimize supercooling. Temperature readings were taken every 60. seconds until the mixture solidified and the temperature stabilized. A graph was drawn of the results (see Figure 5).

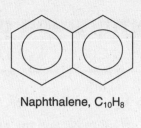

Naphthalene, $C_{10}H_8$

Figure 4

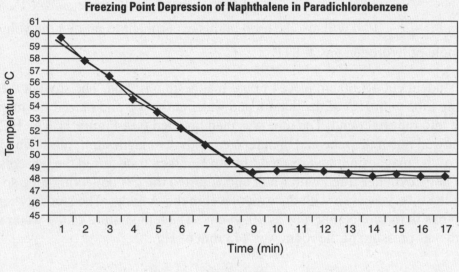

Figure 5

Analysis:

1. What was the freezing point of the solution?

   The intersection of the two lines occurred at approximately 48.4 °C

2. What was the freezing point depression of the solution? The standard freezing point for paradichlorobenzene is 53.0°C.

$$\Delta T_f = T_f^\circ - T_f = 53.0\ ^\circ C - 48.4\ ^\circ C = 4.6\ ^\circ C$$

3. What was the molality of the naphthalene? The freezing point depression constant for paradichlorobenzene is 7.1 °C · m⁻1.

$$m = \frac{\Delta T_f}{k_f} = \frac{4.6^\circ C}{7.1^\circ C \cdot m^{-1}} = 0.65 m$$

4. What was the molecular mass of naphthalene?

$$MM_{solute} = \frac{k_f \cdot g\ solute}{kg\ solvent \cdot \Delta T_f} = \frac{7.1^\circ \cancel{C} \cdot \cancel{kg} \cdot 4.0g}{mol \cdot 0.0520\ \cancel{kg} \cdot 4.6^\circ \cancel{C}} = 120\ g \cdot mol^{-1}$$

5. What was the percent error of the results compared to the actual molecular mass of 128 g · mol⁻¹?

$$\%\ error = \frac{observed - expected}{expected} \times 100\% = \frac{120 - 128}{128} \times 100\% = -6\%$$

The results were about 6% below expected.

Experiment 5: Determination of the Molar Volume of a Gas

Background: Avogadro's law ($V_1n_2 = V_2n_1$), where moles, $n = \dfrac{grams}{MW\,(grams/mole)}$ expresses the relationship between molar mass, the actual mass and the number of moles of a gas. The molar volume of a gas at STP, $\bar{V}_{STP}$ is equal to the volume of the gas measured at STP divided by the number of moles; $\bar{V}_{STP} = \dfrac{V_{STP}}{n}$. Dalton's Law of Partial Pressure ($P_{total} = P_1 + P_2 + P_3 + \dots$) and the derivation, $P_i = \dfrac{n_i}{n_{total}} \cdot P_{total}$ will also be used in this experiment to predict the volume occupied by one mole of hydrogen gas at STP.

Scenario: A student cut a 4.60 cm piece of pure magnesium ribbon. The student then prepared a gas collecting tube with HCl at the bottom and water at the top and placed the magnesium in a copper cage into the water end. The tube was then inverted (see Figure 1). The HCl being more dense flowed down the tube and then reacted with the magnesium, producing bubbles of gas. After all of the magnesium had reacted, the student then transferred the gas collecting tube carefully to a large cylinder filled with water and adjusted the gas collecting tube so that the water level in the tube was even with the water level in the cylinder. The student then read the volume of the gas in the collecting tube as 40.44 cm^3. The temperature of the water was 26.3 °C and the air pressure of the room was 743 mm of Hg.

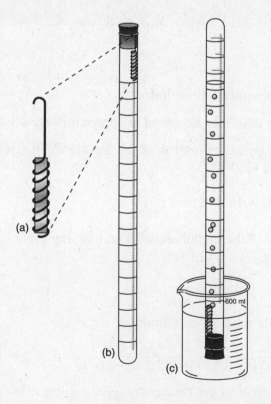

Figure 1

Analysis:

1. Given that 1.00 m of the pure Mg ribbon weighs 0.816 g, determine the mass of Mg used in the experiment.

$$\frac{4.60 \text{ cm Mg}}{1} \times \frac{1.00 \text{ m Mg}}{100. \text{ cm Mg}} \times \frac{0.816 \text{ g Mg}}{1.00 \text{ m Mg}} = 0.0375 \text{ g Mg}$$

2. Calculate the number of moles of Mg used in the experiment.

$$\frac{0.0375 \text{ g Mg}}{1} \times \frac{1 \text{ mol Mg}}{24.3 \text{ g Mg}} = 0.00154 \text{ mol Mg}$$

3. Calculate the partial pressure of the hydrogen gas in the mixture. The vapor pressure of the water at 26.3 °C is 25.7 mm Hg.

$$P_{H_2} = P_{tot} - P_{H_2 0}$$

$$= 743 \text{ mm Hg} - 26 \text{ mm Hg} = 717 \text{ mm Hg}$$

4. Calculate the volume of the hydrogen gas at STP, assuming that the gas temperature equals the temperature of the aqueous HCl.

$$V_2 = V_1 \frac{P_1}{P_2} \times \frac{T_2}{T_1} = 40.44 \text{ cm}^3 \times \frac{717 \text{ mm Hg}}{760 \text{ mm Hg}} \times \frac{273 \text{ K}}{299 \text{ K}} = 34.8 \text{ cm}^3$$

5. Write the balanced equation for the reaction.

$$Mg_{(s)} + 2H^+_{(aq)} \rightarrow H_{2(g)} + Mg^{2+}_{(aq)}$$

6. Determine the number of moles of $H_2(g)$ produced, based on the moles of Mg used.

$$\frac{0.00154 \text{ mol Mg}}{1} \times \frac{1.00 \text{ mol } H_2}{1.00 \text{ mol Mg}} = 0.00154 \text{ mol } H_{2(g)}$$

7. Calculate the molar volume of the $H_2(g)$ at STP.

$$\bar{V}_{STP} = \frac{34.8 \text{ cm}^3 \text{ } H_2}{0.00154 \text{ mol } H_2} \times \frac{1.000 \text{ L}}{1000 \text{ cm}^3} = 22.6 \text{ L} \cdot \text{mol}^{-1}$$

8. What was the percent error in this experiment?

$$\% \text{ error} = \frac{obs - exp}{exp} \times 100\% = \frac{22.6 \text{ L} \cdot \text{mol}^{-1} - 22.4 \text{ L} \cdot \text{mol}^{-1}}{22.4 \text{ L} \cdot \text{mol}^{-1}} \times 100\% = 0.9\%$$

9. Determine the number of moles of hydrogen gas by using the Ideal Gas Law.

$$PV = nRT$$

$$n = \frac{P \cdot V}{R \cdot T} = \frac{717/760 \text{ atm} \cdot 0.04044 \text{ L} \cdot \text{mol} \cdot \text{K}}{0.0821 \text{ L} \cdot \text{atm} \cdot 299 \text{ K}} = 1.55 \times 10^{-3} \text{ mol } H_2$$

Experiment 6: Standardization of a Solution Using a Primary Standard

and

Experiment 7: Determination of Concentration by Acid-Base Titration

Background: A neutralization reaction results when a strong acid reacts with a strong base and is represented by the following net-ionic equation:

$$H^+_{(aq)} + OH^-_{(aq)} \rightarrow H_2O_{(l)}$$

The equilibrium constant for this reaction at room temperature is approximately 10^{14}. The magnitude of the equilibrium constant shows that the reaction proceeds essentially to the product side, using up nearly all of the ion present as the limiting reactant. If both H^+ and OH^- are present in equal quantities as reactants, the resulting solution will be neutral, because water is the only product of the reaction. When weak acids are titrated with strong bases, the resulting solution will be basic because the conjugate base of the weak acid will form. When weak bases are titrated with strong acids, the resulting solution will be acidic because the conjugate acid of the weak base will form.

In Part I of this experiment a titration will be performed by titrating a standardized solution of HCl with a NaOH solution whose concentration is not known. One cannot make a standardized solution of NaOH directly by weighing out an exact amount of NaOH and diluting it with distilled water. This is because solid NaOH absorbs both H_2O and CO_2. The titration will allow the concentration of the OH^- to be calculated accurately. The equivalence point is the point in the titration when enough titrant has been added to react exactly with the substance in solution being titrated and will be determined by using an indicator. The end point, the point in the titration at which the indicator changes color, will be very close to 7. At this point, a drop of acid or base added to the solution will change the pH by several pH units.

In Part II, a weak acid is titrated with the strong base from Part I and the equivalence point will be somewhat higher than 7 (8 – 9). By titrating a known amount of solid acid with the standardized NaOH, it will be possible to determine the number of moles of H^+ the acid furnished. From this information, one can obtain the equivalent mass (EM_a) of the acid:

$$EM_a = \frac{\text{grams acid}}{\text{moles of } H^+}$$

However, the equivalent mass of the acid may or may not be the same as the molecular mass of the acid since some acids produce more than one mole of H^+ per mole of acid. In order to find the molecular mass from the EM_a, the molecular formula is required.

If a graph is drawn of pH versus mL of NaOH added, there will be a significant change in pH in the vicinity of the equivalence point. It is important to understand that the equivalence point will not be at pH 7, but will be slightly higher. The value of the equilibrium constant for the dissociation of the acid can be obtained from this graph. Since a weak acid is being studied in Part II, the dissociation can be represented as

$$HA + H_2O \Leftrightarrow H_3O^+ + A^-$$

which results in the equilibrium expression:

$$K_a = \frac{[H_3O^+][A^-]}{[HA]}$$

This can be rearranged to read $[H_3O^+] = K_a \times \frac{[HA]}{[A^-]}$

When the acid is half neutralized, half of the number of moles of HA originally present will be converted to the conjugate base A^-. So $[HA] = [A^-]$ and K_a will be equal to $[H_3O^+]$ since $[HA] / [A^-] = 1$. Therefore, when the acid is half-neutralized, pH = pK_a. The point at which the pH is equal to pK_a can be seen in Figure 1.

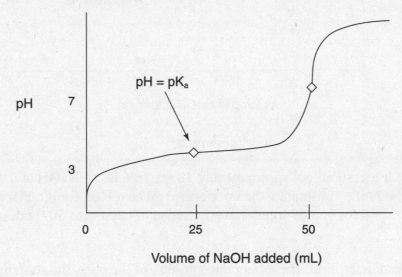

Figure 1

If one were to titrate a weak diprotic acid such as maleic acid with the known NaOH solution, the graph would show two separate inflections — representing the neutralization of each hydrogen — assuming the pK_a values differ by 4 or more pK units (see Figure 2). The dissociation of a diprotic acid occurs in two separate steps:

$$H_2A + H_2O \Leftrightarrow H_3O^+ + HA^-$$
$$HA^- + H_2O \Leftrightarrow H_3O^+ + A^{2-}$$

and results in two separate equilibrium expressions:

$$K_{a1} = \frac{[H_3O^+][HA^-]}{[H_2A]}$$

$$K_{a2} = \frac{[H_3O^+][A^{2-}]}{[HA^-]}$$

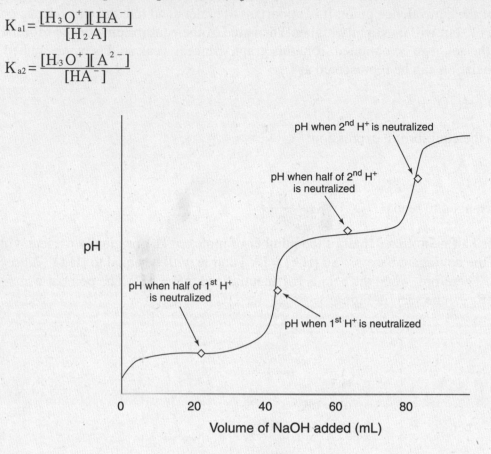

Figure 2

Scenario: A student measured out approximately 10 mL of 6.00 M NaOH and diluted the base to approximately 600 mL. The student then performed an acid-base titration (see Figure 3) and determined that 48.7 mL of NaOH solution were needed to neutralize 50.0 mL of a 0.100 M HCl solution.

Once the student had determined the exact concentration of the base, the student then proceeded to determine the equivalent mass of an unknown acid. To do this, the student measured out 0.500 grams of an unknown solid acid and titrated it with the standardized base, recording pH with a calibrated pH meter as the base was added. The student added 43.2 mL of the base but went too far past the end point and needed to back-titrate with 5.2 mL of the 0.100 M HCl to exactly reach the end point.

Analysis:

1. Calculate the molarity of the NaOH solution from Part I.

$$50.0 \text{ mL} \times \frac{0.100 \text{ mol HCl}}{1000 \text{ mL}} \times \frac{1 \text{ mol NaOH}}{1 \text{ mol HCl}} \times \frac{1}{0.0487 \text{ L}} = 0.103 \text{ M NaOH}$$

2. The graph of the titration from this experiment is presented here. Determine the K_a or K_a's.

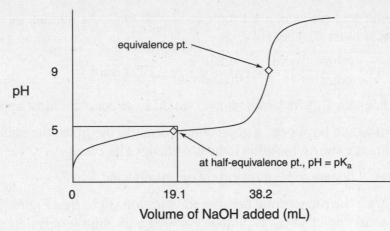

Figure 3

$$K_a = 10^{-pKa} = 10^{-5} = 1 \times 10^{-5}$$

$$K = \frac{(HA)(OH^-)}{(A^-)} \cdot \frac{(H^+)}{(H^+)} = \frac{K_W}{K_A} = \frac{(OH^-)^2}{(A^-)}$$

$$K_A = \frac{K_W(A^-)}{(OH^-)^2}$$

3. What is the pH at the equivalence point?

The pH at the equivalence point appears to be around 9. Therefore, $(OH^-) \approx 10^{-5} M$

$$(A^-) = \frac{(43.2 \text{ mL} \times 0.103 \text{ M}) - (5.2 \text{ mL} \times 0.100 \text{ M})}{43.2 \text{ mL} + 5.2 \text{ mL}} = 0.0812 \text{ M at equiv. point}$$

$$Ka = Kw \cdot \frac{(A^-)}{(OH^-)^2} = \frac{(1 \times 10^{-14})(0.0812)}{(1 \times 10^{-5})^2} = 8.12 \times 10^{-6}$$

4. Why is the equivalence point not a pH of 7?

The neutralization reaction is $HA + OH^- \Rightarrow A^- + H_2O$, where HA is the weak acid and A^- is its conjugate base. This assumes that HA is a monoprotic acid. Because the principal product of this reaction is the weak base, A^-, the resulting solution will be basic with a pH greater than 7.

5. Determine the EM_a of the solid acid.

$$43.2 \text{ mL} \times \frac{0.103 \text{ mol OH}^-}{1000 \text{ mL}} = 0.00445 \text{ mol OH}^- \text{ dispensed from buret}$$

$$5.2 \text{ mL} \times \frac{0.100 \text{ mol H}^+}{1000 \text{ mL}} = 0.00052 \text{ mol H}^+ \text{ used in back-titration}$$

$(0.00445 - 0.00052) \text{ mol} = 0.00393 \text{ mol OH}^- \text{ actually used to neutralize}$

Since H^+ from the acid reacts in a 1:1 mole ratio with OH^-, the number of moles of H^+ furnished by the acid must also be 0.00393.

$$\text{equivalent mass} = \frac{\text{grams of acid}}{\text{moles of } H^+} = \frac{0.500 \text{ g}}{0.00393 \text{ mol}} = 127 \text{ g/mol}$$

6. Referring to Appendix B, which indicator(s) would have been the most appropriate to use?

Phenolphthalein would have been a good choice because the pH at the equivalence point falls within the range over which this indicator changes its color.

7. Does the solid acid appear to be monoprotic or diprotic and why?

This acid appears to be monoprotic because the titration curve (see Figure 3 above) only shows one inflection point so far as plotted. Another end-point (equivalence point) is generally hard to find, using only indicators above pH=10.5 since the indicator equilibrium reaction $HIn \rightleftharpoons H^+ + In^- \left(K_{HIn} = 10^{-12} \right)$ interacts and interferes.

Experiment 8: Determination of Concentration by Oxidation-Reduction Titration and an Actual Student Lab Write-Up*

Background: Commercially available bleaching solutions contain NaOCl, sodium hypochlorite. Sodium hydroxide is reacted with chlorine gas to produce the hypochlorite ion, OCl^-.

$$Cl_{2\,(g)} + 2OH^-_{(aq)} \rightarrow OCl^-_{(aq)} + Cl^- + H_2O_{(l)}$$

In solution, NaOCl dissociates into sodium ions (Na^+) and hypochlorite ions (OCl^-). Bleaching involves an oxidation-reduction reaction in which the Cl in the OCl^- ion (oxidizing agent) is reduced to the chloride ion (Cl^-). The reducing agent is either a dye, which fades, or the stain being removed.

The amount of hypochlorite ion present in bleach can be determined by an oxidation-reduction titration. In this experiment, an iodine-thiosulfate titration will be utilized. The iodide ion is oxidized to form iodine, I_2. This iodine is then titrated with a solution of sodium thiosulfate of known concentration. Three steps are involved:

1. An acidified solution of iodide ion and hypochlorite ion is oxidized to iodine:

    $$2H^+_{(aq)} + OCl^-_{(aq)} + 2I^-_{(aq)} \rightarrow Cl^-_{(aq)} + I_{2\,(aq)} + H_2O_{(l)}$$

2. However, iodine is not very soluble in water. Therefore, an aqueous solution of iodide ion is added to the iodine to form the complex ion, triiodide ion, $I_3^-{}_{(aq)}$. In dilute concentrations, the triiodide ion is yellow and in concentrated solutions it is a dark reddish-brown.

    $$I_{2\,(aq)} + I^-_{(aq)} \rightarrow I_3^-{}_{(aq)}$$

3. Finally, the triiodide ion is titrated with a known solution of thiosulfate ions, which forms iodide ions:

    $$I_3^-{}_{(aq)} + 2S_2O_3^{2-}{}_{(aq)} \rightarrow 3I^-_{(aq)} + S_4O_6^{2-}{}_{(aq)}$$

In this step the reddish brown color of the triiodide begins to fade to yellow and finally to clear, indicating only iodide ions present. However, this is not the best procedure for determining when all of the I_3^- has disappeared since it is not a sensitive reaction and the change from pale yellow to colorless is not distinct. A better procedure is to add a soluble starch solution shortly prior to reaching the end point, since if it is added to soon, too much iodine or triiodide ion may be present forming a complex that may not be reversible in the titration. The amount of thiosulfate is proportional to the amount of hypochlorite ion present.

The experimental error involved in measuring small volumes of liquids is usually greater than the error when measuring larger volumes. Diluted samples will be prepared, called aliquots, to improve the accuracy.

Scenario: A student diluted 50.00 mL of a commercial bleach to 250.00 mL in a volumetric flask and then titrated a 20. mL aliquot. The titration required 35.50 mL of 0.10000 M $Na_2S_2O_3$ solution. The price of the gallon jug of bleach was \$1.00. The density of the bleach was $1.15 \text{ g} \cdot \text{mL}^{-1}$.

Analysis:

1. Calculate the number of moles of $S_2O_3^{2-}$ ion required for titration.

$$= \frac{35.50 \text{ mL Na}_2\text{S}_2\text{O}_3\text{sol'n}}{1} \times \frac{1 \text{ L sol'n}}{1000 \text{ mL Na}_2\text{S}_2\text{O}_3\text{sol'n}} \times$$

$$\frac{0.10000 \text{ mol Na}_2\text{S}_2\text{O}_3}{1 \text{ L sol'n}} \times \frac{1 \text{ mol S}_2\text{O}_3^{2-}}{1 \text{ mol Na}_2\text{S}_2\text{O}_3} = 3.550 \times 10^{-3} \text{ mol S}_2\text{O}_3^{2-}$$

2. Calculate the number of moles of I_2 produced in the titration mixture.

$$\frac{3.550 \times 10^{-3} \text{ mol S}_2\text{O}_3^{2-}}{1} \times \frac{1 \text{ mol I}_2}{2 \text{ mol S}_2\text{O}_3^{2-}} = 1.775 \times 10^{-3} \text{ mol I}_2$$

3. Calculate the number of moles of OCl^- ion present in the diluted bleaching solution that was titrated.

$$\frac{1.775 \times 10^{-3} \text{ mol I}_2}{1} \times \frac{1 \text{ mol OCl}^-}{1 \text{ mol I}_2} = 1.775 \times 10^{-3} \text{ mol OCl}^-$$

4. Calculate the mass of NaOCl present in the diluted bleaching solution titrated.

$$\frac{1.775 \times 10^{-3} \text{ mol OCl}^-}{1} \times \frac{1 \text{ mol NaOCl}}{1 \text{ mol OCl}^-} \times \frac{74.44 \text{ g NaOCl}}{1 \text{ mol NaOCl}} = 0.1321 \text{ g NaOCl}$$

5. Determine the volume of commercial bleach present in the diluted bleaching solution titrated (aliquot).

$$\frac{20.00 \text{ mL diluted bleach sol'n}}{1} \times \frac{50.00 \text{ mL commercial bleach}}{250.00 \text{ mL diluted bleach sol'n}}$$

$$= 4.00 \text{ mL commercial bleach}$$

6. Calculate the mass of commercial bleach titrated.

$$\frac{4.00 \text{ mL commercial bleach}}{1} \times \frac{1.15 \text{ g commerical bleach}}{1.00 \text{ mL commercial bleach}} = 4.60 \text{ g commercial bleach}$$

7. Determine the percent NaOCl in the commercial bleach.

$$\frac{\text{part}}{\text{whole}} \times 100\% = \frac{0.1321 \text{ g NaOCl}}{4.60 \text{ g commercial bleach}} \times 100\% = 2.87\%$$

8. Calculate the mass of 1.00 gallon of the commercial bleach. 1 U.S. gallon = 3785 mL.

$$= \frac{1.00 \text{ gal comm. bleach}}{1} \times \frac{3785 \text{ mL}}{1 \text{ gal}} \times \frac{1.15 \text{ g comm. bleach}}{1.00 \text{ mL comm. bleach}} = 4.35 \times 10^3 \text{ g}$$

9. Calculate the cost of 100. g of the commercial bleach.

$$\frac{100.\ \text{g comm. bleach}}{1} \times \frac{1.00\ \text{gal comm. bleach}}{4.35 \times 10^3\ \text{g comm. bleach}}$$

$$\times \frac{\$1.00}{1.00\ \text{gal comm. bleach}} = \$0.0230$$

10. Determine the cost of the amount of commercial bleach required to supply 100. g of NaOCl.

$$\frac{100.\ \text{g NaOCl}}{1} \times \frac{100.\ \text{g comm. bleach}}{2.87\ \text{g NaOCl}} \times \frac{\$0.0230}{100.\ \text{g comm. bleach}} = \$0.801$$

A Sample Lab Report*

I. Determining The Percent Sodium Hypochlorite in Commercial Bleaching Solutions

II. Date: 1/22/98

III. Purpose: Determine the percent sodium hypochlorite in various commercial bleaching solutions by titration. Compare the cost effectiveness of different brands of commercial bleaching solutions.

IV. Theory: Commercial bleaching solutions are preared by reacting chlorine (Cl_2) with a base. If the base is sodium hydroxide (NaOH), the product is sodium hypochlorite (NaOCl). Commercially available bleaching solutions usually contain NaOCl.

V. Procedure: Refer to Chemical Education Resources, Modular Laboratory Program in Chemistry, 1992**.

VI. Results:

brand name of commercial bleaching solution	Clorox Bleach and Hughes Bleach	
cost of commercial bleaching solution per gallon	Clorox $1.49 per gallon Hughes $1.50 per gallon	
molarity of $Na_2S_2O_3$ solution, M	Clorox 0.1	Hughes 0.1
volume of diluted bleaching solution titrated, mL	Clorox 25	Hughes 25
final buret reading, mL	Clorox 25.9	Hughes 41.85
initial buret reading, mL	Clorox 0	Hughes 0
volume of $Na_2S_2O_3$ solution used, mL	Clorox 25.9	Hughes 41.85
volume of commercial bleaching solution titrated, mL	Clorox 10	Hughes 10
mass of commercial bleaching solution titrated, g	Clorox 10.041	Hughes 10.945

VII. Calculations:

		Determination	
		(1) Clorox	**(2)** Hughes
Density of commercial bleaching solution, $g \cdot mL^{-1}$ **(1)** density $= \dfrac{mass}{volume} = \dfrac{10.041\,g}{10\,mL} = 1.0041\,g \cdot mL^{-1}$ **(2)** density $= \dfrac{mass}{volume} = \dfrac{10.945\,g}{10\,mL} = 1.0945\,g \cdot mL^{-1}$		1.0041	1.0945
Volume of commercial bleaching solution diluted to 100 mL, mL $10\,mL = (25\,mL)\left(\dfrac{x}{100\,mL}\right) \quad 10\,mL = \dfrac{25\,mL\,(x)}{100\,mL} \quad x = \dfrac{100\,mL}{25\,mL} \quad x = 4\,mL$		4	
Number of moles of $S_2O_3^{2-}$ ion required for titration, mol **(1)** $(25.9\,mL\,Na_2S_2O_3)\left(\dfrac{1\,L}{1000\,mL}\right)\left(\dfrac{0.00259\,moles\,Na_2S_2O_3}{0.0256\,L\,of\,sol'n}\right)\left(\dfrac{1\,mol\,S_2O_3^{2-}}{1\,mol\,Na_2S_2O_3}\right)$ $\dfrac{0.1\,mol\,Na_2S_2O_3}{1\,L} \times \dfrac{1\,L}{1000\,mL} \times \dfrac{25.9\,mL}{1} = 0.00259 \quad 25.9\,mL = 0.00259\,L$ **(2)** $(41.85\,mL\,Na_2S_2O_3)\left(\dfrac{1\,L}{1000\,mL}\right)\left(\dfrac{0.004185\,moles\,Na_2S_2O_3}{0.04185\,L\,of\,sol'n}\right)\left(\dfrac{1\,mol\,S_2O_3^{2-}}{1\,mol\,Na_2S_2O_3}\right)$ $\dfrac{0.1\,moles\,Na_2S_2O_3}{1\,L} \times \dfrac{1\,L}{1000\,mL} \times \dfrac{41.85\,mL}{1} = 0.004185 \quad 41.85\,mL = 0.04185\,L$		0.00259	0.00418
Number of moles of I_2 produced in the titration mixture, mol **(1)** $(0.00259\,moles\,S_2O_3^{2-}\,ion)\left(\dfrac{1\,mol\,I_2}{2\,mol\,S_2O_3^{2-}\,ion}\right) = 0.001295$ **(2)** $(0.004185\,moles\,S_2O_3^{2-}\,ion)\left(\dfrac{1\,mol\,I_2}{2\,moles\,S_2O_3^{2-}\,ion}\right) = 0.0020925$		0.001295	0.0020925
Number of moles of OCl^- ion in diluted bleaching solution titrated, mol **(1)** $(0.001295\,moles\,I_2)\left(\dfrac{1\,mol\,OCl^-\,ion}{1\,mol\,I_2}\right) = 0.001295$ **(2)** $(0.0020925\,moles\,I_2)\left(\dfrac{1\,mol\,OCl^-\,ion}{1\,mol\,I_2}\right) = 0.0020925$		0.001295	0.0020925
Mass of NaOCl present in diluted bleaching solution titrated, g **(1)** $(0.001295\,moles\,OCl^-)\left(\dfrac{1\,mol\,NaOCl}{1\,mol\,OCl^-\,ion}\right)\left(\dfrac{74.44\,g\,NaOCl}{1\,mol\,NaOCl}\right) = 0.09305\,g$ **(2)** $(0.0020925\,moles\,OCl^-)\left(\dfrac{1\,mol\,NaOCl}{1\,mol\,OCl^-\,ion}\right)\left(\dfrac{74.44\,g\,NaOCl}{1\,mol\,NaOCl}\right) = 0.15577\,g$		0.09305	0.15577
Percent NaOCl in commercial bleaching solution, % **(1)** $\left(\dfrac{0.09305\,g}{10.041\,g}\right)(100\%) = 0.9267\%$ **(2)** $\left(\dfrac{0.15577\,g}{10.041\,g}\right)(100\%) = 1.423\%$		0.9267	1.423

	Determination	
	(1) Clorox	**(2)** Hughes
Mean percent NaOCl in commercial bleaching solution, % $\dfrac{(1)+(2)}{2} = \dfrac{.9267\% + 1.423\%}{2} = 1.175\%$	1.175%	
Mass of 1 gal commercial bleaching solution, g **(1)** $\left(1\,\text{gal bleaching sol'n}\right)\left(\dfrac{3.785 \times 10^3\,\text{mL}}{1\,\text{gal bleaching sol'n}}\right)\left(\dfrac{1.0041\,\text{g}}{1\,\text{mL}}\right) = 3800.5\,\text{g}$ **(2)** $\left(1\,\text{gal bleaching sol'n}\right)\left(\dfrac{3.785 \times 10^3\,\text{mL}}{1\,\text{gal bleaching sol'n}}\right)\left(\dfrac{1.0945\,\text{g}}{1\,\text{mL}}\right) = 4142.7\,\text{g}$	3800.5	4142.7
Cost of 100 g commercial bleaching solution, $ **(1)** $\left(100\,\text{g}\right)\left(\dfrac{1\,\text{gallon}}{3800.5\,\text{g}}\right)\left(\dfrac{\$1.49}{1\,\text{gallon}}\right) = \$0.0392$ **(2)** $\left(100\,\text{g}\right)\left(\dfrac{1\,\text{gallon}}{4142.7\,\text{g}}\right)\left(\dfrac{\$1.50}{1\,\text{gallon}}\right) = \$0.0362$	$0.0392	$0.0362
Cost of the amount of commercial bleaching solution required to supply 100g of NaOCl, $ **(1)** $\left(100\,\text{g NaOCl}\right)\left(\dfrac{100\,\text{g}}{0.9267\,\text{g of NaOCl}}\right)\left(\dfrac{\$0.0392}{100\,\text{g}}\right) = \$4.23$ **(2)** $\left(100\,\text{g NaOCl}\right)\left(\dfrac{100\,\text{g}}{1.423\,\text{g of NaOCl}}\right)\left(\dfrac{\$0.0362}{100\,\text{g}}\right) = \$2.54$	$4.23	$2.54

Post-Lab:

1. Graduated cylinders are not as precisely calibrated as burets or volumetric pipets. Briefly explain why it is acceptable to measure the Kl and HCl solutions used in the titration with graduated cylinders rather than with pipets or burets.

   It is acceptable to measure the Kl and HCl solutions used in the titration with graduated cylinders rather than with pipets or burets because we are not after a specific amount, we are after excess.

2. Would the following procedural errors result in an incorrectly high or low calculated percent NaOCl in commercial bleaching solution? Briefly explain.

   (1) A student failed to allow the volumetric pipet to drain completely when transferring the diluted bleaching solution to the Erlenmeyer flask.

   If a student failed to drain completely the volumetric pipet when transferring the diluted bleach, this error would result in an incorrectly low percent of NaOCl. NaOCl is the active ingredient in bleach. If incorrect low amounts are transferred to the Erlenmeyer flask, then the NaOCl concentration is also incorrect.

   (2) A student blew the last drops of solution from the pipet into the volumetric flask when transferring commercial bleaching solution to the flask.

   If a student blew drops of commercial bleaching solution into the volumetric flask, then that student has introduced another set of chemicals from the moisture of their mouth into the solution, altering the experiment and altering the results.

   (3) A student began a titration with an air bubble in the buret tip. The bubble came out of the tip after 5 mL of $Na_2S_2O_3$ solution had been released.

   Having a bubble in the buret tip containing 5 mL of $Na_2S_2O_3$ results in an incorrectly high calculated percent NaOCl in the commercial bleaching solution. $Na_2S_2O_3$ is used to determine the amount of I_2 formed by titration. It reconverts I_2 made from the reaction of OCl^- ion and I^- ion into I^- ions again. So, transferring less $Na_2S_2O_3$ to the pipet, affects the concentration of OCl^- ion in the bleaching solution.

3. An overly efficient student simultaneously prepared two titration mixtures, consisting of diluted bleaching solution, KI solution, and HCl solution. The student found that data from the two titrations yielded significantly different percents NaOCl in the commercial bleaching solution. Which determination would give the higher % NaOCl? Briefly explain.

   Based on my results, I believe the first determination would yield higher NaOCl simply because more care is put into one determination than the other, because as you titrate, the diluted bleach solution must be agitated constantly throughout the $Na_2S_2O_3$ titration to notice color change. So, since the student is working simultaneously, he cannot be precise with the $Na_2S_2O_3$ concentration he titrates into the flask.

4. Based on your titration data and calculations for determination 1:

   (1) Calculate the volume of $Na_2S_2O_3$ solution required had you transferred commercial bleaching solution instead of diluted bleaching solution to the Erlenmeyer flask 1.

   25 mL commercial bleaching solution was diluted by a factor of 10 and required 19 mL of $Na_2S_2O_3$. A 25 mL undiluted commercial bleaching solution would require ten times more $Na_2S_2O_3$. Approximately 190 mL of $Na_2S_2O_3$ would be required for titration.

   (2) Assuming you used only the glassware supplied for the original experiment, briefly comment on the procedural change necessary to titrate 25.00 mL of commercial bleaching solution.

   The procedural change would be to titrate the solution a portion at a time and then calculate the total $Na_2S_2O_3$ used.

   (3) Would you anticipate a greater error in the results of an analysis done in this way than is the case with the standard procedure? Briefly explain.

   I would anticipate greater error in the results of an analysis done this way than in the standard procedure due to there being too many opportunities for misreading and miscalculation to occur.

5. Commercial bleaching solutions found on store shelves are usually labeled: "Contains at least 5.25% sodium hypochlorite." Analysis will often show a lower OCl⁻ ion content. Briefly explain.

   Analysis will show a lower OCl⁻ ion content on commercial bleaching solutions because it is exposed to air and the solution becomes diluted.

   * Student from College of the Canyons, Santa Clarita, CA. The lab write-up is provided to show format only. No guarantee is provided for accuracy or quality of answers.

   ** Chemical Education Resources, Inc., "Determining the Percent Solium Hypochlorite in Commercial Bleaching Solutions," Palmyra, PA., 1992.

Experiment 9: Determination of Mass and Mole Relationship in a Chemical Reaction

Background: This experiment uses the concept of continuous variation to determine mass and mole relationships. Continuous variation keeps the total volume of two reactants constant, but varies the ratios in which they combine. The optimum ratio would be the one in which the maximum amount of both reactants of known concentration are consumed and the maximum amount of product(s) is produced. Since the reaction is exothermic, and heat is therefore a product, the ratio of the two reactants that produces the greatest amount of heat is a function of the actual stoichiometric relationship. Other products that could be used to determine actual molar relationships might include color intensity, mass of precipitate formed, amount of gas evolved, and so on.

Scenario: The active ingredient in commercial bleach is NaClO (see Experiment 8). A student was given a sample of commercial bleach which was labeled 5.20 % NaClO by mass. The student was then directed to prepare 300. mL of a 0.500 M NaClO solution. The student was also directed to prepare 300. mL of a 0.500 M solution of sodium thiosulfate, $Na_2S_2O_3$ in 0.200 M sodium hydroxide. Both solutions were then allowed to reach a room temperature of 25.0°C. Keeping a constant final volume of 50.0 mL of solution, the student mixed various amounts of sodium hypochlorite and sodium thiosulfate together, stirring well, and recorded the maximum temperature (±0.2°C) of the solution with a calibrated temperature probe. The change in temperature (ΔT) is equal to $T_{final} - 25.0$°C. A chart and graph of the data is given in Figure 1.

Trial #	ΔT°C	mL NaClO	mL $Na_2S_2O_3$	vol ratio
1	0	0	50	x
2	3.9	5	45	1:9
3	8.1	10	40	1:4
4	11.3	15	35	1:2.3
5	14.7	20	30	1:1.5
6	18.8	25	25	1:1
7	22.6	30	20	1.5:1
0	27.6	35	15	2.3:1
9	31.9	40	10	4:1
10	12	45	5	9:1
11	0.1	50	0	x

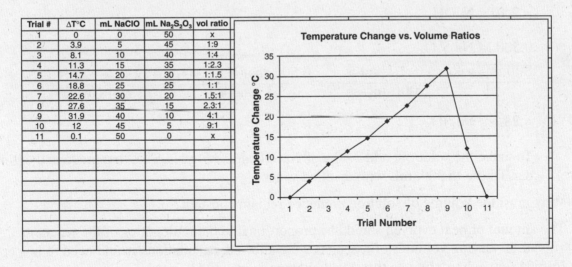

Figure 1

Analysis:

1. Write the net ionic equations involved in this experiment.

$$4\,OCl^-_{(aq)} + S_2O_3^{2-}_{(aq)} + 2\,OH^-_{(aq)} \rightarrow 2\,SO_4^{2-}_{(aq)} + 4\,Cl^-_{(aq)} + H_2O_{(l)}$$

2. Show the steps necessary to produce the 0.5 M sodium hypochlorite solution.

The commercial bleach was 5.20% NaClO. The directions required making a 0.500 M solution. MM of NaClO = 74.44 g · mol^{-1}. 300. mL of solution should contain 11.2 g NaClO:

$$\frac{300.\,\text{mL sol'n}}{1} \times \frac{1\,\text{L sol'n}}{1000\,\text{mL sol'n}} \times \frac{0.500\,\text{mole NaClO}}{1\,\text{L sol'n}} \times \frac{74.44\,\text{g NaClO}}{1\,\text{mole NaClO}} = 11.2\,\text{g NaClO}$$

To obtain 11.2 g of NaClO, using a 5.20% solution, would require

215 g of bleach solution, diluted to 300. mL with distilled water:

$$\frac{11.2\,\text{g NaClO}}{1} \times \frac{100.\,\text{g sol'n}}{5.20\,\text{g NaClO}} = 215\,\text{g bleach sol'n diluted to 300. mL}$$

3. Show the steps necessary to produce the 300. mL of 0.500 M sodium thiosulfite in 0.200 M NaOH.

   (a) 0.200 M NaOH

   $$\frac{300.\,\text{mL sol'n}}{1} \times \frac{1\,\text{L sol'n}}{1000\,\text{mL sol'n}} \times \frac{0.200\,\text{mol NaOH}}{1\,\text{L sol'n}} \times \frac{40.00\,\text{g NaOH}}{1\,\text{mol NaOH}}$$

   $$= 2.40\,\text{g NaOH}$$

   (b) 0.500 M Na$_2$S$_2$O$_3$

   $$\frac{300.\,\text{mL sol'n}}{1} \times \frac{1\,\text{L sol'n}}{1000\,\text{mL sol'n}} \times \frac{0.500\,\text{mol Na}_2\text{S}_2\text{O}_3}{1\,\text{L sol'n}} \times \frac{158.12\,\text{g Na}_2\text{S}_2\text{O}_3}{1\,\text{mol Na}_2\text{SO}_3}$$

   $$= 23.7\,\text{g Na}_2\text{SO}_3$$

   To make the solution, add 2.40 g of NaOH and 23.7 g Na$_2$S$_2$O$_3$ to a graduated cylinder. Dilute to 300. mL with distilled water.

4. Why must the volume of solution produced remain constant?

The amount of heat evolved should be proportional to the mole ratios of the reactants involved. If varying volumes of reactants were allowed, calculations would need to be performed for each mixture so that a valid comparison could be made.

5. Referring to the graph produced, what is the limiting reactant on the line with the positive slope? With the negative slope?

Positive Slope: The mL and mole ratio for equivalency is 4 NaClO to 1 Na$_2$S$_2$O$_3$ (see #1 above). In examining the data table, it can be seen that the Na$_2$S$_2$O$_3$ is in excess, with the NaClO being the limiting reactant; for example, in Trial 7, for 30 mL of 0.500 M NaClO,

$$\frac{30.\,\text{mL NaClO}}{1} \times \frac{1\,\text{mL Na}_2\text{S}_2\text{O}_3}{4\,\text{mL NaClO}} = 7.5\,\text{mL Na}_2\text{S}_2\text{O}_3$$

would be needed for equivalency, but 20. mL of $Na_2S_2O_3$ was used.

Negative Slope: In examining the data table, it can be seen that the NaClO is in excess and that the $Na_2S_2O_3$ is the limiting reactant.

6. Why should the point of intersection of the two lines be used to find the molar ratio rather than ratio associated with the greatest temperature change?

The exact ratios that would have produced the greatest temperature change may not have been used. By drawing regression lines from the data points, and then extrapolating to the x-axis, this allows for an exact determination of the volume ratios that produced the greatest temperature change.

7. If 20. mL of the sodium hypochlorite solution was at 25.0°C and 30. mL of the sodium thiosulfate solution was initially at 35.0°C before they were mixed, and after they were mixed together the final temperature was 34.0 °C, what would be the actual final temperature of the mixture?

A temperature correction factor would be needed:

$$\frac{(20.\,\text{mL} \cdot 25.0°\text{C}) + (30.\,\text{mL} \cdot 35.0°\text{C})}{50.\,\text{mL}} = 31.0°\text{C}$$

If the measured temperature is 34.0°C, ΔT should be reported as 34.0°C − 31.0°C = 3.0°C

8. What is the purpose of having sodium hydroxide in both reactants?

The bleach solution is a base. Diluting the bleach in a base of approximately the same molarity, does not appreciably change the temperature when the bleach is diluted.

9. What conclusions can be drawn from Trial 9?

According to the balanced net ionic equation

$$4OCl^-_{(aq)} + S_2O_3^{2-}{}_{(aq)} + 2OH^-_{(aq)} \rightarrow 2SO_4^{2-}{}_{(aq)} + 4Cl^-_{(aq)} + H_2O_{(l)}$$

4 moles of OCl^- are required for every 1 mole of $S_2O_3^{2-}$. In trial 9, the trial which showed the greatest temperature change, 40. mL of NaClO was used for 10 mL of $Na_2S_2O_3$.

$$\frac{40.\,\text{mL NaClO}}{1} \times \frac{1\,\text{L}}{1000\,\text{mL}} \times \frac{0.500\,\text{mol NaClO}}{1\,\text{L sol'n}} 0.020\,\text{mol ClO}^-$$

$$\frac{10.\,\text{mL Na}_2\text{S}_2\text{O}_3}{1} \times \frac{1\,\text{L}}{1000\,\text{mL}} \times \frac{0.500\,\text{mol Na}_2\text{S}_2\text{O}_3}{1\,\text{L sol'n}} = 0.0050\,\text{mol S}_2\text{O}_3^{2-}$$

$$\frac{0.020\,\text{mol ClO}^-}{0.0050\,\text{mol S}_2\text{O}_3^{2-}} = 4:1\,\text{molar ratio}$$

Experiment 10 A: Determination of the Equilibrium Constant, K_a, for a Chemical Reaction

Background: Weak acids, upon dissolving in water, dissociate slightly according to the following equation:

$$HA + H_2O \rightleftharpoons H_3O^+ + A^-$$

The equilibrium expression for this dissociation is: $K_a = \dfrac{[H_3O^+][A^-]}{[HA]}$

Most weak acids dissociate 5% or less, that is, 95% or more of the acid remains as HA. The smaller the value of K_a, the weaker the acid.

If the acid contains more than one ionizable hydrogen, the acid is known as a polyprotic acid. Specifically, diprotic acids, such as sulfuric acid, H_2SO_4, contain two ionizable hydrogens and triprotics, such as phosphoric acid, H_3PO_4, contain three ionizable hydrogens.

In the case of a diprotic acid, the dissociation occurs in two steps and a separate equilibrium constant exists for each step:

$$H_2A + H_2O \rightleftharpoons H_3O^+ + HA^- \qquad K_{a1} = \dfrac{[H_3O^+][HA^-]}{[H_2A]}$$

$$HA^- + H_2O \rightleftharpoons H_3O^+ + A^{2-} \qquad K_{a2} = \dfrac{[H_3O^+][A^{2-}]}{[HA^-]}$$

The second dissociation generally occurs to a smaller extent than the first dissociation so that K_{a2} is smaller than K_{a1} $(H_3O^+)^2 \times (A^{2-}) / (H_2A) = K_{a1} \times K_{a2} = K_\Sigma$. Exceptions are generally biomolecules in specific solvent media.

This experiment will determine K_a for a monoprotic weak acid.

A common weak acid, acetic acid, dissociates according to the following equation:

$$\underset{\text{acid}}{HC_2H_3O_2} + \underset{\text{base}}{H_2O} \rightleftharpoons \underset{\substack{\text{conjugate}\\\text{acid}}}{H_3O^+} + \underset{\substack{\text{conjugate}\\\text{base}}}{C_2H_3O_2^-}$$

The equilibrium expression for this dissociation is:

$$K_a = \dfrac{[H_3O^+][C_2H_3O_2^-]}{[HC_2H_3O_2]} = 1.8 \times 10^{-5}\,M$$

If $[HC_2H_3O_2] = [C_2H_3O_2^-]$, the two terms cancel, resulting in $H_3O^+ = 1.8 \times 10^{-5}$ M.

$$K_a = \dfrac{[H_3O^+]\cancel{[C_2H_3O_2^-]}}{\cancel{[HC_2H_3O_2]}} = 1.8 \times 10^{-5}\,M$$

Scenario: A student prepared three solutions by adding 12.00, 18.50, and 35.00 mL of a 7.50×10^{-2} M NaOH solution to 50.00 mL of a 0.100 M solution of a weak monoprotic acid, HA. The solutions were labeled X, Y and Z respectively. Each of the solutions were diluted to a total volume of 100.00 mL with distilled water. The pH readings of these solutions, obtained through the use of a calibrated pH meter, were (X) 6.50, (Y) 6.70 and (Z) 7.10.

Analysis:

1. Convert the pH of each solution to an equivalent H_3O^+ concentration.

   $-pH = \log_{10}[H_3O^+]$

   Step 1: Enter pH
   Step 2: Change sign by pressing +/− key
   Step 3: Press "2nd" then "log" or "inv" then "log"

   X: 3.16×10^{-7} M H_3O^+
   Y: 2.00×10^{-7} M H_3O^+
   Z: 7.94×10^{-8} M H_3O^+

2. For each solution, calculate the number of moles of acid added.

   The moles of acid was constant for each solution,

   $$\frac{50.00 \text{ mL}}{1} \times \frac{1 \text{ L}}{1000 \text{ mL}} \times \frac{0.100 \text{ moles HA}}{1 \text{ L}} = 5.00 \times 10^{-3} \text{ mol HA}$$

3. For each solution, calculate the number of moles of OH^- added.

   X: $\dfrac{12.00 \text{ mL}}{1} \times \dfrac{1 \text{ L}}{1000 \text{ mL}} \times \dfrac{7.50 \times 10^{-2} \text{ mol } OH^-}{\text{L}} = 9.00 \times 10^{-4} \text{ mol } OH^-$

   Y: $\dfrac{18.50 \text{ mL}}{1} \times \dfrac{1 \text{ L}}{1000 \text{ mL}} \times \dfrac{7.50 \times 10^{-2} \text{ mol } OH^-}{\text{L}} = 1.39 \times 10^{-3} \text{ mol } OH^-$

   Z: $\dfrac{35.00 \text{ mL}}{1} \times \dfrac{1 \text{ L}}{1000 \text{ mL}} \times \dfrac{7.50 \times 10^{-2} \text{ mol } OH^-}{\text{L}} = 2.69 \times 10^{-3} \text{ mol } OH^-$

4. For each solution, determine the concentration of HA and of A^- ion present at equilibrium.

   X: $[HA] = \dfrac{\text{initial moles HA} - \text{moles } OH^- \text{ added}}{\text{volume sol'n}}$

   $[HA] = \dfrac{5.00 \times 10^{-3} - 9.00 \times 10^{-4}}{0.10000 \text{ L}} = 4.10 \times 10^{-2}$ M

   moles A^- formed = moles OH^- added

   $[A^-] = \dfrac{9.00 \times 10^{-4} \text{ mol } A^-}{0.10000 \text{ L}} = 9.00 \times 10^{-3}$ M

   Y: $[HA] = \dfrac{\text{initial moles HA} - \text{moles } OH^- \text{ added}}{\text{volume sol'n}}$

   $[HA] = \dfrac{5.00 \times 10^{-3} - 1.39 \times 10^{-3}}{0.10000 \text{ L}} = 3.61 \times 10^{-2}$ M

   moles A^- formed = moles OH^- added

   $[A^-] = \dfrac{1.39 \times 10^{-3} \text{ mol } A^-}{0.10000 \text{ L}} = 1.39 \times 10^{-2}$ M

Z: $[HA] = \dfrac{\text{initial moles HA} - \text{moles OH}^- \text{ added}}{\text{volume sol'n}}$

$[HA] = \dfrac{5.00 \times 10^{-3} - 2.63 \times 10^{-3}}{0.10000 \text{ L}} = 2.37 \times 10^{-2} \text{ M}$

moles A^- formed = moles OH^- added

$[A^-] = \dfrac{2.63 \times 10^{-3} \text{ mol A}^-}{0.10000 \text{ L}} = 2.63 \times 10^{-2} \text{ M}$

5. Determine the reciprocal of the A^- ion concentration for each solution.

X: $\dfrac{1}{9.00 \times 10^{-3}} = 1.11 \times 10^{-2} \text{ M}^{-1}$

Y: $\dfrac{1}{1.39 \times 10^{-2}} = 7.19 \times 10^{1} \text{ M}^{-1}$

Z: $\dfrac{1}{2.63 \times 10^{-2}} = 3.80 \times 10^{1} \text{ M}^{-1}$

6. Plot the reciprocal of the A^- ion concentration on the ordinate against the H_3O^+ ion concentration on the abscissa. Draw a straight "best-fit" line (see Figure 1).

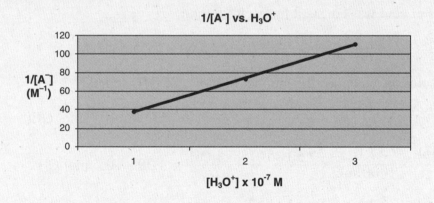

Figure 1

7. Determine the slope of the line.

$\text{Slope} = \dfrac{\Delta y}{\Delta x} = \dfrac{\Delta[1/A^-]}{\Delta[H_3O^+]} = \dfrac{1.11 \times 10^{2} - 3.80 \times 10^{1}}{(3.16 \times 10^{-7}) - (7.94 \times 10^{-8})} = 3.08 \times 10^{8}$

8. Determine the initial concentration of the acid, HA.

$= \dfrac{5.00 \times 10^{-3} \text{ mol HA}}{0.10000 \text{ L}} = 5.00 \times 10^{-2} \text{ M}$

9. Determine the K_a for the weak acid.

$\dfrac{1}{\text{slope}} = \dfrac{\Delta[H_3O^+]}{\Delta[1/A^-]} = (H_3O^+) \cdot (A^-)$

$K_a = \dfrac{1}{HA^* \cdot \text{slope}} = \dfrac{1}{5.00 \times 10^{-2} \cdot (3.08 \times 10^{8})} = 6.49 \times 10^{-8} \text{ M}$

*= concentration of HA prior to dissociation or neutralization

Experiment 10 B: Determination of the Equilibrium Constant, K_{sp}, for a Chemical Reaction

Background: Lead iodide is relatively insoluble, with a solubility of less than 0.002 M at 20°C. Lead iodide dissolved in water is represented as:

$$PbI_{2\,(s)} \rightleftharpoons Pb^{2+}_{(aq)} + 2I^-_{(aq)}$$

The solubility product expression (K_{sp}) for this reaction is:

$$K_{sp} = \left[Pb^{2+}\right]\left[I^-\right]^2$$

If lead iodide is in a state of equilibrium in solution, the product of the molar concentration of lead ions times the square of the molar concentration of the iodide ions will be equal to a constant. This constant does not depend upon how the state of equilibrium was achieved.

When standard solutions of lead nitrate, $Pb(NO_3)_2$ and potassium iodide, KI, are mixed in the presence of KNO_3, a yellow precipitate of lead (II) iodide, PbI_2, forms. The potassium nitrate helps to keep the solution at a nearly constant ionic molarity and promotes the formation of well-defined crystals of lead iodide. The lead iodide will be allowed to come to equilibrium in the solution by mixing the solution very well and then allowing the precipitate to settle completely followed by centrifuging. The iodide ion, I^-, is colorless and not able to be directly measured in a spectrophotometer (see Figure 1). Therefore, the I^- will be oxidized to I_2 which is brown in water by an acidified potassium nitrite, KNO_2, solution. The concentration of I_2 is rather small, however, the absorption of light at 525 nm is sufficient to allow detection of the I_2 molecule. $[Pb^{2+}]$ in solution will be determined from the initial state of the system, the measured $[I^-]$ and the relationship between $[Pb^{+2}]$ and $[I^-]$ in the equilibrium expression.

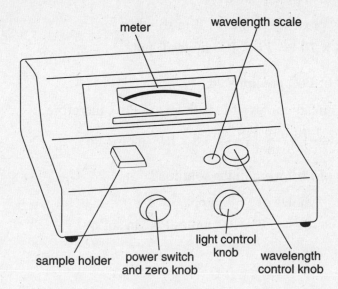

Figure 1

Scenario: A student mixed 5.0 mL of 0.015 M $Pb(NO_3)_2$ with 3.0 mL of 0.030 M KI in 2.0 mL of 0.200 M KNO_3. The test tube was shaken for 15 minutes and the solid precipitate of lead (II) iodide was allowed to settle. The tube was centrifuged to remove any excess lead (II) iodide from the supernatant. To the supernatant, she added KNO_2 to oxidize the I^- to I_2. The supernatant was then analyzed for I^- by using a spectrophotometer. She then made known dilutions of a 0.10 M potassium iodide solution in acidified KNO_2 to create a calibration curve to be able to determine the I^- in the supernatant.

Analysis:

1. How many moles of Pb^{2+} were initially present?

$$\frac{5.0 \text{ mL sol'n}}{1} \times \frac{1 \text{ L sol'n}}{1000 \text{ mL sol'n}} \times \frac{0.015 \text{ mol } Pb(NO_3)_2}{1 \text{ L sol'n}}$$

$= 7.5 \times 10^{-5}$ mol $Pb(NO_3)_2$ yielding 7.5×10^{-5} mol Pb^{2+}

2. How many moles of I^- were initially present?

$$\frac{3.0 \text{ mL sol'n}}{1} \times \frac{1 \text{ L sol'n}}{1000 \text{ mL sol'n}} \times \frac{0.030 \text{ mol KI}}{1 \text{ L sol'n}}$$

$= 9.0 \times 10^{-5}$ mol KI yielding 9.0×10^{-5} mol I^-

3. From known dilutions of KI in acidified KNO_2, a calibration curve was created. From the calibration curve, it was determined that the concentration of I^- at equilibrium in the supernatant was 1.4×10^{-3} mole/liter. How many moles of I^- were present in 10. mL of the final solution?

$$\frac{1.4 \times 10^{-3} \text{ moles}}{L} \times \frac{0.010 \text{ L}}{1} = 1.4 \times 10^{-5} \text{ mol } I^- \text{ in sol'n at equil.}$$

4. How many moles of I^- precipitated?

moles I^- originally present − moles of I^- in solution

$= 9.0 \times 10^{-5} - 1.4 \times 10^{-5} = 7.6 \times 10^{-5}$ moles I^- ppt.

5. How many moles of Pb^{2+} precipitated?

There are twice as many moles of I^- as there are Pb^{2+}, therefore,

$$\frac{7.6 \times 10^{-5} \text{ mol } I^-}{1} \times \frac{1.00 \text{ mol } Pb^{2+}}{2.00 \text{ mol } I^-} = 3.8 \times 10^{-5} \text{ moles } Pb^{2+}$$

6. How many moles of Pb^{2+} were in the solution?

total moles of Pb^{2+} − moles of Pb^{2+} in ppt

$= 7.5 \times 10^{-5} - 3.8 \times 10^{-5} = 3.7 \times 10^{-5}$ moles Pb^{2+} in sol'n

7. What was the molar concentration of Pb^{2+} in the solution at equilibrium?

$$\frac{3.7 \times 10^{-5} \text{ mol } Pb^{2+}}{0.010 \text{ L}} = 3.7 \times 10^{-3} \text{ M}$$

8. Determine K_{sp}

$$K_{sp} = [Pb^{2+}][I^-]^2 = (3.7 \times 10^{-3} \text{ M})(1.4 \times 10^{-3} \text{ M})^2 = 7.3 \times 10^{-9} \text{ M}^3$$

9. The published value of K_{sp} at 25°C for PbI_2 is 7.1×10^{-9}. Determine the percent error.

$$\% \text{ error} = \frac{\text{obs} - \text{exp}}{\text{exp}} \times 100\% = \frac{(7.3 \times 10^{-9} - 7.1 \times 10^{-9})}{7.1 \times 10^{-9}} \times 100\% = 2.8 \%$$

Experiment 10 C: Determination of the Equilibrium Constant, K_c, for a Chemical Reaction

Background: Many reactions do not go to completion. Rather, they reach an intermediate state in which both reactants and products are present simultaneously. When the concentrations of all species remain constant at a particular temperature over time, a state of equilibrium has been achieved. The equilibrium constant, K_c, relates the concentrations of products to reactants at equilibrium. K_c values significantly greater than 1 indicate that the products are favored, while K_c values significantly less that 1 indicate that the reactants are more predominant.

In this experiment the equilibrium between iron (III) ions, Fe^{3+}, and thiocyanate ions, SCN^-, will be investigated. The equilibrium equation is:

$$Fe^{3+}_{(aq)} + SCN^-_{(aq)} \rightleftharpoons FeSCN^{2+}_{(aq)}$$

From the equation, an equilibrium expression can be created:

$$K_c = \frac{\left[FeSCN^{2+}\right]}{\left[Fe^{3+}\right]\left[SCN^-\right]}$$

The $FeSCN^{2+}$ ion is red. The degree of color is proportional to the concentration. The concentration of $FeSCN^{2+}$ will be found through the use of a calibrated spectrophotometer whose most optimal wavelength was determined in which a known concentration of $FeSCN^{2+}$ is made up and the wavelength varied for maximum absorbance. Since the $FeSCN^{2+}$ is red, the wavelength of the light in the spectrophotometer should be the complement color and appear blue-violet. Then, with the spectrophotometer set at this optimal wavelength, the absorbances of various known concentration of $FeSCN^{2+}$ will be recorded and graphed. From this graph the concentration of the $FeSCN^{2+}$ at equilibrium will be extrapolated through Beer's law.

Scenario: A student determined that the optimal wavelength for the absorbance of $FeSCN^{2+}$ experiment was 445 nm. Then the student prepared samples of known concentrations of $FeSCN^{2+}$ ranging from 4.0×10^{-5} M to 1.4×10^{-4} M. The samples were then examined by means of a spectrophotometer and their transmittances recorded. From the transmittance, the absorbance was calculated and graphed. Next, he mixed 5.0 mL of 2.0×10^{-3} M $Fe(NO_3)_3$ with 5.0 mL of 2.0×10^{-3} M KSCN. This solution was then analyzed in the spectrophotometer and through extrapolation, he was able to determine that the concentration of $FeSCN^{2+}$ at equilibrium was 1.3×10^{-4} M.

Analysis:

1. How many moles of Fe^{3+} and SCN^- were initially present?

$$\frac{0.0050 \; \text{L sol'n}}{1} \times \frac{2.0 \times 10^{-3} \; \text{moles } Fe^{3+}}{\text{L sol'n}} = 1.0 \times 10^{-5} \; \text{moles } Fe^{3+}$$

$$\frac{0.0050 \; \text{L sol'n}}{1} \times \frac{2.0 \times 10^{-3} \; \text{moles } SCN^-}{\text{L sol'n}} = 1.0 \times 10^{-5} \; \text{moles } SCN^-$$

2. How many moles of FeSCN^{2+} were in the mixture at equilibrium?

Total volume of sol'n = 5.0 mL + 5.0 mL = 10.0 mL

$$\frac{1.3 \times 10^{-4} \text{ moles FeSCN}^{2+}}{\text{L sol'n}} \times \frac{0.0100 \text{ L sol'n}}{1} = 1.3 \times 10^{-6} \text{ moles FeSCN}^{2+}$$

3. How many moles of Fe^{3+} and SCN$^-$ were used up in making the FeSCN^{2+}?

The balanced equation at equilibrium is:

$$Fe^{3+}_{(aq)} + SCN^-_{(aq)} \rightleftharpoons FeSCN^{2+}_{(aq)}$$

The ratio of Fe^{3+} to SCN$^-$ is 1:1 and the ratio of Fe^{3+} and SCN$^-$ to FeSCN^{2+} is 1:1

Therefore, 1.3×10^{-6} moles of Fe^{3+} and 1.3×10^{-6} moles of SCN$^-$ were used in making the FeSCN^{2+}.

4. How many moles of Fe^{3+} and SCN$^-$ remain in the solution at equilibrium?

initial # of moles − # of moles used up in making FeSCN^{2+}

$= 1.0 \times 10^{-5}$ moles Fe^{3+} − 1.3×10^{-6} moles Fe^{3+} = 8.7×10^{-6} moles Fe^{3+}

$= 1.0 \times 10^{-5}$ moles SCN$^-$ − 1.3×10^{-6} moles SCN$^-$ = 8.7×10^{-6} moles SCN$^-$

5. What are the concentrations of Fe^{3+}, SCN$^-$, and FeSCN^{2+} at equilibrium?

$$\left[Fe^{3+}\right] = \frac{8.7 \times 10^{-6} \text{ moles Fe}^{3+}}{0.01 \text{ L sol'n}} = 8.7 \times 10^{-4} \text{ M Fe}^{3+}$$

$$\left[SCN^-\right] = \frac{8.7 \times 10^{-6} \text{ moles SCN}^-}{0.01 \text{ L sol'n}} = 8.7 \times 10^{-4} \text{ M SCN}^-$$

[FeSCN^{2+}] = 1.3 x 10^{-4} M FeSCN^{2+}

6. Determine K$_c$ for the reaction.

$$Fe^{3+}_{(aq)} + SCN^-_{(aq)} \rightleftharpoons FeSCN^{2+}_{(aq)}$$

$$K_c = \frac{\left[FeSCN^{2+}\right]}{\left[Fe^{3+}\right]\left[SCN^-\right]} = \frac{1.3 \times 10^{-4} \text{ M}}{\left(8.7 \times 10^{-4}\right)^2 \text{M}^2} = 1.7 \times 10^2 \text{ M}^{-1}$$

Experiment 11: Determination of Appropriate Indicators for Various Acid-Base Titrations

and

Experiment 19: Preparation and Properties of Buffer Solutions

Background: Strong acids generally dissociate into ions nearly completely. The hydronium ion concentration of a strong monoprotic acid solution is therefore essentially equal to the concentration of the acid in solution*. In mono-hydroxy strong bases, the hydroxide ion generally dissociates nearly completely and the hydroxide ion concentration is equivalent to the concentration of the base in solution**. In 1909, Sören Sörenson proposed the system that we use today in measuring the concentration of acids and bases. This system is based on exponents to overcome the difficulties of dealing with small numbers.

$$pH = \log\frac{1}{[H^+]} = -\log[H^+] \qquad [H_3O^+] = 10^{-pH}$$

$$pOH = \log\frac{1}{[OH^-]} = -\log[OH^-] \qquad [OH^-] = 10^{-pOH}$$

Since $K_w = 1.0 \times 10^{-14} = (H^+) \cdot (OH^-)$; $\log K_w = -14.00 = \log(H^+) + \log(OH^-)$; $[-\log(H^+)] + [-\log(OH^-)] = -[-14.00]$, then $pH + pOH = 14.00$***

For weak acids, those that do not dissociate completely (usually 5% dissociation or less), hydronium ion concentration is not equal to the concentration of the anion of the acid in solution (it is less). In weak bases, as well, the concentration of the hydroxide ion is not equal to the concentration of the cation of the base (it is less).

In the case of a weak acid, acetic acid, which is represented by HAc, the dissociation reaction can be written as:

$$HAc_{(aq)} + H_2O_{(l)} \Leftrightarrow Ac^-_{(aq)} + H_3O^+_{(aq)}$$

The equilibrium expression is therefore:

$$K_a = \frac{[H_3O^+][Ac^-]}{[HAc]} = 1.8 \times 10^{-5}\ M$$

The concentration of the acetic acid that has not dissociated is considered to be approximately equal to the initial concentration of the acid because the extent of the dissociation of weak acids is small. However, often it is not negligible and then to solve for the equilibrium concentrations, use of the quadratic equation or other special mathematics is needed.

* more accurately as the anion of the acid

** more accurately as the cation of the base

*** temperature dependent

In hydrolysis, a salt reacts with water. The ions that hydrolyze do so because a weak acid or a weak base is formed. The process of hydrolysis removes ions from the solution and is the driving force for the reaction. The reaction may produce a solution that is acidic, basic or neutral according to the following chart:

Strong base + strong acid	No hydrolysis — neutral
NaOH + HCl	Salt formed — NaCl $H^+ + OH^- \Leftrightarrow H_2O$ $K = \dfrac{1}{[H^+][OH^-]} = \dfrac{1}{K_w}$ Methyl Red (end point = pH 5) Bromthymol Blue (end point = pH 7) Phenolphthalein (end point = pH 9)
Strong base + weak acid	Basic — only anion hydrolyzes
NaOH + HAc	Salt formed — NaAc $Ac^- + H_2O \Leftrightarrow HAc + OH^-$ $K_b = \dfrac{[HAc][OH^-]}{[Ac^-]} \times \dfrac{[H^+]}{[H^+]} = \dfrac{K_w}{K_{HAc}}$ Phenolphthalein (end point = pH 9)
Weak base + strong acid	Acidic — only cation hydrolyzes
NH₃ + HCl	Salt formed — NH₄Cl $NH_4^+ + H_2O \Leftrightarrow NH_3 + H_3O^+$ $K_a = \dfrac{[NH_3][H_3O^+]}{[NH_4^+]} \times \dfrac{[OH^-]}{[OH^-]} = \dfrac{K_w}{K_{NH_3}}$ Methyl Red (end point = pH 5) or Methyl Orange (end point = pH 4)
Weak base + weak acid	Variable pH— both ions hydrolyze. pH depends upon extent of hydrolysis of each ion involved.
NH₃ + HCN	Salt formed — NH₄CN $NH_4^+ + H_2O \Leftrightarrow NH_3 + H_3O^+$ $CN^- + H_2O \Leftrightarrow HCN + OH^-$ $K_a = \dfrac{[NH_3][HCN]}{[NH_4^+][CN^-]} \times \dfrac{[H^+][OH^-]}{[H^+][OH^-]} = \dfrac{K_w}{K_{NH_3} \times K_{HCN}}$ pH of NH₄CN (aq) is greater than 7 because CN^- ($K_b = 2.0 \times 10^{-5}$) is more basic than NH_4^+ ($K_a = 5.6 \times 10^{-10}$) is acidic.

If solutions undergo only very small changes in pH after small amounts of strong acids or bases are added, the solution is called a buffer. Buffers may be prepared by either combining a weak acid and a salt of the acid or by adding a weak base to a salt of the base. Where both ions of a salt can hydrolyze, that salt may also act as a buffer (for example, NH₄AC, pH=7.0). The solutions that result may have a common ion and resist changes in pH.

The simplest quantitative method for determining pH is with the use of indicators. An indicator is a colored substance usually derived from plant material that can exist in either an acid or base form. The two forms have different colors. If one knows the pH at which the indicator turns from one form to the other, one can then determine from the observed color whether the solution has a pH higher or lower than this value. Methyl orange changes color over the pH interval from 3.1 to 4.4. Below pH 3.1 it is in the acid form, which is red. In the interval from 3.1 to 4.4 it is gradually converted to the basic form, which is yellow. By pH 4.4 the conversion is complete.

See Appendix B for a chart of acid-base indicators.

Scenario: Part I: pH of a Strong Acid

A student obtains 100.0 mL of 0.100 M HCl. The student prepares serial dilutions of the acid to obtain solutions of the following concentrations: 0.0500 M, 0.0100 M, 0.00500 M and 0.00100 M. The student determines the pH of each of the solutions with a pH meter and then from an assortment of indicators available and using Appendix B, adds the appropriate indicator to samples of the solutions produced.

Part II: pH of a Weak Acid

This time the student obtains 100.0 mL of 0.100 M acetic acid solution and repeats the steps from Part I.

Part III: pH of Various Salt Solutions

0.100 M solutions of NH_4Ac, $NaCl$, $NaAc$, NH_4Cl, $NaHCO_3$ and Na_2CO_3 are available. The student measures each solution with the calibrated pH meter and also adds appropriate indicators to samples of each solution and observes color changes.

Part IV: pH of Buffer Solutions

The student prepares a buffer solution by adding 4.00 g of sodium acetate to 200 mL of 0.250 M acetic acid. The solution is then diluted to 100.0 mL with distilled water. The student then obtains 4 beakers labeled 1, 2, 3 and 4. To beakers 1 and 2 she adds 50.0 mL of the buffer solution. To beakers 3 and 4 she adds 50.0 mL of distilled water. She then uses a calibrated pH meter and obtains the pH of the solutions in the beakers. To beakers 1 and 3, she pipets 25.0 mL of 0.250 M HCl, mixes well and determines the pH using the calibrated pH meter. The pH of each beaker was as follows: (1) 4.32, (2) 1.01, (3) 5.25, (4) 12.88. She also selects an appropriate indicator and records the color produced. To beakers 2 and 4 she adds 25.0 mL of 0.250 M NaOH and repeats the process as she did with the HCl.

Analysis: Because the data obtained from this lab is quite extensive, sample questions will be asked to determine knowledge of the processes involved.

1. A solution from Part I was determined to have a pH of 1.00. Determine the pOH, the $[H_3O^+]$ and the $[OH^-]$.

   $pOH = 14.00 - pH = 14.00 - 1.00 = 13.00$

   $[H^+] = 10^{-pH} = 10^{-1.00} = 0.100$ M

   $[OH^-] = 10^{-pOH} = 10^{-13.00} = 1.00 \times 10^{-13}$ M

2. A solution from Part I had an $[OH^-]$ of 2.00×10^{-12} M. Determine the pH.

   $pOH = -\log [OH^-] = -\log (2.00 \times 10^{-12}) = 11.70$

   $pH = 14.00 - pOH = 14.00 - 11.70 = 2.30$

3. The pH of a 0.100 M acetic acid solution in Part II was found to be 2.87. Calculate the K_a of acetic acid.

   $[H_3O^+] = 10^{-2.87} = 1.35 \times 10^{-3}$ M at equilibrium

   $$HAc_{(aq)} + H_2O_{(l)} \rightleftharpoons Ac^-_{(aq)} + H_3O^+_{(aq)}$$

	HAc	Ac⁻	H₃O⁺
I	0.100 M	0	~0
C	−0.00135	+0.00135	+0.00135
E	~0.099	0.00135	0.00135

   $$K_a = \frac{[Ac^-][H_3O^+]}{[HAc]} = \frac{(0.00135)(0.00135)}{0.099} = 1.8 \times 10^{-5}$$

4. Calculate the hydrolysis constant K_h for the ammonium chloride solution from Part III. Assume that $[NH_4^+]$ is the same as the initial concentration of ammonium chloride.

   From collected data, pH = 5.13, so $[H_3O^+] = 10^{-5.13} = 7.4 \times 10^{-6}$ M

   $$NH_4^+ + H_2O \rightleftharpoons NH_3 + H_3O^+$$

	NH₄⁺	NH₃	H₃O⁺
I	0.100 M	0	~0
C	−7.4 × 10⁻⁶	+7.4 × 10⁻⁶	+7.4 × 10⁻⁶
E	~0.100	7.4 × 10⁻⁶	7.4 × 10⁻⁶

   $$K_h = \frac{[NH_3][H_3O^+]}{[NH_4^+]} = \frac{(7.4 \times 10^{-6})(7.4 \times 10^{-6})}{0.100} = 5.5 \times 10^{-10}$$

5. Calculate the theoretical pH of the original buffer solution, assuming the volume change due to the addition of the solid is negligible.

   $$0.200 \text{ L} \times \frac{0.250 \text{ mol HAc}}{\text{L}} = 0.0500 \text{ mol HAc}$$

   $$\frac{4.00 \text{ g NaAc}}{1} \times \frac{1 \text{ mol NaAc}}{82.034 \text{ g NaAc}} \times \frac{1 \text{ mol Ac}^-}{1 \text{ mol NaAc}} = 0.0488 \text{ mol Ac}^-$$

Since, $\dfrac{[HAc]}{[Ac^-]} = \dfrac{n_{HAc}/V_{sol'n}}{n_{Ac^-}/V_{sol'n}}$

Then, $\left[H^+\right] = K_a \cdot \dfrac{n_{HAc}}{n_{Ac^-}} = 1.8 \times 10^{-5} \cdot \dfrac{0.0500 - x \text{ mol HAc}}{0.0488 - x \text{ mol Ac}^-} = 1.84 \times 10^{-5} \text{ M}^*$

where x is negligible

*extra significant figure

$pH = -\log [H^+] = -\log (1.84 \times 10^{-5}) = 4.73$

6. Calculate the theoretical pH of the buffer solution with the added HCl (beaker 1).

$0.0500 \text{ L} \times \dfrac{0.250 \text{ mol HAc}}{L} = 0.0125 \text{ mol HAc}$

$0.0500 \text{ L} \times \dfrac{0.0488 \text{ mol Ac}^-}{0.200 \text{ L}} = 0.0122 \text{ mol Ac}^-$

$0.0250 \text{ L} \times \dfrac{0.250 \text{ mol H}^+}{L} = 0.00625 \text{ mol H}^+$

$Ac^- + H^+ \rightleftharpoons HAc$

	Ac⁻	*H⁺*	*HAc*
I	0.0122	0.00625	0.0125
C	−0.00625	−0.00625	+0.00625
E	0.00595	~0	0.01875

$\left[H^+\right] = K_a \times \dfrac{n_{HAc}}{n_{Ac^-}} = 1.8 \times 10^{-5} \times \dfrac{0.01875 \text{ mol HAc}}{0.00595 \text{ mol Ac}^-} = 5.67 \times 10^{-5} \text{ M}$

$pH = -\log [H^+] = -\log (5.67 \times 10^{-5}) = 4.25$

7. Calculate the theoretical pH of the buffer solution with the added NaOH (beaker 2).

$0.0500 \text{ L} \times \dfrac{0.250 \text{ mol HAc}}{L} = 0.0122 \text{ mol HAc}$

$0.0500 \text{ L} \times \dfrac{0.488 \text{ mol Ac}^-}{0.200 \text{ L}} = 0.0122 \text{ mol Ac}^-$

$0.0250 \text{ L} \times \dfrac{0.250 \text{ mol OH}^-}{L} = 0.00625 \text{ mol OH}^-$

$HAc + OH^- \rightleftharpoons Ac^- + H_2O$

	HAc	*OH⁻*	*Ac⁻*
I	0.0125	0.00625	0.0122
C	−0.00625	−0.00625	+0.00625
E	0.00625	~0	0.01845

$\left[H^+\right] = K_a \times \dfrac{n_{HAc}}{n_{Ac^-}} = 1.8 \times 10^{-5} \times \dfrac{0.00625 \text{ mol HAc}}{0.01845 \text{ mol Ac}^-} = 6.10 \times 10^{-6} \text{ M}$

$pH = -\log [H^+] = -\log (6.10 \times 10^{-6}) = 5.21$

8. Calculate the theoretical pH of the distilled water and added HCl (beaker 3).

$$\left[H^+\right] = 0.0250 \;\cancel{L} \times \frac{0.250 \text{ mol } H^+}{\cancel{L}} \times \frac{1}{(0.0250 + 0.0500)\,L} = 8.33 \times 10^{-2} \text{ M}$$

$$pH = -\log\,[H^+] = -\log\,(8.33 \times 10^{-2}) = 1.08$$

9. Calculate the theoretical pH of the distilled water and added NaOH (beaker 4).

$$\left[OH^-\right] = 0.0250 \;\cancel{L} \times \frac{0.250 \text{ mol } OH^-}{\cancel{L}} \times \frac{1}{(0.0250 + 0.0500)\,L} = 8.33 \times 10^{-2} \text{ M}$$

$$pOH = -\log\,[OH^-] = -\log\,(8.33 \times 10^{-2}) = 1.08$$

$$pH = 14.00 - pH = 14.00 - 1.08 = 12.92$$

10. A 0.1 M solution of an acid was tested with the following indicators and the following colors observed:

Phenolphthalein: clear	Bromthymol blue: blue
Thymol blue: yellow	Alizarin yellow: yellow
Methyl orange: yellow	Methyl red: yellow

Using Appendix B, what is the approximate pH of the solution? What two indicators would you use to find the pH more precisely and why?

From the indicators that were used, the following is a summary of their properties:

Phenolphthalein: range 8.3 to 10.0	Lower color-colorless; Upper color-dark pink
Bromthymol blue: range 6.0 to 7.6	Lower color-yellow; Upper color-blue
Thymol blue: range 8.1 to 9.5	Lower color-yellow; Upper color-blue
Alizarin yellow: range 9.9 to 11.8	Lower color-yellow; Upper color-dark orange
Methyl orange: range 2.5 to 4.4	Lower color-red; Upper color- yellow
Methyl red: range 4.4 to 6.2	Lower color-red; Upper color-yellow
m-Cresol purple: range 7.5 to 9.0	Lower color-yellow; Upper color-violet

Using Appendix B, you would choose bromthymol blue (range 6.0 to 7.6, Lower color-yellow; Upper color-blue) and m-Cresol purple (range 7.5 to 9.0, Lower color-yellow; Upper color-violet) because bromthymol blue's high-end color (blue) ends where m-cresol purple's low-end color (yellow) begins. The pH would be between 7.5–8.5.

Experiment 12: Determination of the Rate of a Reaction and Its Order and an Actual Student Lab Write-Up*

Background: Reaction rates depend on several criteria: the concentration of the reactants, the nature of the reaction, temperature, and presence of catalysts. The rate of most reactions increases when the concentration of any reactant increases. For the reaction

$$aA + bB \rightarrow cC$$

the rate can be expressed by the rate equation, rate = $k(A)^x(B)^y$. The values of x and y are usually whole number integers ranging from 0 to 3 and can only be determined by examining lab data. These numbers represent the order of the reactant. If the order of the reactant is 0, then increasing the concentration of that reactant has no effect on the rate. If the reactant order is 1, then doubling the concentration of that reactant will double the rate of the reaction. If the reactant order is 2, then doubling the concentration of that reactant will increase the rate 4 times. And if the reactant order is 3, then doubling the concentration of that reactant will increase the reaction rate eight times. It is possible to have reactant orders that are fractions or that are negative. To obtain the overall order of the reaction, simply add the reactant orders.

k is known as the rate-specific constant. It is a value, unique for each reaction, that allows the equality to exist. It is constant for the reaction as long as the temperature does not change. It depends principally upon the nature of the reactants and the temperature at which the reaction occurs. For reactions between ions in aqueous solution, it is affected by the total concentration of ions in the solution.

Svante Arrhenius proposed that a minimum amount of energy was necessary for a reaction to proceed. This minimum amount of energy is called the activation energy, E_a. The Arrhenius equation

$$k = Ae^{-Ea/RT}$$

relates the temperature to the rate specific constant, k. A more useful derivation of this equation is

$$\ln k = \frac{-E_a}{RT} + \ln A$$

where A is called the collision frequency factor constant and considers the collision frequency and the geometry of colliding species, which in this experiment will be ignored. R is the universal gas constant. This equation follows the straight line relationship: $y = mx + b$. A plot of the natural logarithm of k versus 1/T gives a straight line. The slope of the graph will be used to determine the activation energy.

This experiment studies the kinetics or reaction mechanisms, and their rates, when iodine is added to acetone:

$$CH_3 \overset{\overset{\textstyle O}{\|}}{-C}-CH_3(aq) + I_2(aq) \rightarrow CH_3 \overset{\overset{\textstyle O}{\|}}{-C}-CH_2I(aq) + H^+(aq) + I^-(aq)$$

Therefore a rate law can be created:

$$\text{rate} = k \, (\text{acetone})^x \cdot (I_2)^y \cdot (H^+)^z$$

It is known that the amount of iodine does not affect the rate of the reaction. Therefore, the rate order of I_2 is 0. Investigating the reaction in terms of the change in the concentration of iodine over time gives the relationship

$$\text{rate} = \frac{-\Delta(I_2)}{\Delta t}$$

The negative sign cancels the negative value of the change in I_2 which is due to the disappearance of I_2, making the rate a positive value. Since the rate of the overall reaction does not depend on the iodine, the reaction rate can be studied by making the iodine the limiting reactant present in excess acetone and H^+, concentrations high enough so that their concentrations do not change significantly through the course of the reaction. One simply then measures the time required for the I_2 to be consumed by varying the concentrations of H^+ and acetone — easily determined since I_2 is yellow in solution.

Scenario: A student prepared a data table for the results that he got in doing this experiment:

Mixture	Acetone	H^+	$(I_2)_0$	Time (sec)	Temp. °C	rate = $(I_2)_0 \cdot \text{sec}^{-1}$
I	0.80 M	0.20 M	0.0010 M	240	25°C	4.2×10^{-6}
II	1.60 M	0.20 M	0.0010 M	120	25°C	8.3×10^{-6}
III	0.80 M	0.40 M	0.0010 M	120	25°C	8.3×10^{-6}
IV	0.80 M	0.20 M	0.00050 M	120	25°C	4.2×10^{-6}

Analysis:

1. Determine the rate order for each reactant.

$$\text{rate} = k \cdot (\text{acetone})^x \cdot ((I_2)_0)^y \cdot (H^+)^z$$
$$\text{rate I} = 4.2 \times 10^{-6} = k \cdot (0.80)^x \cdot (0.0010)^y \cdot (0.20)^z$$
$$\text{rate II} = 8.3 \times 10^{-6} = k \cdot (1.60)^x \cdot (0.0010)^y \cdot (0.20)^z$$

Ratio of Rate II to Rate I:

$$\frac{\text{rate II}}{\text{rate I}} = \frac{8.3 \times 10^{-6}}{4.2 \times 10^{-6}} = 2.0 = \frac{k \cdot (1.60)^x \cdot \cancel{(0.0010)^y} \cdot \cancel{(0.20)^z}}{k \cdot (0.80)^x \cdot \cancel{(0.0010)^y} \cdot \cancel{(0.20)^z}} =$$

$$= \log 2 = x \cdot \log 2; \; \mathbf{x = 1}$$

$$\text{rate III} = 8.3 \times 10^{-6} = k \cdot (0.80)^x \cdot (0.0010)^y \cdot (0.40)^z$$
$$\text{rate IV} = 4.2 \times 10^{-6} = k \cdot (0.80)^x \cdot (0.00050)^y \cdot (0.20)^z$$

Ratio of rate III to rate IV:

$$\frac{\text{rate III}}{\text{rate IV}} = \frac{8.3 \times 10^{-6}}{4.2 \times 10^{-6}} = 2.0 = \frac{k \cdot \cancel{(0.80)^x} \cdot (0.00050)^y \cdot (0.40)^z}{k \cdot \cancel{(0.80)^x} \cdot (0.00050)^y \cdot (0.20)^z}$$

$$= \log 2 = z \cdot \log 2; \; \mathbf{z = 1}$$

Ratio of rate IV to rate I:

$$\frac{\text{rate IV}}{\text{rate I}} = \frac{4.2 \times 10^{-6}}{4.2 \times 10^{-6}} = 1.0 = \frac{\cancel{k} \cdot \cancel{(0.80)^x} \cdot (0.00050)^y \cdot \cancel{(0.20)^z}}{\cancel{k} \cdot \cancel{(0.80)^x} \cdot (0.0010)^y \cdot \cancel{(0.20)^z}}$$

$$= \log 1 = y \log 0.5; \; \mathbf{y = 0}$$

As the concentration of the acetone doubles, the rate doubles; $x = 1$.

As the concentration of the H^+ doubles, the rate doubles; $z = 1$.

The concentration of the I_2 has no affect on the rate; $y = 0$.

2. Determine the rate specific constant, k, for the reaction.

$$k = \frac{\text{rate}}{(\text{acetone})(H^+)}$$

Mixture 1: $k = \dfrac{4.2 \times 10^{-6}}{(0.80)(0.20)} = 2.6 \times 10^{-5}$

Mixture 2: $k = \dfrac{8.3 \times 10^{-6}}{(1.60)(0.20)} = 2.6 \times 10^{-5}$

Mixture 3: $k = \dfrac{8.3 \times 10^{-6}}{(0.80)(0.20)} = 2.6 \times 10^{-5}$

Mixture 4: $k = \dfrac{4.2 \times 10^{-6}}{(0.80)(0.20)} = 2.6 \times 10^{-5}$

Average: $k = 2.6 \times 10^{-5}$

3. Given the following information, predict the time required for the reaction to reach completion:

Acetone	HCl	I_2	H_2O
25 mL	25 mL	25 mL	25 mL
3.2 M	4.0 M	0.020 M	—

Total volume of mixture = 25 mL + 25 mL + 25 mL + 25 mL = 1.0×10^2 mL

Concentration of each after mixing:

Acetone: $\dfrac{0.025 \; \cancel{L}}{1} \times \dfrac{3.2 \; \text{moles}}{\cancel{L}} \times \dfrac{1}{0.10 \; \text{L sol'n}} = 0.80$ M

HCl: $\dfrac{0.025 \; \cancel{L}}{1} \times \dfrac{4.0 \; \text{moles}}{\cancel{L}} \times \dfrac{1}{0.10 \; \text{L sol'n}} = 1.0$ M

I_2: $\dfrac{0.025 \; \cancel{L}}{1} \times \dfrac{0.020 \; \text{moles}}{\cancel{L}} \times \dfrac{1}{0.10 \; \text{L sol'n}} = 0.0050$ M

rate $= k \cdot (\text{acetone})^x \cdot ((I_2)_o)^y \cdot (H^+)^z$

$= 2.6 \times 10^{-5} \cdot (0.80)^1 \cdot (0.0050)^0 \cdot (1.0)^1 = \mathbf{2.1 \times 10^{-5} \; moles \cdot L^{-1} \cdot sec^{-1}}$

4. The student used the ratios in Mixture I and ran the experiment at two different temperatures. Calculate the rate, the rate constant, log k and 1/T for each temperature studied. From the data, plot k versus 1/T and determine the activation energy. Given that the activation energy for the reaction is 8.6×10^4 Joules, calculate the % error.

25 °C   240 seconds

40 °C   45 seconds

Begin by creating a table summarizing the known information and leaving room for calculated results:

Acetone	H$^+$	(I$_2$)$_o$	Temp °C	Time (sec)	rate	k	log k	1/T (K)
0.80	0.20	0.0010	25	240	4.2×10^{-6}	2.6×10^{-5}	−4.58	3.36×10^{-3}
0.80	0.20	0.0010	40	45	2.2×10^{-5}	1.4×10^{-5}	−3.86	3.19×10^{-3}

$rate = k \cdot (acetone)^1 \cdot ((I_2)_o)^0 \cdot (H^+)^1$

$rate = (I_2)_0 \cdot time^{-1}$

At 25°C = 298 K:

$$rate = \frac{0.0010 \ M}{240 \ sec} = 4.2 \times 10^{-6}$$

$$k = \frac{4.2 \times 10^{-6}}{(0.80)(0.20)} = 2.6 \times 10^{-5}$$

At 40°C = 313 K:

$$rate = \frac{0.0010 \ M}{45 \ sec} = 2.2 \times 10^{-5}$$

$$k = \frac{2.2 \times 10^{-5}}{(0.80)(0.20)} = 1.4 \times 10^{-4}$$

$$\text{Slope log k vs. } 1/T = \frac{y_2 - y_1}{x_2 - x_1} = \frac{-3.86 + 4.58}{3.19 \times 10^{-3} - 3.36 \times 10^{-3}} = -4.24 \times 10^3$$

$$E_a = -2.30 \log k \cdot R \cdot T = -2.30 \cdot \frac{\log k}{T^{-1}} \cdot R$$

$$E_a = -2.30 \times 8.31 \times (-4240) = 8.10 \times 10^4 \ Joules$$

$$\% \ error = \frac{obs - exp}{exp} \times 100\% = \frac{8.1 \times 10^4 - 8.6 \times 10^4}{8.6 \times 10^4} \times 100\%$$

$$= -5.8\%$$

A Sample Lab Report*

I. Title: Chemical Kinetics: An Iodine Clock Reaction

II. Date: 2/3/98

III. Purpose: Determine the initial rate of a chemical reaction using an internal indicator.

Examine the dependence of the initial reaction rate upon the initial concentrations of the reactants.

Find the reaction order, the partial orders, and the experimental rate constant of the reaction.

IV. Theory: The rate of the reaction:

$$3I^- + H_2O_2 + 2H_3O^+ \rightarrow I_3^- + 3H_2O \qquad (1)$$

may be determined by measuring the time required for a fixed amount of $S_2O_3^{2-}$ to react with the I_3^- formed in (1)

$$2S_2O_3^{2-} + I_3^- \rightarrow S_4O_6^{2-} + 3I^- \qquad (2)$$

The moment the $S_2O_2^{2-}$ is all used up, excess I_3^- is free to react with starch to form a dark-blue complex. The coefficients in the rate equation.

$$Rate = k[H_2O_2]^x[H_3O^+]^z[I^-]^y \qquad (3)$$

May be determined by the initial rate method, where the

$$rate = \frac{\Delta[I_3^-]}{\Delta t} = \frac{1}{2}\frac{[S_2O_3^{2-}]}{\Delta t} = \frac{5 \times 10^{-4}\,M}{sec}$$

V. Procedure: Chemical Education Resources, Inc., "Chemical Kinetics: An Iodine Clock Reaction," 1988.**

VI. Results:

Part I Standardization of Hydrogen Peroxide	
Average concentration of H_2O_2	0.10 M

Part II. Reaction Order with Respect to Hydrogen Peroxide
Beaker A   Beaker B

0.05 M KI (mL)	[I⁻] (M)	0.05 M Na₂S₂O₃ (mL)	Buffer1 (mL)	0.1 M H₂O₂ (mL)	[H₂O₂] (M)	Time (s)	Initial rate (M/s)
10.0	0.010	1.0	19.0	20.0	0.040	47	1.0638×10^{-5}
10.0	0.010	1.0	24.0	15.0	0.030	51	9.8039×10^{-6}
10.0	0.010	1.0	28.0	11.0	0.022	60	8.3333×10^{-6}
10.0	0.010	1.0	30.00	9.00	0.018	70	7.1429×10^{-6}
10.0	0.010	1.0	32.0	7.00	0.014	90	5.5556×10^{-6}

Order with respect to H_2O_2 = First order according to Graph for Part 2

X = 1st order = 0.6107   *slope = order = x

$[H_2O_2]$	log $[H_2O_2]$	Time, sec	Rate	Log rate
0.04	−1.397940009	47	1.06383×10^{-05}	−4.973127854
0.03	−1.522878745	51	9.80392×10^{-06}	−5.008600172
0.022	−1.657577319	60	8.33333×10^{-06}	−5.079181246
0.018	−1.744727495	70	7.14286×10^{-06}	−5.146128036
0.014	−1.853871964	90	5.55556×10^{-06}	−5.255272505

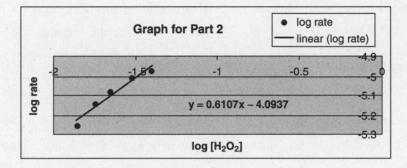

Part III. Reaction Order with Respect to Hydronium Ion								
Beaker A				Beaker B				
0.05 KI (mL)	$[I^-]$ (M)	0.05M $Na_2S_2O_3$ (mL)	Buffer 2 (mL)	0.1 M H_2O_2 (mL)	$[H_2O_2]$ (M)	Time (s)	Initial rate (M/s)	Partial Order
10.0	0.010	1.0	19.0	20.0	0.040	45	1.11111×10^{-5}	0.037761
10.0	0.010	1.0	24.0	15.0	0.030	50	1.00000×10^{-5}	0.0172
10.0	0.010	1.0	28.0	11.0	0.022	75	6.66667×10^{-6}	−0.193819
10.0	0.010	1.0	30.0	9.0	0.018	104	4.80769×10^{-6}	−0.34387
10.0	0.010	1.0	32.0	7.0	0.014	130	3.84615×10^{-6}	−0.31940
Average order with respect to H_3O^+ = 0.1604256								Zero order

Part IV. Reaction Order with Respect to Iodide Ion							
Beaker A				Beaker B			
0.05 MKI (mL)	**[I⁻] (M)**	**0.05 M $Na_2S_2O_3$ (mL)**	**Buffer1 (mL)**	**0.1 M H_2O_2 (mL)**	**[H_2O_2] (M)**	**Time (sec)**	**Initial rate (M/s)**
10.0	0.0100	1.0	19.0	20.0	0.040	35	1.42857×10^{-5}
8.0	0.0080	1.0	21.0	20.0	0.040	52	9.61538×10^{-6}
6.0	0.0060	1.0	23.0	20.0	0.040	62	8.06452×10^{-6}
4.0	0.0040	1.0	25.0	20.0	0.040	71	7.04225×10^{-6}
3.0	0.0030	1.0	26.0	20.0	0.040	140	3.57143×10^{-6}

Partial order with respect to iodide ion = First order according to Graph for Part 4

Y = 1st order = 0.9789 *slope = order = y

[I⁻]	***log [I⁻]***	***time (sec)***	***rate***	***log rate***
0.01	−2	35	1.42857×10^{-05}	−4.84509804
0.008	−2.096910013	52	9.61538×10^{-06}	−5.017033339
0.006	−2.22184875	62	8.06452×10^{-06}	−5.093421685
0.004	−2.397940009	71	7.04225×10^{-06}	−5.152288344
0.003	−2.522878745	140	3.57143×10^{-06}	−5.447158031

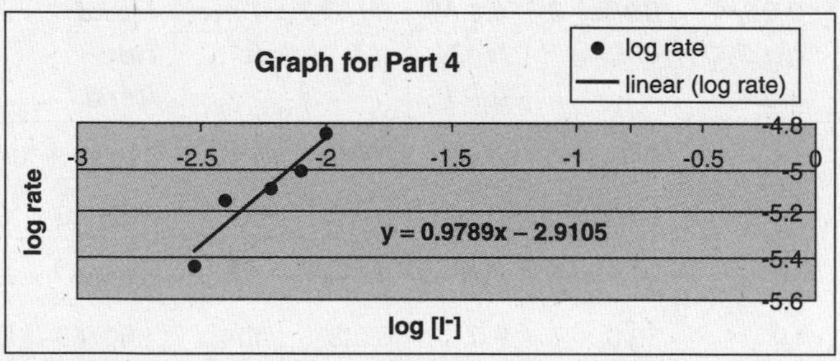

Part V. Overall Reaction Order and the Rate Constant	
Overall reaction order:	Second Order
Average experimental rate constant	7.1467×10^{-4}

VII. Calculations:

II. Reaction Order with Respect to Hydrogen Peroxide

2. Calculate the initial reaction rate for each trial.

$$rate = \frac{\Delta[I_3^-]}{time\,(s)} = \frac{5 \times 10^{-4}M}{s}$$

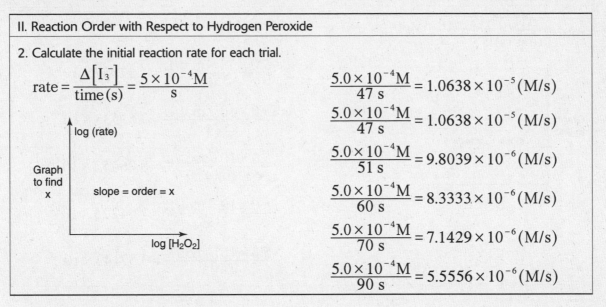

Graph to find x

log (rate)

slope = order = x

log [H$_2$O$_2$]

$$\frac{5.0 \times 10^{-4}M}{47\,s} = 1.0638 \times 10^{-5}\,(M/s)$$

$$\frac{5.0 \times 10^{-4}M}{47\,s} = 1.0638 \times 10^{-5}\,(M/s)$$

$$\frac{5.0 \times 10^{-4}M}{51\,s} = 9.8039 \times 10^{-6}\,(M/s)$$

$$\frac{5.0 \times 10^{-4}M}{60\,s} = 8.3333 \times 10^{-6}\,(M/s)$$

$$\frac{5.0 \times 10^{-4}M}{70\,s} = 7.1429 \times 10^{-6}\,(M/s)$$

$$\frac{5.0 \times 10^{-4}M}{90\,s} = 5.5556 \times 10^{-6}\,(M/s)$$

III. Reaction Order with Respect to Hydronium Ion

1. Calculate the reaction rates as in step 2 of Part II.

$$rate = \frac{\Delta[I_3^-]}{time\,(s)} = \frac{5 \times 10^{-4}M}{s}$$

$$\frac{5.0 \times 10^{-4}M}{45\,s} = 1.1111 \times 10^{-5}$$

$$\frac{5.0 \times 10^{-4}M}{50\,s} = 1.0000 \times 10^{-5}$$

$$\frac{5.0 \times 10^{-4}M}{75\,s} = 6.66667 \times 10^{-6}$$

$$\frac{5.0 \times 10^{-4}M}{104\,s} = 4.80769 \times 10^{-6}$$

$$\frac{5.0 \times 10^{-4}M}{130\,s} = 3.84615 \times 10^{-6}$$

Find z for each of the five reaction mixtures and average the results

0.037761+0.0172+(−.193819)+(−.34387)+ (−.31940) = −.802128

−.802128/5 = −.1604256

zero order

Algebraic solution of order z.

log (rate$_0$ Part II) = log k+z log (1.00 × 10^{-5})

log (rate$_0$ Part III) = log k+z log (3.16 × 10^{-5})

log(1.00 × 10^{-5}) = −5.00

log (3.16 × 10^{-5}) = −4.50

log k = log rate (II) − z (5.0)

log k = log rate (III) − z (4.50)

log rate (II) − z (5.0) = log rate (III) − z (4.50)

$$z(0.5) = \log\left(\frac{rate\ III}{rate\ II}\right)$$

*formula to calcuate z

$$z = \frac{\log\left(\dfrac{rate\ III}{rate\ II}\right)}{0.5}$$

$$^*z = \frac{\log\left(\dfrac{1.1111 \times 10^{-5}}{1.0638 \times 10^{-5}}\right)}{0.5} = 0.037761$$

$$^*z = \frac{\log\left(\dfrac{1.0000 \times 10^{-5}}{9.8039 \times 10^{-6}}\right)}{0.5} = 0.0172$$

$$^*z = \frac{\log\left(\dfrac{6.66667 \times 10^{-6}}{8.3333 \times 10^{-6}}\right)}{0.5} = -0.193819$$

$$^*z = \frac{\log\left(\dfrac{4.80769 \times 10^{-6}}{7.14286 \times 10^{-6}}\right)}{0.5} = -0.34387$$

$$^*z = \frac{\log\left(\dfrac{3.84615 \times 10^{-6}}{5.55556 \times 10^{-6}}\right)}{0.5} = -0.31940$$

IV. Reaction Order with Respect to Iodide Ion

$$\text{rate} = \frac{\Delta\left[I_3^-\right]}{\text{time}\,(s)} = \frac{5 \times 10^{-4}\,M}{s}$$

log (rate)

Graph to find y

slope = order = y

log [I⁻]

$$\frac{5.0 \times 10^{-4}\,\text{mol/L}}{35\,s} = 1.42857 \times 10^{-5}$$

$$\frac{5.0 \times 10^{-4}\,\text{mol/L}}{52\,s} = 9.61538 \times 10^{-6}$$

$$\frac{5.0 \times 10^{-4}\,\text{mol/L}}{62\,s} = 8.06452 \times 10^{-6}$$

$$\frac{5.0 \times 10^{-4}\,\text{mol/L}}{71\,s} = 7.04225 \times 10^{-6}$$

$$\frac{5.0 \times 10^{-4}\,\text{mol/L}}{140\,s} = 3.57143 \times 10^{-6}$$

V. Overall Reaction Order and the Rate Constant

1. Add all the partial orders to obtain the total reaction order.

Total reaction order: $1 + 0 + 1 = 2$

2. Average experimental rate constant using the experimental rate equation:

   $\text{rate} = k[H_2O_2]^1 [H_3O^+]^0 [I^-]^1$ for parts II, III, and IV.

Part II. $\text{rate} = k[H_2O_2]^1$

(1) $1.06383 \times 10^{-5} = k(0.04)$ $k = \dfrac{1.06383 \times 10^{-5}}{0.04} = 2.6595 \times 10^{-4}$ 2.6595×10^{-4}

(2) $9.80392 \times 10^{-6} = k(0.03)$ $k = \dfrac{9.80392 \times 10^{-6}}{0.03} = 3.2680 \times 10^{-4}$ 3.2680×10^{-4}

(3) $8.33333 \times 10^{-6} = k(0.022)$ $k = \dfrac{8.33333 \times 10^{-6}}{0.022} = 3.7878 \times 10^{-4}$ 3.7878×10^{-4}

(4) $7.14286 \times 10^{-6} = k(0.018)$ $k = \dfrac{7.14286 \times 10^{-6}}{0.018} = 3.9682 \times 10^{-4}$

3.9682×10^{-4}

$+3.9682 \times 10^{-4}$

$\overline{0.00176517 \div 5 = 3.53 \times 10^{-4}}$

(5) $5.5556 \times 10^{-6} = k(0.014)$ $k = \dfrac{5.5556 \times 10^{-6}}{0.014} = 3.9682 \times 10^{-4}$ Average k for Part II

3.53×10^{-4}

Part III. $\text{rate} = k[H_3O^+]^0$

(1) $1.11111 \times 10^{-5} = k(0.04)$ $k = \dfrac{1.11111 \times 10^{-5}}{0.04} = 2.7777 \times 10^{-4}$ 2.7777×10^{-4}

(2) $1.0000 \times 10^{-5} = k(0.03)$ $k = \dfrac{1.0000 \times 10^{-5}}{0.03} = 3.3333 \times 10^{-4}$ 3.3333×10^{-4}

(3) $6.66667 \times 10^{-6} = k(0.022)$ $k = \dfrac{6.66667 \times 10^{-6}}{0.022} = 3.0303 \times 10^{-4}$ 3.0303×10^{-4}

(4) $4.80769 \times 10^{-6} = k(0.018)$ $k = \dfrac{4.80769 \times 10^{-6}}{0.018} = 2.6710 \times 10^{-4}$

2.6710×10^{-4}

$+2.7472 \times 10^{-4}$

$\overline{0.00145595 \div 5 = 2.91 \times 10^{-4}}$

(5) $3.84615 \times 10^{-6} = k(0.014)$ $k = \dfrac{3.84615 \times 10^{-6}}{0.014} = 2.7472 \times 10^{-4}$ Average k for Part III

2.91×10^{-4}

Part IV

(1) $1.42857 \times 10^{-5} = k(0.0100)$ $k = \dfrac{1.42857 \times 10^{-5}}{0.0100} = 2.7472 \times 10^{-3}$ 1.4286×10^{-3}

                                          1.2020×10^{-3}

(2) $9.61538 \times 10^{-6} = k(0.0080)$ $k = \dfrac{9.61538 \times 10^{-6}}{0.0080} = 1.2020 \times 10^{-3}$ 1.3441×10^{-3}

                                          1.7606×10^{-3}

(3) $8.06452 \times 10^{-6} = k(0.0060)$ $k = \dfrac{8.06452 \times 10^{-6}}{0.0060} = 1.3441 \times 10^{-3}$ $\underline{+1.7857 \times 10^{-3}}$

                                          7.521×10^{-3}

(4) $7.04225 \times 10^{-6} = k(0.0040)$ $k = \dfrac{7.04225 \times 10^{-6}}{0.0040} = 1.7606 \times 10^{-3}$

(5) $3.57143 \times 10^{-6} = k(0.0020)$ $k = \dfrac{3.57143 \times 10^{-6}}{0.0020} = 1.7857 \times 10^{-3}$ Average k for Part IV

                                          1.50×10^{-3}

The average experimental rate constant:

$$k = \frac{(3.53 \times 10^{-4}) + (2.91 \times 10^{-4}) + (1.5 \times 10^{-3})}{3} = 7.1467 \times 10^{-4}$$

Student from College of the Canyons, Santa Clarita, CA. The lab write-up is provided to show format only. No guarantee is provided for accuracy or quality of answers.

**Chemical Education Resources, Inc., "Chemical Kinetics: Iodine Clock Reaction," Palmyra, PA, 1988.*

Experiment 13: Determination of Enthalpy Changes Associated with a Reaction and Hess's Law

Background: Thermochemistry is the study of heat changes and transfers associated with chemical reactions. In this thermochemical laboratory study, you will determine the enthalpy change that occurs when a strong base, sodium hydroxide, reacts with a strong acid, hydrochloric acid. Other mixtures studied will include ammonium chloride mixed with sodium hydroxide and ammonia mixed with hydrochloric acid. These three reactions are represented as:

$$NaOH_{(aq)} + HCl_{(aq)} \rightarrow NaCl_{(aq)} + H_2O_{(l)}$$

$$NH_4Cl_{(aq)} + NaOH_{(aq)} \rightarrow NH_3_{(aq)} + NaCl_{(aq)} + H_2O_{(l)}$$

$$NH_3_{(aq)} + HCl_{(aq)} \rightarrow NH_4Cl_{(aq)}$$

So that there is no gain nor loss of heat, thermochemical reactions that are called "adiabatic" are carried out in well-insulated containers called calorimeters. The reactants and products in the calorimeter are called the chemical system. Everything else is called the surroundings, and includes the air above the calorimeter, the water in the calorimeter, the thermometer and stirrer, etc.

Hess's law, or the law of constant heat summation, states that at constant pressure, the enthalpy change for a process is not dependent on the reaction pathway, but is dependent only upon the initial and final states of the system. The enthalpy changes of individual steps in a reaction can be added or subtracted to obtain the net enthalpy change for the overall reaction.

An exothermic reaction occurs when the enthalpy of the reactants is greater than the enthalpy of the products, $-\Delta H$ ($\Delta H = \Sigma\Delta H_{f\,products} - \Sigma\Delta H_{f\,reactants}$). In endothermic reactions, the enthalpy of the reactants is lower than that of the products, $+\Delta H$.

Some of the heat transferred to the surroundings during an exothermic reaction are absorbed by the calorimeter and its parts. In order to account for this heat, a calorimeter constant or heat capacity of the calorimeter is required and usually expressed in $J \cdot °C^{-1}$.

Scenario: A student constructed a 'coffee cup calorimeter' (see Figure 1). To determine the heat capacity of the calorimeter, the student placed 50.0 mL of room temperature distilled water in the calorimeter. A calibrated temperature probe recorded the temperature as 23.0°C. The student then added 50.0 mL of warm distilled water (61.0°C) to the calorimeter and recorded the temperature every 30 seconds for the next three minutes. The calorimeter was then emptied and dried. Next, the student measured the temperature change when 50.0 mL of 2.00 M HCl was added to 50.0 mL of 2.00 M NaOH. Again, temperature change was recorded every 30 seconds for three minutes. The same procedure was followed for 2.00 M NH₄Cl with 2.00 M NaOH and finally, 2.00 M NH₃ with 2.00 M HCl.

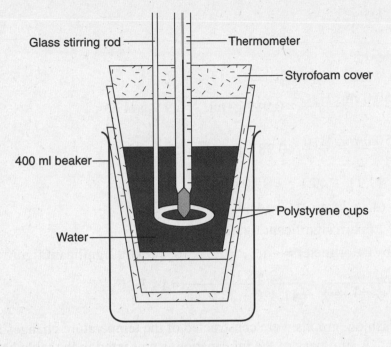

Glass stirring rod — — Thermometer

— Styrofoam cover

400 ml beaker —

— Polystyrene cups

Water —

Figure 1

Analysis:

1. A graph was constructed to determine the heat capacity of the calorimeter (see Figure 2). From the graph, determine the heat capacity of the calorimeter. The specific gravity of water at 23.0°C is 0.998 and at 61.0°C it is 0.983.

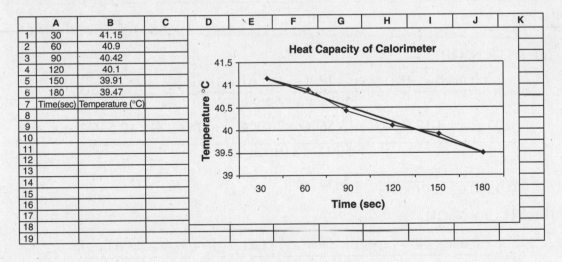

	A	B	C	D	E	F	G	H	I	J	K
1	30	41.15									
2	60	40.9									
3	90	40.42									
4	120	40.1									
5	150	39.91									
6	180	39.47									
7	Time(sec)	Temperature (°C)									
8											
9											
10											
11											
12											
13											
14											
15											
16											
17											
18											
19											

Heat Capacity of Calorimeter

Figure 2

Extrapolating the regression line to the Y axis (0 seconds) gives a temperature of 41.4°C at the moment the room temperature and warm water were mixed.

Average temperature of room temperature and warm water:

$$= \frac{23.0°C + 61.0°C}{2} = 42.0°C$$

$$C_{calorimeter} = \frac{q_{calorimeter}}{(T_{mix} - T_{initial})}$$

$$q_{calorimeter} = -q_{water}$$

$$q_{water} = (\text{mass water}) \cdot (\text{specific heat}) \cdot (T_{mix} - T_{avg})$$

At 23.0°C: $\dfrac{50.0 \text{ mL } H_2O}{1} \times 0.983 \text{ g} \cdot mL^{-1} = 49.9 \text{ g } H_2O$

At 61.0°C: $\dfrac{50.0 \text{ mL } H_2O}{1} \times 0.983 \text{ g} \cdot mL^{-1} = 49.1 \text{ g } H_2O$

Total mass = 49.9 g + 49.1 g = 99.0 g H_2O

$= (99.0 \text{ g}) \cdot (4.18 \text{ J} / \text{g} \cdot °C) \cdot (41.4 °C - 42.0 °C)$

$= -2.5 \times 10^2 \text{ J}$ (extra significant figure carried)

Heat gained by calorimeter $= -q_{water} = 2.5 \times 10^2 \text{ J}$ (extra significant figure carried)

$$C_{calorimeter} = \frac{q_{calorimeter}}{(T_{mix} - T_{initial})} = \frac{2.5 \times 10^2 \text{ J}}{(41.4°C - 23.0°C)} = 14 \text{ J} \cdot °C^{-1}$$

2. In the same fashion, graphs were constructed of the temperature changes for each of the three reactions. A summary of the information is presented in the table below. Calculate the heat evolved in each reaction (kJ/mol of product). Assume the density of each solution = 1.00 g · mL^{-1}.

	$T_{initial}$ (°C)	$T_{at\ mixing}$ (°C)
HCl + NaOH	23.0	35.6
NH₄Cl + NaOH	22.9	24.1
NH₃ + HCl	23.0	33.1

(a) HCl + NaOH

$$q_{rxn} = \frac{-[(\text{mass}_{sol'n}) \times (\text{specific heat}_{sol'n}) \times (\Delta T_{sol'n})] + (C_{calorimeter} \times \Delta T_{sol'n})}{\text{volume}_{sol'n} \times \text{molarity}}$$

$$= \frac{\left[100 \text{ g}_{sol'n} \times 4.18 \text{ J/g} \cdot °C \times (3.56°C - 23.0°C)\right] + (14 \text{ J}/°C \times 12.5°C)}{0.500 \text{ L} \times 2.0 \text{ mole/L}}$$

$$= \frac{-5.4 \times 10^3 \text{ J}}{0.0500 \text{ L} \times 2.0 \text{ mol/L}} = -54 \text{ kJ/mol}$$

(b) NH₄Cl + NaOH

$$q_{rxn} = \frac{-[(\text{mass}_{sol'n}) \times (\text{specific heat}_{sol'n}) \times (\Delta T_{sol'n})] + (C_{calorimeter} \times \Delta T_{sol'n})}{\text{volume}_{sol'n} \times \text{molarity}}$$

$$= \frac{-\left[100. \text{ g}_{sol'n} \times 4.18 \text{ J/g} \cdot °C \times (24.1°C - 22.9°C)\right] + (14 \text{ J/g} \cdot °C \times 1.2°C)}{0.0500 \text{ L} \times 2.0 \text{ mol/L}}$$

$$= \frac{-5.2 \times 10^2 \text{ J}}{0.0500 \text{ L} \times 2.0 \text{ mol/L}} = -5.2 \text{ kJ/mol}$$

(c) $NH_3 + HCl$

$$q_{rxn} = \frac{-[(mass_{sol'n}) \times (specific\ heat_{sol'n}) \times (\Delta T_{sol'n})] + (C_{calorimeter} \times \Delta T_{sol'n})}{volume_{sol'n} \times molarity}$$

$$= \frac{\left[100\,g_{sol'n} \times 4.18\,J/g°C \times (33.1°C - 23.0°C)\right] + (13.6\,J/°C \times 10.1°C)}{0.0500\,L \times 2.0\,mole/L}$$

$$= \frac{-4.4 \times 10^3\,J}{.00500\,L \times 2.0\,mol/L} = -44\,kJ/mol$$

3. Write the net ionic equation, including the ΔH's, for the first two reactions studied and rearrange the equation(s) where necessary to produce the third reaction and its ΔH.

$H^+_{(aq)} + OH^-_{(aq)} \rightarrow H_2O_{(l)}$ $\hspace{2cm}$ $\Delta H = -54\ kJ \cdot mol^{-1}$

$NH_{3(aq)} + H_2O_{(l)} \rightarrow NH_4^+{}_{(aq)} + OH^-_{(aq)}$ $\hspace{1cm}$ $\Delta H = +5.2\ kJ \cdot mol^{-1}$

$NH_{3(aq)} + H^+_{(aq)} \rightarrow NH_4^+{}_{(aq)}$ $\hspace{2.5cm}$ $\Delta H = -49\ kJ \cdot mol^{-1}$

4. Calculate the percent error between the measured ΔH and the calculated ΔH.

$$\%\ error = \frac{obs - exp}{exp} \times 100\% = \frac{-44 - (-49)}{-49} \times 100\% = -10\%$$

Experiment 14: Separation and Qualitative Analysis of Cations and Anions

Background: Qualitative analysis answers the question "What is and what is not present?". Although technology has replaced many of the "wet chemistry" techniques that are employed in these lab procedures, nevertheless, the technology does not replace the understanding that comes from knowing the reactions. The techniques employed in "wet chemistry" such as decanting, filtering, centrifuging, proper methods of determining pH, washing, and so on. are beyond the scope of this review. Furthermore, the detailed qualitative schemes for each reaction, other than being presented in a condensed flowchart style (see Appendix D), are also beyond the scope of a review. Instead, a "Test Tube Mystery" problem will be proposed. The solution of this "mystery" will require you to have mastered the techniques and understanding of some of the reactions in qualitative analysis.

Scenario: Following is a list of solutions used in this "mystery".

> 0.1 M $Ni(NO_3)_2$
>
> 0.1 M $SnCl_4$ in 3 M HCl
>
> 0.1 M $Cu(NO_3)_2$
>
> 0.1 M $Ca(NO_3)_2$
>
> 0.1 M $Al(NO_3)_3$
>
> 0.1 M $AgNO_3$
>
> 3 M H_2SO_4
>
> 6 M NH_3
>
> 6 M NaOH
>
> 6 M HCl

A student was given 10 test tubes labeled 1–10. Each tube contained a different solution from the list above. Samples of the solutions in each tube were mixed in all possible combinations and the results presented in the following table. From the information presented and the use of Appendix D, logically determine the contents of each test tube, write formulas of precipitates where necessary and support your reasoning.

	1	2	3	4	5	6	7	8	9	10
1	X	H	H, S AMMONIA SMELL*	NR	NR	NR	NR	GREEN SOL'N.	NR	P
2	X	X	NR	H	P, D	P GREEN	P, D, H	P BLUE	P	P BROWN
3	X	X	X	H	P WHITE	P, D BLUE	P, D, H	P, D BLUE	NR	P, D BROWN
4	X	X	X	X	NR	NR	NR	NR	NR	P (small P)
5	X	X	X	X	X	NR	NR	NR	NR	NR
6	X	X	X	X	X	X	NR	NR	NR	NR
7	X	X	X	X	X	X	X	NR	NR	P
8	X	X	X	X	X	X	X	X	NR	NR
9	X	X	X	X	X	X	X	X	X	NR
10	X	X	X	X	X	X	X	X	X	X

P = precipitate

D = precipitate dissolves in excess reagent

G = gas

S = smoke

NR = no reaction

H = heat evolved

observed odor in tube 3 before mixing

Analysis:

Test Tube 1: 6 M HCl: Had to be a strong acid since when mixed with tube 3 it got hot and I had already identified tube 3 as NH_3 by odor. When mixed with samples from test tube 2, the reaction produced a great deal of heat, therefore tube 1 had to contain a strong acid. Could be either 6 M HCl or 3 M H_2SO_4. When mixed with test tube 3, a white smoke was produced. I concluded that the smoke was NH_4Cl, therefore, by default, test tube 1 had to contain 6 M HCl. To confirm that this test tube contained HCl, I mixed samples from test tube 1 with test tube 10, which I believed to contain Ag^+ from prior data in the table, and a white precipitate of AgCl was formed.

Test Tube 2: 6 M NaOH: When samples of this test tube were mixed with samples from test tube 1, a great deal of heat was produced. The same occurred when I mixed samples of this tube with test tube 4. Therefore, test tube 2 had to contain a strong base since I had already identified test tube 1 as a strong acid. The only strong base among the choices was 6 M NaOH. When I mixed samples from test tube 2 with samples from test tube 10 which I thought contained Ag^+, a brownish-gray precipitate was formed, indicative of AgOH. When I mixed samples of test tube 2 with samples from test tube 7, which I believed contained Sn^{4+}, a precipitate formed, which I believed to be $Sn(OH)_4$. Upon adding more solution from test tube 2, the precipitate dissolved, consistent with a shift in equilibrium. When I mixed samples from test tube 2 with samples from test tube 6, which I believed contained Ni^{2+}, a green precipitate was produced, $Ni(OH)_2$.

Test Tube 3: 6 M NH₃: I *carefully* smelled all tubes to begin with and this tube was definitely ammonia. Since I was able to rule out this tube initially, I then mixed a small amount of known ammonia with all other tubes. As I mixed the ammonia with the other tubes, I noticed considerable heat being produced in tubes 1, 4 and 7, which I suspected was due to an acid present. With tube 1, smoke appeared, which led me to believe that tube 1 was the HCl which produced a white cloud (smoke) of NH_4Cl. Tubes 1 and 3 were now ruled out. I then noticed precipitates formed in tubes 8 (blue), tube 6 (blue), tube 5 (white), tube 10 (brownish-grey) and tube 7 (white). When excess ammonia was added to tube 10, the precipitate dissolved. When excess ammonia was added to test tube 8, the precipitate dissolved and the solution turned dark blue indicative of Cu^{2+}. When I added excess ammonia to test tube 6, the precipitate dissolved and the solution turned a medium blue color, indicative of the presence of $Ni(NH_3)_6^{2+}$.

Test Tube 4: 3 M H₂SO₄: Test tube 4 had to contain an acid since when I mixed it with test tube 3 which I had concluded contained a base, it got hot. Since I had already identified test tube 1 as 6 M HCl I was left with the choice of this tube containing either the 3 M H_2SO_4 or $SnCl_4$ in 3 M HCl. When I mixed test tube 4 with tube 2 the reaction produced a great amount of heat. Therefore, I concluded that this tube had to contain a strong acid. I was able to rule out the $SnCl_4$ earlier (see test tube 7), therefore this tube must have contained the H_2SO_4.

Test Tube 5: 0.1 M Al(NO₃)₃: When I mixed samples of test tube 5 with a sample from test tube 3, a white precipitate formed. No heat was evident. The white precipitate could have been either $Al(OH)_3$ or $Sn(OH)_4$, however, since no heat was produced and the $Al(NO_3)_3$ did not contain an acid, it can be concluded that test tube 5 contained $Al(NO_3)_3$. When I mixed samples of test tube 5 which I believed contained Al^{3+} with samples of test tube 2 which I believed contained OH^-, a white precipitate formed which I believed to be $Al(OH)_3$. When I added more solution from test tube 2, increasing the OH^- concentration, the precipitate dissolved, consistent with a shift in equilibrium.

Test Tube 6: 0.1 M Ni(NO₃)₂: I was able to identify this tube through observation since the Ni^{2+} ion is green. No other solution in the list should have had a green color. I was able to confirm this when I mixed samples of test tube 6 with test tube 3 (ammonia) producing $Ni(NH_3)_6^{2+}$, a known light blue solution. When I mixed samples from test tube 6 with test tube 4 which I believed to contain OH^-, a green precipitate formed consistent with the color of $Ni(OH)_2$.

Test Tube 7: 0.1 M $SnCl_4$ in 3 M HCl: When I mixed samples from tube 3 (6 M NH_3) with samples from test tube 7, a white precipitate formed and heat was produced. Tube 7 could contain Al^{3+} or Sn^{4+}. Since heat was produced, an acid must have been present. Between the two choices, only $SnCl_4$ was originally mixed with acid, therefore I concluded that test tube 7 contained $SnCl_4$. When I mixed samples of test tube 7 with samples from test tube 2 that I believed contained OH^-, a white precipitate formed, consistent with $Sn(OH)_4$. Upon adding more NaOH, the precipitate dissolved resulting from a shift in equilibrium.

Test Tube 8: 0.1 $Cu(NO_3)_2$: I was able to identify this tube by observation, knowing that the Cu^{2+} ion is blue. No other solution in the list should have a blue color. I confirmed this by forming $Cu(NH_3)_4^{2+}$ (dark blue solution) with samples from test tube 2 that I had initially identified as NH_3.

Test Tube 9: 0.1 M $Ca(NO_3)_2$: When I mixed samples from test tube 9 with samples from test tube 2 that I had already identified as OH^-, a white precipitate was produced consistent with $Ca(OH)_2$.

Test Tube 10: 0.1 M $AgNO_3$: When I mixed samples from test tube 10 with samples from test tube 2 that I had already identified as OH^-, a brownish-gray precipitate consistent with the color of AgOH. Since I had already ruled out test tube 9 and this was the last tube that needed identification, I can conclude that test tube 10 contained Ag^+. Finally, when I mixed the samples of test tube 10, which I believed to contain Ag^+, with samples from test tube 1 which contained what I believed to be HCl, a white precipitate formed consistent with the color of AgCl.

Experiment 15: Synthesis of a Coordination Compound and Its Chemical Analysis

and

Experiment 17: Colorimetric or Spectrophotometric Analysis

Background: Coordination compounds or complexes are an exciting branch of chemistry and area where much research is focused today. There are several types of coordination compounds. One type of coordination compound consists of a cation bonded to a complex ion, often highly colored, i.e., $K_2[Ni(CN)_4]$, potassium tetracyanonickelate(II). The complex ion, the section in square brackets, consists of a centrally located metallic ion, surrounded by ligands. Ligands may be either polar molecules or simple anions. The bonding within this complex ion is through coordinate covalent bonds. Another type of coordination compound forms when neither the ligands nor the central atom has a charge, i.e., $[Cr(CO)_6]$, hexacarbonylchromium (0), or when the central atom does have a charge, i.e. $[Ni(H_2O)_2(NH_3)_4](NO_3)_2$, tetraamminediaquanickel(II) nitrate.

The rate of reaction involving the formation of complex ions is often very rapid. Given this fact, the ion that is formed is usually the most stable, consistent with the principles of equilibrium and subject to shift. Labile reactions are reactions that involve very fast reversible complex ion formation. However, complex ions may not be labile, that is they do not exchange ligands rapidly and are referred to as nonlabile. These complex ions exchange ligands slowly in substitution reactions and may be favored more kinetically than thermodynamically. Adding a catalyst and thereby changing the rate of formation of the product(s) may change the nature of the complex ion produced. An easy method to determine whether a complex ion is labile or not is to note whether a color change occurs in a solution containing the ion when a good complexing ligand is added.

Chemists are involved in both synthesis, that is creating compounds, and analysis, determing the nature of compounds. In this lab, a coordination compound containing Ni^{2+}, NH_3 and Cl^- will be synthesized. Once it has been synthesized, the next step will be to analyze the coordination compound to determine its exact formula.

When the compound that is synthesized is irradiated with white light, some of the light at particular wavelengths will be selectively absorbed. To determine the wavelength(s) of light absorbed, we can expose a solution of the compound to varying wavelengths of monochromatic light and record the responses. If a particular wavelength of light is not absorbed, the intensity of the light directed at the solution (I_o) will match the intensity of the light transmitted by the solution (I_t). For wavelengths of light absorbed, the intensity will be measurably less intense, therefore, $I_t < I_o$.

$$\%T = \frac{I_t}{I_o} \times 100\%$$

The ratio of I_t to I_o can be helpful in identification when unknown concentrations and characteristics are compared to known values. The lowest percent transmittance (%T) is found at the wavelength to which the sample is most sensitive. This wavelength is known as the analytical wavelength, λ_{max}. Once the analytical wavelength has been determined, then it is possible to investigate or determine: 1) concentration of the solution (c); 2) pathway of light through the solution (b); and 3) the molar absorptivity (ε) of the solution. When multiplied together, these three variables provide the absorbance (A) of the solution, as expressed by Beer's law:

$$A = \varepsilon bc$$

The spectrophotometer used has two scales–absorbance (log scale, 0–2) and percent transmittance (linear scale, 0–100). Most readings are taken from the transmittance scale and then converted to absorbance through the relationship

$$A = 2.000 - \log(\%T)$$

Beer's law allows us to determine concentration since absorbance is directly proportional to it. To negate influences due to cuvette differences, or differences between sample tubes, a reference solution is made containing all of the components except the species being analyzed. From this reference point, accurate absorbances can then be determined. This data is then plotted (absorbance vs. known concentrations) and the unknown concentration then extrapolated from the Beer's law plot. Factors that limit this technique include: 1) sensitivity of the instrument being used — usually best between 10 and 90%T, 2) the magnitude of the molar absorptivity (ε), 3) fluctuations due to pH changes, and 4) temperature changes.

Scenario: A student reacted green nickel chloride hexahydrate ($NiCl_2 \cdot 6\,H_2O$) with NH_3. A solid bluish purple solid was produced:

$$Ni^{2+}_{(aq.\ green)} + 2Cl^-_{(aq)} + nNH_{3(aq)} \rightarrow Ni(NH_3)_n Cl_{2\,(s,\ bluish\ purple)}$$

However, NH_3 in water produces a small amount of hydroxide ion:

$$NH_{3(aq)} + H_2O_{(l)} \rightleftharpoons NH^+_{4(aq)} + OH^-_{(aq)} \quad K_b = 1.75 \times 10^{-5}$$

So the student simultaneously produced green $Ni(OH)_2$

$$Ni^{2+}_{(aq)} + 2OH^-_{(aq)} \rightleftharpoons Ni(OH)_{2\,(s)}$$

resulting in an impure synthesis. To maximize the yield of the coordination compound, the products were separated based on their solubilities by heating the products in a small amount of water at 60°C. Further treatment to ensure that no $Ni(OH)_2$ remained involved cooling the solution mixture to 0°C and washing the product with cold ethanol, filtering out the crystals and finally washing them with cold, concentrated NH_3. The crystals were then dried and weighed to determine yield of product.

After the coordination compound was made, it was then analyzed to determine (1) the mass percent of NH_3 in the compound and (2) the mass percent of Ni^{2+} in the compound.

The analysis of NH_3 began by adding excess HCl to the synthesized compound according to the following reaction:

$$Ni(NH_3)_n Cl_{2(s)} + nH_3O^+_{(aq)} \rightleftharpoons nNH_4^+_{(aq)} + Ni^{2+}_{(aq)} + (nH_2O) + 2Cl^-_{(aq)}$$

The student was not told the coordination number of the compound, so "n" was used to represent the number of ammonia ligands surrounding the nickel(II) ion. After the reaction has reached completion, additional HCl was then titrated back with a standardized NaOH solution:

$$H_3O^+_{(aq)} + OH^-_{(aq)} \rightarrow 2H_2O_{(l)}$$

At the equivalence point, the point at which the moles of OH^- equals the moles of H_3O^+, an indicator mixture of bromcresol green and methyl red with a suitable end point, was used to determine the equivalence point (pH of 5.1). The reason that the equivalence point is at 5.1 is that the ammonium ion present as a product hydrolyzes resulting in an acidic solution:

$$NH_4^+_{(aq)} + H_2O_{(l)} \rightleftharpoons NH_3_{(aq)} + H_3O^+_{(aq)}$$

To determine the mass percent of Ni^{2+} in the compound one takes advantage of the fact that the $[Ni(NH_3)_n]^{2+}$ ion absorbs light. To determine the wavelength of light, which this ion absorbs, the analytical wavelength, λ_{max}, was determined. Fifty mL aliquots of standardized solutions of $[Ni(NH_3)_n]^{2+}$ were made up by mixing nickel(II) sulfate hexahydrate in water with excess NH_3 and the absorbances plotted. The reason that $NiSO_4 \cdot 6 H_2O$ was used to provide the $[Ni(NH_3)_n]^{2+}$ was that its molecular weight is known. Furthermore, SO_4^{2-} is less likely to disturb the $[Ni(NH_3)_n]^{2+}$ coordination than would Cl^- and the $NiCl_2 \cdot 6 H_2O$ deliquesces. The absorbing species, $[Ni(NH_3)_n]^{2+}$, is identical in both the known and unknown solutions, therefore, the molar absorptivity, ε, would be the same. The molar ratio of Ni^{2+} between $Ni(NH_3)_n Cl_2$ and $NiSO_4 \cdot 6 H_2O$ is 1:1, making direct comparisons possible.

Data:

I. Synthesis

    a. Mass of $NiCl_2 \cdot 6 H_3O$ used = 7.00 g

    b. Mass of (impure) synthesized $Ni(NH_3)_n Cl_2$ produced = 5.50 g

II. Mass Percent of NH_3 in $Ni(NH_3)_n Cl_2$

    a. molarity of NaOH solution = 0.100 M

    b. molarity of HCl solution = 0.250 M

    c. volume of HCl solution added = 25.0 mL

    d. volume of NaOH solution required = 29.6 mL

    e. mass of $Ni(NH_3)_n Cl_2$ used in titration = 0.130 g

III. Mass Percent of Ni^{2+} in $Ni(NH_3)_nCl_2$

   a. molar mass of $NiSO_4 \cdot 6\,H_2O = 262.88\ g \cdot mol^{-1}$

   b. mass of $NiSO_4 \cdot 6\,H_2O = 0.3000\ g$

   c. mass of $Ni(NH_3)_nCl_2 = 0.3500\ g$

   d. initial color of $NiSO_4$ solution = blue-green

   e. initial color of $Ni(NH_3)_nCl_2$ solution = bluish-purple

   f. final color of $NiSO_4$ after NH_3 added = bluish-purple

   g. final color of $Ni(NH_3)_nCl_2$ after NH_3 added = bluish-purple

IV. %T for the standard $[Ni(NH_3)_n]^{2+}$

Wavelength (nm)	%T of the standard $[Ni(NH_3)_n]^{2+}$ sol'n
540	36.4
560	15.5
580	7.5
600	5.2
620	8.1
640	16.3

Approximate λ_{max} of $[Ni(NH_3)_n]^{2+}$ ion (nm) = 600 nm

%T of the standard $[Ni(NH_3)_n]^{2+}$ solution at 600 nm = 5.2

%T of the unknown $[Ni(NH_3)_n]^{2+}$ = 22.4

V. Analysis

Mass Percent of NH_3 in $Ni(NH_3)_nCl_2$

Recap: Excess HCl was added to the synthesized $Ni(NH_3)_nCl_2$ to form NH_4^+. The solution was then titrated with NaOH to reach an endpoint of 5.1.

1. Determine the number of moles of HCl originally.

$$\frac{25.0\ \cancel{mL\ HCl}}{1} \times \frac{1\ \cancel{L}}{1000\ \cancel{mL}} \times \frac{0.250\ mol\ HCl}{1\ \cancel{L\ sol'n}} = 0.00625\ moles\ HCl$$

2. Determine the number of moles of NaOH added.

$$\frac{29.6\ \cancel{mL\ NaOH}}{1} \times \frac{1\ \cancel{L}}{1000\ \cancel{mL}} \times \frac{0.100\ mol\ NaOH}{1\ \cancel{L\ sol'n}} = 0.00296\ moles\ NaOH$$

3. Determine the number of moles of HCl that remained after the NH_3 was neutralized.

$$\frac{29.6\ \cancel{mL\ NaOH}}{1} \times \frac{1\ \cancel{L}}{1000\ \cancel{mL}} \times \frac{0.100\ \cancel{mol\ NaOH}}{1\ \cancel{L\ sol'n}} \times \frac{1\ mol\ HCl}{1\ \cancel{mol\ NaOH}} = \mathbf{0.00296\ mol\ HCl}$$

4. Determine the number of moles of NH_3 in the $Ni(NH_3)_nCl_2$ sample.

$= 0.00625\ mol\ HCl - 0.00296\ mol\ NaOH = 0.00329\ mol\ HCl$ used to react with the NH_3

$$\frac{0.00329\ \cancel{mol\ HCl}}{1} \times \frac{1\ mol\ NH_3}{1\ \cancel{mol\ HCl}} = 0.00329\ mol\ NH_3$$

5. Calculate the mass of NH_3 in the $Ni(NH_3)_nCl_2$ sample.

$$\frac{0.00329 \text{ mol } NH_3}{1} \times \frac{17.04 \text{ g } NH_3}{1 \text{ mol } NH_3} = 0.0561 \text{ g } NH_3$$

6. Calculate the mass percent of NH_3 in $Ni(NH_3)_nCl_2$.

$$= \frac{0.0561 \text{ g } NH_3}{0.130 \text{ g } Ni(NH_3)_n Cl_2} \times 100\% = 43.2\%$$

Mass Percent of Ni^{2+} in $Ni(NH_3)_nCl_2$

Recap: 50. mL samples of known concentration of $[Ni(NH_3)_n]^{2+}$ were made up by using $NiSO_4 \cdot 6\ H_2O$. The concentrations were then plotted versus their absorbance. The unknown solution's absorbance was then extrapolated from this graph and used in the calculation for the mass percent of Ni^{2+}.

7. Calculate the number of moles of $NiSO_4 \cdot 6\ H_2O$ added to form the standard solution.

$$\frac{0.3000 \text{ g } NiSO_4 \cdot 6\ H_2O}{1} \times \frac{1 \text{ mol } NiSO_4 \cdot 6\ H_2O}{262.88 \text{ g } NiSO_4 \cdot 6\ H_2O} = 0.0041 \text{ mol of } NiSO_4 \cdot 6\ H_2O$$

8. Calculate the number of moles of $Ni(NH_3)_nCl_2$ in the standard solution.

$$\frac{0.001141 \text{ mol of } NiSO_4 \cdot 6\ H_2O}{1} \times \frac{1 \text{ mol } Ni(NH_3)_n Cl_2}{1 \text{ mol of } NiSO_4 \cdot 6\ H_2O}$$

$$= 0.001141 \text{ mol } Ni(NH_3)_nCl_2$$

9. Calculate the concentration of $Ni(NH_3)_nCl_2$ in the standard solution.

$$\frac{0.001141 \text{ mol } Ni(NH_3)_n Cl_2}{50.0 \text{ mL}} \times \frac{1000. \text{ mL}}{1 \text{ L}} = 0.0228 \text{ M}$$

10. Calculate the absorbance of the standard (known) and unknown solutions of $[Ni(NH_3)_n]^{2+}$ from their measured transmittance.

absorbance of standard solution $= 2.000 - \log(32.6) = 0.487$

absorbance of unknown solution of $Ni(NH_3)_nCl_2 = 2.000 - \log(22.4) = 0.650$

11. Calculate the concentration of the unknown solution of $Ni(NH_3)_nCl_2$, given that

$$\frac{A_s}{c_s} = \varepsilon b = \frac{A_{syn}}{c_{syn}}$$

where A_s = absorbance of $[Ni(NH_3)_n]^{2+}$ from standard solution, A_{syn} = absorbance of $[Ni(NH_3)_n]^{2+}$ from synthesized $Ni(NH_3)_nCl_2$, c_s = concentrations of standard solution and c_{syn} = concentrations of synthesized solution.

$$\frac{0.650}{1} \times \frac{0.0228 \text{ mol}}{L} \times \frac{1}{0.487} = 0.0304 \text{ M}$$

12. Calculate the concentration of the $Ni^{2+}_{(aq)}$ ion in the unknown solution.

$$\frac{0.0304 \text{ mol } Ni(NH_3)_n Cl_2}{L} \times \frac{1 \text{ mol } Ni^{2+}}{1 \text{ mol } Ni(NH_3)_n Cl_2} = 0.0304 \text{ M}$$

13. Calculate the mass of $Ni^{2+}_{(aq)}$ ion in 50.0 mL of the unknown solution.

$$\frac{0.0304 \text{ mol } Ni^{2+}}{L} \times \frac{58.69 \text{ g } Ni^{2+}}{1 \text{ mol } Ni^{2+}} \times \frac{50.0 \text{ mL sol'n}}{1} \times \frac{1 \text{ L}}{1000 \text{ mL}} = 0.0892 \text{ g } Ni^{2+}$$

14. Calculate the mass percent of $Ni^{2+}_{(aq)}$ in the unknown solution.

$$\frac{0.0892 \text{ g } Ni^{2+}_{(aq)}}{0.350 \text{ g } Ni(NH_3)_n Cl_2} \times 100\% = 25.5\%$$

Empirical Formula and Percent Yield of $Ni(NH_3)_n Cl_2$

15. Determine the mass percent of Cl^- in $Ni(NH_3)_n Cl_2$

$= 100\% - (\text{mass \% } Ni^{2+}_{(aq)}) (\text{mass \% } NH_3)$

$= 100\% - (25.5\% + 43.1\%) = 31.4\%$

16. Determine the empirical formula of $Ni(NH_3)_n Cl_2$.

mass % Ni^{2+} = 25.5%

mass % NH_3 = 43.1%

mass % Cl^- = 31.4%

Assuming 100. g $Ni(NH_3)_n Cl_2$

$$\frac{25.5 \text{ g } Ni^{2+}}{1} \times \frac{1 \text{ mol } Ni^{2+}}{58.69 \text{ g } Ni^{2+}} = 0.434 \text{ mol } Ni^{2+}$$

$$\frac{43.1 \text{ g } NH_3}{1} \times \frac{1 \text{ mol } NH_3}{17.04 \text{ g } NH_3} = 2.53 \text{ mol } NH_3$$

$$\frac{31.4 \text{ g } Cl^-}{1} \times \frac{1 \text{ mol } Cl^-}{35.45 \text{ g } Cl^-} = 0.886 \text{ mol } Cl^-$$

The Lowest Common Multiplier (LCM) is 0.434, which gives an actual ratio of

$Ni^{2+} \frac{0.434}{0.434} : NH_3 \frac{2.53}{0.434} : Cl^- \frac{0.886}{0.434} = 1.00 : 5.83 : 2.04$

Lowest whole number ration = 1 $Ni^{2+} : 6NH_3 : 2 Cl^-$

Therefore, the empirical formula is $Ni(NH_3)_6 Cl_2$

MM of $Ni(NH_3)_6 Cl_2 = 231.83 \text{ g} \cdot \text{mol}^{-1}$

17. Determine the theoretical yield of product in grams.

$$\frac{7.00 \text{ g } NiCl_2 \cdot 6 H_2O}{1} \times \frac{1 \text{ mol } NiCl_2 \cdot 6 H_2O}{237.71 \text{ g } NiCl_2 \cdot 6 H_2O} = 0.0294 \text{ mol}$$

mols of $NiCl_2 \cdot 6 H_2O$ = moles $Ni(NH_3)_6 Cl_2 = 0.0294$ mol

$$\frac{0.0294 \text{ mol } Ni(NH_3)_6 Cl_2}{1} \times \frac{231.83 \text{ g } Ni(NH_3)_6 Cl_2}{1 \text{ mol } Ni(NH_3)_6 Cl_2} = 6.82 \text{ g } Ni(NH_3)_6 Cl_2$$

18. Determine the actual % yield of the impure product.

$$= \frac{\text{actual yield of product}}{\text{theoretical yield of product}} \times 100\%$$

$$= \frac{5.50 \text{ g } Ni(NH_3)_6 Cl_{2 \text{ impure}}}{6.83 \text{ g } Ni(NH_3)_6 Cl_{2 \text{ pure}}} \times 100\% = 80.5\%$$

Experiment 16: Analytical Gravimetric Determination

Background: Analytical gravimetric analysis is a method for determining the amount of a given substance in a solution by precipitation, filtration, drying, and weighing. The steps generally follow the pattern of weighing the sample, dissolving the sample in an appropriate solvent, forming a precipitate, filtering the precipitate, drying the precipitate and then finally weighing the precipitate. From the data obtained and through mass-mass stoichiometry, the nature of the sample can be determined. In this experiment, a sample of alum, $KAl(SO_4)_2 \cdot 12\ H_2O$, will be analyzed for the sulfate content and compared to the theoretical percent found from the formula.

The precipitate that will be formed in this experiment is barium sulfate. The precipitate is neither typically "curdy" nor gelatinous, but rather forms very fine crystals.

Scenario: A student obtained a Büchner funnel, a piece of Whatman No. 42 filter paper, and a filter flask. She then weighed out 1.059 grams of alum and dissolved it in 50.0 mL of distilled water. She then calculated how much 0.200 M $Ba(NO_3)_2$ she would need to totally precipitate all of the sulfate ion present in solution. She carefully added twice this amount to the alum solution, stirring constantly. She then heated the solution to just under the boiling point for 15 minutes and then allowed the solution to stand overnight. The next day she discovered that fine crystals had appeared in the solution. She weighed the filter paper and determined its mass on an analytical balance as 1.675 g. She then filtered the solution through the Büchner funnel containing the filter paper and collected the crystals. The filter paper was removed and allowed to dry in a drying oven set at 50°C to avoid charring of the paper. She returned later and weighed the paper and crystals and determined their total mass to be 2.715 grams.

Analysis:

1. Calculate how much 0.200 M $Ba(NO_3)_2$ would be needed to totally precipitate all of the sulfate ion present in the alum solution.

$$\frac{1.059\ \text{g KAl(SO}_4)_2 \cdot 12\ \text{H}_2\text{O}}{1} \times \frac{1\ \text{mol KAl(SO}_4)_2 \cdot 12\ \text{H}_2\text{O}}{474.46\ \text{g KAl(SO}_4)_2 \cdot 12\ \text{H}_2\text{O}}$$

$$\times \frac{2\ \text{mol Ba(NO}_3)_2}{1\ \text{g KAl(SO}_4)_2 \cdot 12\ \text{H}_2\text{O}} \times \frac{1\ \text{L Ba(NO}_3)_2}{0.200\ \text{mol Ba(NO}_3)_2} \times \frac{1000\ \text{mL}}{1\ \text{L}}$$

$$= 22.3\ \text{mL}$$

2. What was the percent sulfate ion in the alum based upon the experiment?

2.715 g paper + barium sulfate − 1.675 g paper = 1.040 g $BaSO_4$

$$\frac{1.040 \text{ g } BaSO_4}{1} \times \frac{1 \text{ mol } BaSO_4}{233.37 \text{ g } BaSO_4} \times \frac{1 \text{ mol } SO_4^{2-}}{1 \text{ mol } BaSO_4} \times \frac{96.04 \text{ g } SO_4^{2-}}{1 \text{ mol } SO_4^{2-}} = 0.4281 \text{ g } SO_4^{2-}$$

$$\% \text{ sulfate} = \frac{\text{part}}{\text{whole}} \times 100\% = \frac{0.4281 \text{ g } SO_4^{2-}}{1.059 \text{ g } KAl(SO_4)_2 \cdot 12 \text{ } H_2O} \times 100\% = 40.43\%$$

3. What would the theoretical % sulfate be in the alum?

$$= \frac{192.08 \text{ g } SO_4^{2-}}{474.46 \text{ g } KAl(SO_4)_2 \cdot 12 H_2O} \times 100\% = 40.48\%$$

4. What was the % error in this investigation?

$$\% \text{ error} = \frac{\text{obs} - \text{exp}}{\text{exp}} \times 100\% = \frac{40.43 - 40.48}{40.48} \times 100\% = -0.15\%$$

Experiment 18: Separation by Chromatography

Background: Methods to separate mixtures can be based on many principles such as differences in solubility, boiling and freezing points, polarity, density, and so on. In chromatography, differences in polarity can be used to separate mixtures and identify their components. Molecular sizes and shapes are also used in sieving (gel) and recognition (host-guest) chromatography. Common methods include: 1) paper; 2) thin layer; 3) column; 4) high performance liquid (HPLC); and 5) gas (GC) or vapor phase chromatography (VPC). No matter what method of chromatography is employed, it is a powerful analytical tool that involves a stationary phase (either a liquid or solid) and a mobile phase (either a liquid or gas). The stationary phase attracts the components of the mixture. The medium of the stationary phase may be polar, attracting polar components of the mixture, or it may be nonpolar, attracting nonpolar and excluding polar components. In paper chromatography, the mobile phase or solvent travels through the stationary phase (the paper) depositing components of the mixture along the way on the stationary phase based upon differences in intermolecular interactions. The leading edge of the solvent is known as the solvent front. The movement of the mobile phase, due to capillary action, through the stationary phase and selectively depositing components is characterized by determining a retention or retardation factor (R_f):

$$R_f = \frac{\text{Distance traveled by component (cm)}}{\text{Distance traveled by solvent front (cm)}}$$

R_f values can range from 0.0 which indicates that the component did not travel at all, to a maximum of 1.0, which indicates that the component traveled to the solvent front, resulting in no or little retention. It is best to determine the solvent characteristics for a good separation prior to determining the R_f.

Scenario: A student was given a sample containing a mixture of food dyes. The sample appeared green. The student was also given four commercial grade samples of FDC food dyes including Blue 1, Blue 2, Red 3 and Yellow 5. The student spotted the chromatography paper with the known dyes and the unknown mixture, making sure the paper dried before each application, and then placed the paper into a developing chamber that contained a small amount of eluting solvent (see Figure 1). After the solvent front had reached a point near the top of the paper, the paper was removed, a line was marked to show the solvent front and the paper dried. The chromatogram is shown in Figure 2.

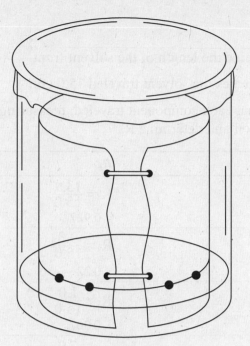

Figure 1

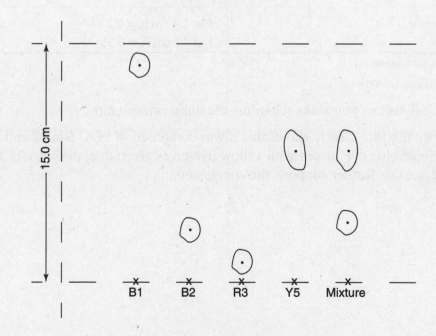

15.0 cm

B1   B2   R3   Y5   Mixture

Figure 2 (not to scale)

Analysis:

1. Using a cm rule, determine the length of the solvent front.

    The chromatogram indicates the solvent traveled 15.0 cm.

2. Measure the distance that each component traveled, measuring from the middle of the spot to the point of application and determine R_f.

Dye	R_f
FDC* Blue 1	$R_f = \dfrac{13.9}{15.0}$ $= 0.927$
FDC Blue 2	$R_f = \dfrac{3.3}{15.0}$ $= 0.22$
FDC Red 3	$R_f = \dfrac{1.0}{15.0}$ $= 0.067$
FDC Yellow 5 **	$R_f = \dfrac{7.9}{15.0}$ $= 0.52$
Unknown Mixture	1 at 3.4 cm $R_f = 0.23$ 1 at 7.8 cm $R_f = 0.53$

*FDC — Food, Drug and Cosmetic Act

**Causes hyperactivity in some children

3. What conclusions can you make regarding the unknown mixture?

    The unknown mixture clearly shows that it was composed of FDC Blue 2 and FDC Yellow 5. Since blue dye mixed with yellow dye gives green dye, the color of the unknown sample, this further supports the conclusion.

Experiment 20: Determination of Electrochemical Series

and

Experiment 21: Measurements Using Electrochemical Cells and Electroplating

Background: When electrons are transferred during the course of a reaction, the reaction is called an oxidation-reduction reaction, or redox reaction for short. The reactant that donates electrons is said to be oxidized and the species that gains electrons is said to be reduced (OIL RIG- Oxidation is Losing, Reduction is Gaining). To illustrate oxidation, examine the half-reaction

$$\text{ox: } Cu_{(s)} \rightarrow Cu^{2+}_{(aq)} + 2e^- \qquad E^o_{ox} = -0.34V$$

Here you can see that the reactant, solid copper, has a 0 oxidation state. It changes to the copper (II) ion by losing 2 electrons and is said to be oxidized. The mass of the solid copper will decrease as the reaction proceeds.

In the reduction reaction

$$\text{red: } Ag^+_{(aq)} + e^- \rightarrow Ag_{(s)} \qquad E^o_{red} = 0.799V$$

the Ag^+ ion is reduced to the silver atom. If the summation of the voltages is positive, the reaction is said to be spontaneous and work can be done. If the summation of the voltages is negative, then work must be done on the system, through the use of a battery, to cause the reaction to occur. In the example above, if a copper wire was placed in a solution of silver nitrate, the copper wire would oxidize to Cu^{2+} ions and silver ions in solution would be reduced and begin to collect on the copper wire as silver atoms.

In Part I of this experiment, different metals will be added to solutions of different aqueous ions to determine whether a spontaneous redox reaction occurs. Also tested will be the colors of three halogens and their corresponding halide ion in mineral oil. Halogens such as $Br_{2(aq)}$, $Cl_2(aq)$ and $I_2(aq)$ dissolve in nonpolar solvents such as mineral oil to give a characteristic color. And finally in Part I, various combinations of these halogens and halides will be mixed and observations recorded.

In Part II, cells will be constructed to determine the relative magnitudes of E and E^0, the determination of a solubility product and a formation constant. In both parts, the relative activity of elements will be determined.

Scenario: Part I

A student obtained a 24-well plate and placed samples of solid metals into solutions of various metallic ions. The results are presented in Table 1 below:

Table 1			
	$Cu_{(s)}$	$Mg_{(s)}$	$Pb_{(s)}$
$Cu^{2+}_{(aq)}$	X	rxn	rxn
$Mg^{2+}_{(aq)}$	no rxn	X	no rxn
$Pb^{2+}_{(aq)}$	no rxn	rxn	X

rxn=reaction occurs

Next, drops of bromine water, chlorine water and iodine water were mixed in samples of mineral oil and the color of the oil and evidence of a reaction were recorded in the Table 2 below. Also tested was the halide ion mixed with mineral oil.

Table 2	
Bromine water $(Br_2)_{(aq)}$ + mineral oil	Orange-brown
Chlorine water $(Cl_2)_{(aq)}$ + mineral oil	Faint greenish-yellow
Iodine water $(I_2)_{(aq)}$ + mineral oil	Pink
$Br^-_{(aq)}$ + mineral oil	Colorless
$Cl^-_{(aq)}$ + mineral oil	Colorless
$I^-_{(aq)}$ + mineral oil	Colorless

Another plate was set up with bromine water, chlorine water and iodine water mixed with samples of $Br^-_{(aq)}$, $Cl^-_{(aq)}$ and $I^-_{(aq)}$ to see if the halogens could be reduced by any of the halide ions. The results of such mixtures are presented in the Table 3 below:

Table 3			
	$Br^-_{(aq)}$	$Cl^-_{(aq)}$	$I^-_{(aq)}$
$Br_{2(aq)}$	no rxn	mineral oil was orangish-brown	mineral oil was pink
$Cl_{2(aq)}$	mineral oil was orangish-brown	no rxn	mineral oil was pink
$I_{2(aq)}$	mineral oil was pink	mineral oil was pink	no rxn

Scenerio: Part II

A student constructed an electrochemical cell as shown in Figure 1:

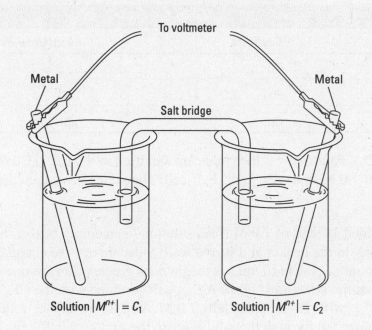

Figure 1

The student began by constructing a half-cell using Zn as a reference electrode in 1.0 M $Zn^{2+}_{(aq)}$. The voltages between the Zn electrode and other electrodes in 1.0 M solutions of their metallic ion at 25°C are presented in Table 4 below. A salt bridge of 1.0 M KNO_3 connected the two half-cells. With the voltmeter set to read a positive voltage, the wire connected to the + terminal on the meter, usually a red wire, was connected to the cathode, the location where reduction was supposedly occurring (RED CAT). The other wire coming from the meter labeled negative and usually black in color, was connected to the anode, the location where oxidation was supposedly occurring (AN OX). However, anode and cathode definitions are for the electrodes in solutions. The anodes produce e^- and the cathodes receive them, so that the signs outside the solutions reverse.

Table 4			
	Voltage	*Anode*	*Cathode*
Zn–Ag	1.40	Zn	Ag
Zn–Cu	0.99	Zn	Cu
Zn–Fe	0.55	Zn	Fe
Zn–Mg	0.60	Mg	Zn
Zn–-Pb	0.48	Zn	Pb

Next, the student used three different combinations of metals for both anodes and cathodes and measured the potential difference. The results are presented in Table 5.

Table 5		
Anode	*Cathode*	*Measured Potential Difference*
Fe	Cu	0.50
Mg	Pb	1.00
Pb	Cu	0.51

A 0.0100 M $Cu(NO_3)_2$ solution was then made up and used in the $Cu^{2+} \mid Cu$ half-cell. The $[Zn^{2+}_{(aq)}]$ was kept at 1.0 M in the $Zn \mid Zn^{2+}$ half-cell. The voltage was then determined and measured as 1.01 volts.

The student then added 10. mL of 1.0 M NaCl solution to an empty beaker. He then added one drop of 1.0 M $AgNO_3$ to the beaker and stirred well. Since there is an abundance of Cl^- in the beaker, and the amount of NaCl (10. mL) is magnitudes greater than the one drop of 1.0 M $AgNO_3$, it can be assumed that most of the $Ag^+_{(aq)}$ will combine with the $Cl^-_{(aq)}$ and that the concentration of the $Cl^-_{(aq)}$ will remain essentially 1.0 M. A silver metal electrode was immersed in this solution and connected through the salt bridge to the $Zn \mid Zn^{2+}$ half-cell. The potential difference was measured as 0.91 volts.

And finally, the same procedure was followed as above, but instead of obtaining an unknown concentration of silver ion, the student prepared a solution containing 2 drops of 1.00 M $Cu(NO_3)_2$ in 10. mL of 6.00 M NH_3. The student determined that it took 20. drops to equal 1 mL. This solution was then added to the cell containing the copper electrode. The voltage was read as 0.56 V. The cell can be represented as:

$$Zn_{(s)} \mid Zn^{2+}_{(aq)} (1.00 \text{ M}) \parallel Cu^{2+}_{(aq)}(? \text{ M}) \mid Cu_{(s)}$$

Analysis:

1. Examine Table 1. Write balanced net ionic equations where reactions occurred and identify the oxidizing agent (OA) and the reducing agent (RA) for each equation.

   $Cu^{2+}_{(aq)} + Mg_{(s)} \rightarrow Cu_{(s)} + Mg^{2+}_{(aq)}$ OA = Cu^{2+}; RA = $Mg_{(s)}$

   $Cu^{2+}_{(aq)} + Pb_{(s)} \rightarrow Cu_{(s)} + Pb^{2+}_{(aq)}$ OA = Cu^{2+}; RA = $Pb_{(s)}$

   $Cu^{2+}_{(aq)} + Zn_{(s)} \rightarrow Cu_{(s)} + Zn^{2+}_{(aq)}$ OA = Cu^{2+}; RA = $Zn_{(s)}$

   $Pb^{2+}_{(aq)} + Mg_{(s)} \rightarrow Pb_{(s)} + Mg^{2+}_{(aq)}$ OA = Pb^{2+}; RA = $Mg_{(s)}$

2. Referring to Table 1, list the metals studied in order of increasing ease of oxidation.

   Copper < Lead < Magnesium

3. Referring to Table 1, list the metallic ions studied in order of increasing ease of reduction.

   $Mg^{2+}_{(aq)} < Pb^{2+}_{(aq)} < Cu^{2+}_{(aq)}$

4. Explain briefly what was occurring in Table 2 and write the balanced net ionic equations for reactions which occurred.

"Like dissolves like" is a fundamental rule of solubility. Mineral oil is nonpolar. Therefore, one would expect nonpolar solutes to dissolve in it. Halide ions are charged particles, therefore one would not expect them to dissolve in a nonpolar solvent like mineral oil, but should dissolve in water, a polar solvent. From the data, it appears that the nonpolar halogens dissolved in the nonpolar mineral oil. When aqueous solutions of halide ions and halogens are mixed together, the color of the mineral oil layer should indicate whether a reaction had occurred or not. If the color of the halogen appeared in the mineral oil from what initially began as the halide ion, then one can conclude that the halide ion was oxidized to the halogen; i.e. $2Br^-_{(aq)}$ (colorless) $\rightarrow Br_{2(\ell)} + 2e^-$ (orangish-brown). The mixtures in which reactions occurred were:

$$Br_{2(aq)} + 2I^-_{(aq)} \rightarrow 2Br^-_{(aq)} + I_{2(aq)}$$
$$Cl_{2(aq)} + 2Br^-_{(aq)} \rightarrow 2Cl^-_{(aq)} + Br_{2(aq)}$$
$$Cl_{2(aq)} + 2I^-_{(aq)} \rightarrow 2Cl^-_{(aq)} + I_{2(aq)}$$

The most reactive halogen was chlorine, followed by bromine and then iodine which appeared to not react. This order agrees with their order of reactivity in the periodic table.

5. In reference to Table 4 above and using a chart of E^0_{red} potentials, determine the potentials had hydrogen been used as a reference electrode instead of the Zn electrode and calculate the differences.

Reduction Rxn	Voltages Using Zn	E^0_{red} Voltages Using H_2 (Chart)	Difference
$Ag^+ + e^- \rightarrow Ag$	1.40	0.80	0.60
$Cu^{2+} + 2e^- \rightarrow Cu$	0.99	0.34	0.65
$Fe^{3+} + 3e^- \rightarrow Fe$	0.55	−0.04	0.59
$Mg^{2+} + 2e^- \rightarrow Mg$	0.60	−2.37	2.97
$Pb^{2+} + 2e^- \rightarrow Pb$	0.48	−0.13	0.61
$Zn^{2+} + 2e^- \rightarrow Zn$	0.00	−0.76	0.76

6. For each reaction studied in Part II, Table 5, write a balanced redox equation.

$$Cu^{2+}_{(aq)} + Fe_{(s)} \rightarrow Cu_{(s)} + Fe^{2+}_{(aq)}$$
$$Pb^{2+}_{(aq)} + Mg_{(s)} \rightarrow Pb_{(s)} + Mg^{2+}_{(aq)}$$
$$Cu^{2+}_{(aq)} + Pb_{(s)} \rightarrow Cu_{(s)} + Pb^{2+}_{(aq)}$$

7. For each reaction in #3 above, predict the potentials from the data obtained and found in Table 4 and compare it to the measured values found in Table 5.

Redox Rxn	Predicted Voltages	Measured Voltage
$Cu^{2+}_{(aq)} + Fe_{(s)} \rightarrow Cu_{(s)} + Fe^{2+}_{(aq)}$	0.99 + (−0.55) = 0.44	0.50
$Pb^{2+}_{(aq)} + Mg_{(s)} \rightarrow Pb_{(s)} + Mg^{2+}_{(aq)}$	0.48 + 0.60 = 1.08	1.00
$Cu^{2+}_{(aq)} + Pb_{(s)} \rightarrow Cu_{(s)} + Pb^{2+}_{(aq)}$	0.99 + (−0.48) = 0.51	0.51

8. Using the Nernst equation, predict the expected voltage of the cell which contained the 0.0100 M $Cu^{2+}_{(aq)}$ solution and compare it to the actual voltage obtained.

$$Zn_{(s)} + Cu^{2+}_{(aq)} \rightarrow Zn^{2+}_{(aq)} + Cu_{(s)} \qquad E° = 1.00\ V$$

$$E = E^0 - \frac{0.0592}{n} \log Q = 1.00V - \frac{0.0592}{2} \log \frac{[Zn^{2+}]}{[Cu^{2+}]}$$

$$= 1.00V - \frac{0.0592}{2} \log \frac{1}{0.0100} = 0.94V$$

This compares favorably to the 1.01 V that was actually measured.

9. Using the data from the cell containing the unknown $Ag^+_{(aq)}$ solution, calculate the K_{sp} of AgCl.

$$Zn_{(s)} + 2Ag^{2+}_{(aq)} \rightarrow Zn^{2+}_{(aq)} + 2Ag_{(s)}\ E° = 1.56V$$

$$ox:\ Zn_{(s)} \rightarrow Zn^{2+}_{(aq)} + 2e^- \qquad E°_{ox} = 0.76\ V$$

$$red:\ Ag^+_{(aq)} + e^- \rightarrow Ag_{(s)} \qquad E°_{red} = 0.80\ V$$

$$E° = E°_{ox} + E°_{red} = 0.76\ V + 0.80\ V + 1.56\ V$$

$$E = E^0 - \frac{0.0592}{n} \log Q = E^0 - \frac{0.0592}{2} \log \frac{[Zn^{2+}]}{[Ag^+]^2}$$

$$0.91 = 1.56 - \frac{0.0592}{2} \log \frac{1}{[Ag^+]^2}$$

$$\log \frac{1}{[Ag^+]^2} = 21.959 \rightarrow \frac{1}{[Ag^+]^2} = 9.1 \times 10^{21} \qquad [Ag^+] = 1.0 \times 10^{-11}$$

$$K_{sp} = [Ag^+][Cl^-] = (1.0 \times 10^{-11}) \times (1) = 1.0 \times 10^{-11}$$

The actual K_{sp} for AgCl is 1.56×10^{-10} at 25°C.

10. From the data on the cell containing the $Cu^{2+}_{(aq)}$ and $NH_{3(aq)}$ mixture, write the reaction and determine the formation constant, K_f and compare it to the accepted value for K_f of $2 \times 10^{13}\ M^{-4}$.

$$Zn_{(s)} + Cu^{2+}_{(aq)} \rightarrow Zn^{2+}_{(aq)} + Cu_{(s)} \quad E = 0.56\ V; \quad E° = 1.10\ V$$

In a 6.0 M NH_3 solution, most of the $Cu^{2+}_{(aq)}$ will combine with NH_3 to form $Cu(NH_3)_4^{2+}{}_{(aq)}$ as evidenced by the magnitude of K_f.

$$Cu^{2+}_{(aq)} + 4NH_{3(aq)} \rightleftharpoons Cu(NH_3)_4^{2+}{}_{(aq)}$$

It is the uncomplexed $Cu^{2+}_{(aq)}$ ion concentration that needs to be determined through the Nernst equation

$$E = E^0 - \frac{0.0592}{n} \log Q = E^0 - \frac{0.0592}{2} \log \frac{[Zn^{2+}]}{[Cu^{2+}]}$$

$$0.56V = 1.10V - \frac{0.0592}{2} \log \frac{1}{[Cu^{2+}]}$$

$$\log \frac{1}{[Cu^{2+}]} = 18.2$$

$$\frac{1}{[Cu^{2+}]} = 1.75 \times 10^{18} M^{-1}$$

$$[Cu^{2+}] = 5.71 \times 10^{-19} M$$

Since 20. drops = 1 mL, 1 drop = 0.050 mL; and 2 drops were added, $V_1 = 0.10$ mL

$$[Cu(NH_3)_4^{2+}] = \frac{V_1 \times M_1}{V_2} = \frac{0.10mL \times 1.0M}{10.1mL} = 0.010M$$

The formation constant, K_f, is derived from the following equation:

$$Cu^{2+} + 4NH_3 \rightleftharpoons Cu(NH_3)_4^{2+}$$

$$K_f = \frac{[Cu(NH_3)_4^{2+}]}{[Cu^{2+}][NH_3]^4} = \frac{0.0099}{(5.71 \times 10^{-19}) \times (6.0)^4} = 1.4 \times 10^{13}$$

$$\% \text{ error} = \frac{obs - exp}{exp} \times 100\% = \frac{1.3 \times 10^{13} - 2 \times 10^{13}}{2 \times 10^{13}} \times 100\% = -3\%$$

Experiment 22: Synthesis, Purification and Analysis of an Organic Compound

Background: When acids react with alcohols, an ester and water is formed:

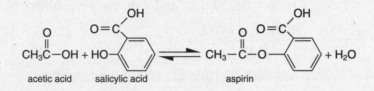

Aspirin, also known as acetylsalicylic acid (ASA), can be synthesized when the carboxyl group in acetic acid (——COOH) reacts with the (—— OH group in the salicylic acid molecule:

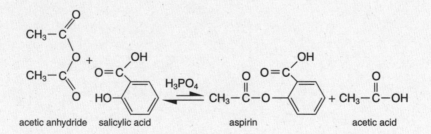

However, the driving force in this reaction is not large, so one usually ends up with an equilibrium mixture of water, salicylic acid, ASA and acetic acid. A better approach to produce ASA (aspirin) is to react acetic anhydride with salicylic acid in the presence of phosphoric or sulfuric acid acting as a catalyst:

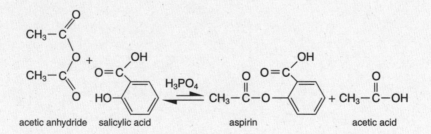

The driving forces in this reaction produce a much higher yield of ASA. One method to determine purity involves determining the melting point of the aspirin produced and comparing it with the known value. If the product is pure, it will have a very distinct melting point. If the sample is impure, there will be a resulting melting range. The final melting point will be lower than the known value by an amount roughly proportional to the amount of impurity present.

A much more definitive method of determining the purity of the aspirin is to analyze the sample through colorimetry. Salicylic acid, not being very soluble in water, is probably the major impurity in the ASA produced. In the presence of Fe^{3+} ion, salicylic acid forms a highly colored magenta or purple complex. By measuring the absorbance of this light and comparing it to known absorbances, it is possible to determine the percent salicylic acid present in the ASA. For a discussion of colorimetry and Beer's law, see Experiments 15 and 17.

Scenario: A student prepared ASA starting with 2.00 g of salicylic acid and 5.00 mL of acetic anhydride (density = 1.08 g · mL^{-1}). After the product was dried, it weighed 1.90 grams. The student then hydrolyzed the ASA with sodium hydroxide and heated the mixture to produce the salicylate dianion:

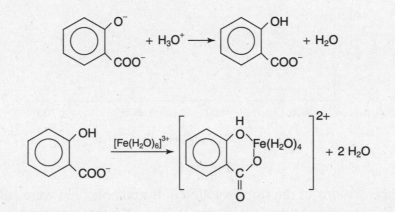

Next, the student then acidified the mixture with a $FeCl_3$–KCl–HCl solution to produce the magenta complex of tetraaquosalicylatoiron (III) ion:

Since the ratio of ASA to the complex ion produced is 1:1, the concentration of the complex ion as determined through colorimetry is the same concentration as that of the ASA. The complex ion is sensitive to pH, therefore care was taken to keep the pH in the range of 0.5–2.0 to avoid formation of di- and trisalicylate complexes of iron (III).

Before the concentration could be determined through colorimetry, the student needed to know the wavelength of light that was most absorbed by the complex ion in order to set the spectrophotometer properly. The student calibrated the spectrophotometer by setting the transmittance to 100% with the $FeCl_3$–KCl–HCl solution as a reference. The optimal wavelength was found to be 525 nm (see Figure 1).

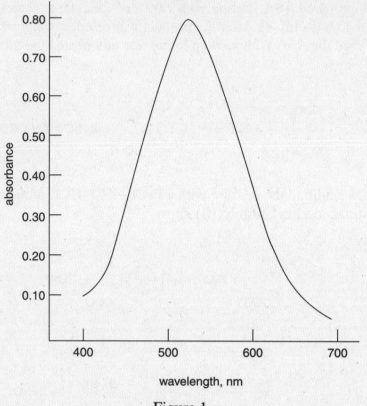

Figure 1

Various known concentrations of the stock solution of the complex ion were colormetrically analyzed and a Beer's law plot drawn (see Figure 2). The ASA that was produced was then treated as described earlier and then colorimetrically analyzed with the concentration determined through extrapolation of the Beer's law plot.

Finally, a small sample of the ASA was determined to have a final melting point of 134°C.

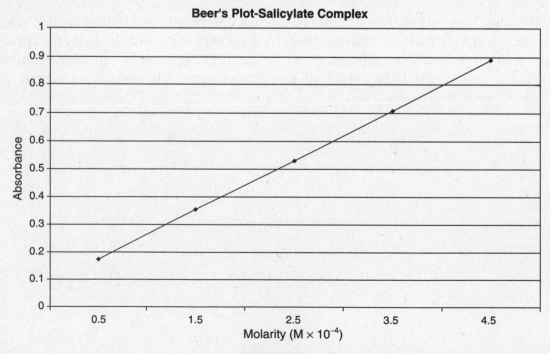

Figure 2

Analysis:

1. Calculate the theoretical yield of ASA to be obtained in this synthesis.

$$\frac{2.00 \text{ g salicylic acid}}{1} \times \frac{1 \text{ mol salicylic acid}}{138.1 \text{ g salicylic acid}} = 0.0145 \text{ mol salicylic acid}$$

$$\frac{5.00 \text{ mL acetic anhydride}}{1} \times \frac{1.08 \text{ g}}{1 \text{ mL}} \times \frac{1 \text{ mol acetic anhydride}}{102.1 \text{ g acetic anhydride}}$$

$$= 0.053 \text{ mol acetic anhydride}$$

Since 1 mol of salicylic acid is required for each mole of acetic anhydride, salicylic acid is the limiting reagent.

$$\frac{0.0145 \text{ mol salicylic acid}}{1} \times \frac{1 \text{ mol ASA}}{1 \text{ mol salicylic acid}} \times \frac{180.17 \text{ g ASA}}{1 \text{ mol ASA}} = 2.61 \text{ g ASA}$$

2. What was the actual percentage yield of the impure product?

$$\% \text{ yield} = \frac{1.90 \text{ g}}{2.61 \text{ g}} \times 100\% = 72.8\% \text{ yield}$$

3. To analyze the purity of the ASA he produced, the student measured out 0.400 g of pure, analytical reagent grade ASA and then treated it with NaOH to create sodium salicylate. He then added a $FeCl_3$–KCl–HCl solution to create a purple salicylate complex. He then diluted this solution to 250. mL with distilled water. Determine the molarity of the stock solution.

$$\frac{0.400 \text{ g ASA}}{0.250 \text{ L}} \times \frac{1 \text{ mol ASA}}{180.17 \text{ g ASA}} = 8.88 \times 10^{-3} \text{ M}$$

4. 5.00, 4.00, 3.00, 2.00 and 1.00 mL samples of the stock solution were then diluted to 100. mL with distilled water. Determine the molarity of each aliquot.

$$\frac{(5.00 \text{ mL stock sol'n}) \cdot (8.88 \times 10^{-3} \text{ M})}{100. \text{ mL standard sol'n}} = 4.44 \times 10^{-4} \text{ M}$$

$$\frac{(4.00 \text{ mL stock sol'n}) \cdot (8.88 \times 10^{-3} \text{ M})}{100. \text{ mL standard sol'n}} = 3.55 \times 10^{-4} \text{ M}$$

$$\frac{(3.00 \text{ mL stock sol'n}) \cdot (8.88 \times 10^{-3} \text{ M})}{100. \text{ mL standard sol'n}} = 2.66 \times 10^{-4} \text{ M}$$

$$\frac{(2.00 \text{ mL stock sol'n}) \cdot (8.88 \times 10^{-3} \text{ M})}{100. \text{ mL standard sol'n}} = 1.78 \times 10^{-4} \text{ M}$$

$$\frac{(1.00 \text{ mL stock sol'n}) \cdot (8.88 \times 10^{-3} \text{ M})}{100. \text{ mL standard sol'n}} = 8.88 \times 10^{-5} \text{ M}$$

5. Samples from these aliquots were then colormetrically analyzed with the spectrophotometer and a Beer's plot drawn (see Figure 2).

To determine the purity of the ASA that was produced, the student then measured out 0.400 g of the ASA that he produced and treated it with NaOH followed by the $Fe-Cl_3-KCl-HCl$ solution and then diluted as before. The transmittance of the Fe^{3+} complex produced from a 5 mL aliquot of ASA synthesized in this experiment was 14%. The reason that the transmittance was taken was because transmittance is a linear scale and the readings were more precisely obtained. Convert the %T to absorbance.

$A = 2.000 - \log (\%T) = 2.000 - \log(14) = 0.854$

6. From the Beer's law plot, determine the concentration of the ASA produced.

From Figure 2, an absorbance of 0.854 corresponds to a 4.27×10^{-4} M concentration of the salicylate complex.

7. Calculate the mass of ASA in the sample produced.

$$= \frac{4.27 \times 10^{-4} \text{ mol ASA}}{L} \times \frac{180.17 \text{ g ASA}}{1 \text{ mol ASA}} \times \frac{100. \text{ mL standard sol'n}}{1}$$

$$\times \frac{1 \text{ L}}{1000 \text{ mL}} \times \frac{250. \text{ mL}}{5.00 \text{ mL}} = 0.385 \text{ g}$$

8. Calculate the percent ASA in the sample produced.

$$\frac{0.385 \text{ g ASA}}{0.400 \text{ g sample}} \times 100\% = 96.3\%$$

9. The melting point of the ASA produced was determined as 134 °C as compared to the known value of 135 °C. Comment on the disparity.

The melting point of the ASA is very close to known literature values. The melting occurred over a very narrow range and was distinct. This confirms a fairly pure synthesis of 96.3%.

Laboratory Manual Resources

1. Vonderbrink, SallyAnn, *Laboratory Experiments for Advanced Placement Chemistry*, Flinn Scientific, Inc., Publishers, Batavia, IL, 1995.

2. Chemical Education Resources, Inc., *Modular Laboratory Program in Chemistry*, Palmyra, PA.

PART IV

AP CHEMISTRY PRACTICE TEST

Answer Sheet for the Practice Test

(Remove This Sheet and Use It to Mark Your Answers)

Section I
Multiple-Choice Questions

1. Ⓐ Ⓑ Ⓒ Ⓓ Ⓔ	26. Ⓐ Ⓑ Ⓒ Ⓓ Ⓔ	51. Ⓐ Ⓑ Ⓒ Ⓓ Ⓔ
2. Ⓐ Ⓑ Ⓒ Ⓓ Ⓔ	27. Ⓐ Ⓑ Ⓒ Ⓓ Ⓔ	52. Ⓐ Ⓑ Ⓒ Ⓓ Ⓔ
3. Ⓐ Ⓑ Ⓒ Ⓓ Ⓔ	28. Ⓐ Ⓑ Ⓒ Ⓓ Ⓔ	53. Ⓐ Ⓑ Ⓒ Ⓓ Ⓔ
4. Ⓐ Ⓑ Ⓒ Ⓓ Ⓔ	29. Ⓐ Ⓑ Ⓒ Ⓓ Ⓔ	54. Ⓐ Ⓑ Ⓒ Ⓓ Ⓔ
5. Ⓐ Ⓑ Ⓒ Ⓓ Ⓔ	30. Ⓐ Ⓑ Ⓒ Ⓓ Ⓔ	55. Ⓐ Ⓑ Ⓒ Ⓓ Ⓔ
6. Ⓐ Ⓑ Ⓒ Ⓓ Ⓔ	31. Ⓐ Ⓑ Ⓒ Ⓓ Ⓔ	56. Ⓐ Ⓑ Ⓒ Ⓓ Ⓔ
7. Ⓐ Ⓑ Ⓒ Ⓓ Ⓔ	32. Ⓐ Ⓑ Ⓒ Ⓓ Ⓔ	57. Ⓐ Ⓑ Ⓒ Ⓓ Ⓔ
8. Ⓐ Ⓑ Ⓒ Ⓓ Ⓔ	33. Ⓐ Ⓑ Ⓒ Ⓓ Ⓔ	58. Ⓐ Ⓑ Ⓒ Ⓓ Ⓔ
9. Ⓐ Ⓑ Ⓒ Ⓓ Ⓔ	34. Ⓐ Ⓑ Ⓒ Ⓓ Ⓔ	59. Ⓐ Ⓑ Ⓒ Ⓓ Ⓔ
10. Ⓐ Ⓑ Ⓒ Ⓓ Ⓔ	35. Ⓐ Ⓑ Ⓒ Ⓓ Ⓔ	60. Ⓐ Ⓑ Ⓒ Ⓓ Ⓔ
11. Ⓐ Ⓑ Ⓒ Ⓓ Ⓔ	36. Ⓐ Ⓑ Ⓒ Ⓓ Ⓔ	61. Ⓐ Ⓑ Ⓒ Ⓓ Ⓔ
12. Ⓐ Ⓑ Ⓒ Ⓓ Ⓔ	37. Ⓐ Ⓑ Ⓒ Ⓓ Ⓔ	62. Ⓐ Ⓑ Ⓒ Ⓓ Ⓔ
13. Ⓐ Ⓑ Ⓒ Ⓓ Ⓔ	38. Ⓐ Ⓑ Ⓒ Ⓓ Ⓔ	63. Ⓐ Ⓑ Ⓒ Ⓓ Ⓔ
14. Ⓐ Ⓑ Ⓒ Ⓓ Ⓔ	39. Ⓐ Ⓑ Ⓒ Ⓓ Ⓔ	64. Ⓐ Ⓑ Ⓒ Ⓓ Ⓔ
15. Ⓐ Ⓑ Ⓒ Ⓓ Ⓔ	40. Ⓐ Ⓑ Ⓒ Ⓓ Ⓔ	65. Ⓐ Ⓑ Ⓒ Ⓓ Ⓔ
16. Ⓐ Ⓑ Ⓒ Ⓓ Ⓔ	41. Ⓐ Ⓑ Ⓒ Ⓓ Ⓔ	66. Ⓐ Ⓑ Ⓒ Ⓓ Ⓔ
17. Ⓐ Ⓑ Ⓒ Ⓓ Ⓔ	42. Ⓐ Ⓑ Ⓒ Ⓓ Ⓔ	67. Ⓐ Ⓑ Ⓒ Ⓓ Ⓔ
18. Ⓐ Ⓑ Ⓒ Ⓓ Ⓔ	43. Ⓐ Ⓑ Ⓒ Ⓓ Ⓔ	68. Ⓐ Ⓑ Ⓒ Ⓓ Ⓔ
19. Ⓐ Ⓑ Ⓒ Ⓓ Ⓔ	44. Ⓐ Ⓑ Ⓒ Ⓓ Ⓔ	69. Ⓐ Ⓑ Ⓒ Ⓓ Ⓔ
20. Ⓐ Ⓑ Ⓒ Ⓓ Ⓔ	45. Ⓐ Ⓑ Ⓒ Ⓓ Ⓔ	70. Ⓐ Ⓑ Ⓒ Ⓓ Ⓔ
21. Ⓐ Ⓑ Ⓒ Ⓓ Ⓔ	46. Ⓐ Ⓑ Ⓒ Ⓓ Ⓔ	71. Ⓐ Ⓑ Ⓒ Ⓓ Ⓔ
22. Ⓐ Ⓑ Ⓒ Ⓓ Ⓔ	47. Ⓐ Ⓑ Ⓒ Ⓓ Ⓔ	72. Ⓐ Ⓑ Ⓒ Ⓓ Ⓔ
23. Ⓐ Ⓑ Ⓒ Ⓓ Ⓔ	48. Ⓐ Ⓑ Ⓒ Ⓓ Ⓔ	73. Ⓐ Ⓑ Ⓒ Ⓓ Ⓔ
24. Ⓐ Ⓑ Ⓒ Ⓓ Ⓔ	49. Ⓐ Ⓑ Ⓒ Ⓓ Ⓔ	74. Ⓐ Ⓑ Ⓒ Ⓓ Ⓔ
25. Ⓐ Ⓑ Ⓒ Ⓓ Ⓔ	50. Ⓐ Ⓑ Ⓒ Ⓓ Ⓔ	75. Ⓐ Ⓑ Ⓒ Ⓓ Ⓔ

CUT HERE

Practice Test

Section I (Multiple-Choice Questions)

Time: 90 minutes

75 questions

45% of total grade

No calculators allowed

This section consists of 75 multiple-choice questions. Mark your answers carefully on the answer sheet.

General Instructions

Do not open this booklet until you are told to do so by the proctor.

Be sure to write your answers for Section I on the separate answer sheet. Use the test booklet for your scratchwork or notes, but remember that no credit will be given for work, notes, or answers written only in the test booklet. Once you have selected an answer, blacken thoroughly the corresponding circle on the answer sheet. To change an answer, erase your previous mark completely, and then record your new answer. Mark only one answer for each question.

Example   Sample Answer

The Pacific is   Ⓐ Ⓑ ● Ⓓ Ⓔ

   (A) a river

   (B) a lake

   (C) an ocean

   (D) a sea

   (E) a gulf

To discourage haphazard guessing on this section of the exam, a quarter of a point is subtracted for every wrong answer, but no points are subtracted if you leave the answer blank. Even so, if you can eliminate one or more of the choices for a question, it may be to your advantage to guess.

Because it is not expected that all test takers will complete this section, do not spend too much time on difficult questions. Answer first the questions you can answer readily, and then, if you have time, return to the difficult questions later. Don't get stuck on one question. Work quickly but accurately. Use your time effectively. The following table is provided for your use in answering questions in Section I.

GO ON TO THE NEXT PAGE

PERIODIC TABLE OF THE ELEMENTS

1 H Hydrogen 1.00797							
3 Li Lithium 6.939	4 Be Beryllium 9.0122						
11 Na Sodium 22.9898	12 Mg Magnesium 24.312						
19 K Potassium 39.102	20 Ca Calcium 40.08	21 Sc Scandium 44.956	22 Ti Titanium 47.90	23 V Vanadium 50.942	24 Cr Chromium 51.996	25 Mn Manganese 54.9380	26 Fe Iron 55.847	27 Co Cobalt 58.9332
37 Rb Rubidium 85.47	38 Sr Strontium 87.62	39 Y Yttrium 88.905	40 Zr Zirconium 91.22	41 Nb Niobium 92.906	42 Mo Molybdenum 95.94	43 Tc Technetium (99)	44 Ru Ruthenium 101.07	45 Rh Rhodium 102.905
55 Cs Cesium 132.905	56 Ba Barium 137.34	57 La Lanthanum 138.91	72 Hf Hafnium 179.49	73 Ta Tantalum 180.948	74 W Tungsten 183.85	75 Re Rhenium 186.2	76 Os Osmium 190.2	77 Ir Iridium 192.2
87 Fr Francium (223)	88 Ra Radium (226)	89 Ac Actinium (227)	104 Rf Rutherfordium (261)	105 Db Dubnium (262)	106 Sg Seaborgium (266)	107 Bh Bohrium (264)	108 Hs Hassium (269)	109 Mt Meitnerium (268)

Lanthanide Series

58 Ce Cerium 140.12	59 Pr Praseodymium 140.907	60 Nd Neodymium 144.24	61 Pm Promethium (145)	62 Sm Samarium 150.35	63 Eu Europium 151.96

Actinide Series

90 Th Thorium 232.038	91 Pa Protactinium (231)	92 U Uranium 238.03	93 Np Neptunium (237)	94 Pu Plutonium (242)	95 Am Americium (243)

						He Helium 4.0026

B Boron 10.811	C Carbon 12.01115	N Nitrogen 14.0067	O Oxygen 15.9994	F Flourine 18.9984	10 Ne Neon 20.183
13 Al Aluminum 26.9815	14 Si Silicon 28.086	15 P Phosphorus 30.9738	16 S Sulfur 32.064	17 Cl Chlorine 35.453	18 Ar Argon 39.948

28 Ni Nickel 58.71	29 Cu Copper 63.546	30 Zn Zinc 65.37	31 Ga Gallium 69.72	32 Ge Germanium 72.59	33 As Arsenic 74.9216	34 Se Selenium 78.96	35 Br Bromine 79.904	36 Kr Krypton 83.80
46 Pd Palladium 106.4	47 Ag Silver 107.868	48 Cd Cadmium 112.40	49 In Indium 114.82	50 Sn Tin 118.69	51 Sb Antimony 121.75	52 Te Tellurium 127.60	53 I Iodine 126.9044	54 Xe Xenon 131.30
78 Pt Platinum 195.09	79 Au Gold 196.967	80 Hg Mercury 200.59	81 Tl Thallium 204.37	82 Pb Lead 207.19	83 Bi Bismuth 208.980	84 Po Polonium (210)	85 At Astatine (210)	86 Rn Radon (222)
110 Uun Unununilium (269)	111 Uuu Unununium (272)	112 Uub Ununbium (277)	113 Uut §	114 Uuq Ununquadium (285)	115 Uup §	116 Uuh Ununhexium (289)	117 Uus §	118 Uuo Ununoctium (293)

64 Gd Gadolinium 157.25	65 Tb Terbium 158.924	66 Dy Dysprosium 162.50	67 Ho Holmium 164.930	68 Er Erbium 167.26	69 Tm Thulium 168.934	70 Yb Ytterbium 173.04	71 Lu Lutetium 174.97
96 Cm Curium (247)	97 Bk Berkelium (247)	98 Cf Californium (251)	99 Es Einsteinium (254)	100 Fm Fermium (257)	101 Md Mendelevium (258)	102 No Nobelium (259)	103 Lr Lawrencium (260)

§ Note: Elements 113, 115, and 117 are not known at this time, but are included in the table to show their expected positions.

Half-reaction	$E°$(V)	Half-reaction	$E°$(V)
$Li^+ + e^- \longrightarrow Li(s)$	−3.05	$Sn^{2+} + 2e^- \longrightarrow Sn(s)$	−0.14
$Cs^+ + e^- \longrightarrow Cs(s)$	−2.92	$Pb^{2+} + 2e^- \longrightarrow Pb(s)$	−0.13
$K^+ + e^- \longrightarrow K(s)$	−2.92	$2H^+ + 2e^- \longrightarrow H_2(g)$	0.00
$Rb^+ + e^- \longrightarrow Rb(s)$	−2.92	$S(s) + 2H^+ + 2e^- \longrightarrow H_2S(g)$	0.14
$Ba^{2+} + 2e^- \longrightarrow Ba(s)$	−2.90	$Sn^{4+} + 2e^- \longrightarrow Sn^{2+}$	0.15
$Sr^{2+} + 2e^- \longrightarrow Sr(s)$	−2.89	$Cu^{2+} + e^- \longrightarrow Cu^+$	0.15
$Ca^{2+} + 2e^- \longrightarrow Ca(s)$	−2.87	$Cu^{2+} + 2e^- \longrightarrow Cu(s)$	0.34
$Na^+ + e^- \longrightarrow Na(s)$	−2.71	$Cu^+ + e^- \longrightarrow Cu(s)$	0.52
$Mg^{2+} + 2e^- \longrightarrow Mg(s)$	−2.37	$I_2(s) + 2e^- \longrightarrow 2 I^-$	0.53
$Be^{2+} + 2e^- \longrightarrow Be(s)$	−1.70	$Fe^{3+} + e^- \longrightarrow Fe^{2+}$	0.77
$Al^{3+} + 3e^- \longrightarrow Al(s)$	−1.66	$Hg_2^{2+} + 2e^- \longrightarrow 2Hg(l)$	0.79
$Mn^{2+} + 2e^- \longrightarrow Mn(s)$	−1.18	$Ag^+ + e^- \longrightarrow Ag(s)$	0.80
$Zn^{2+} + 2e^- \longrightarrow Zn(s)$	−0.76	$Hg^{2+} + 2e^- \longrightarrow Hg(l)$	0.85
$Cr^{3+} + 3e^- \longrightarrow Cr(s)$	−0.74	$2Hg^{2+} + 2e^- \longrightarrow Hg_2^{2+}$	0.92
$Fe^{2+} + 2e^- \longrightarrow Fe(s)$	−0.44	$Br_2(l) + 2e^- \longrightarrow 2 Br^-$	1.07
$Cr^{3+} + e^- \longrightarrow Cr^{2+}$	−0.41	$O_2(g) + 4H^+ + 4e^- \longrightarrow 2 H_2O(l)$	1.23
$Cd^{2+} + 2e^- \longrightarrow Cd(s)$	−0.40	$Cl_2(g) + 2e^- \longrightarrow 2 Cl^-$	1.36
$Tl^+ + e^- \longrightarrow Tl(s)$	−0.34	$Au^{3+} + 3e^- \longrightarrow Au(s)$	1.50
$Co^{2+} + 2e^- \longrightarrow Co(s)$	−0.28	$Co^{3+} + e^- \longrightarrow Co^{2-}$	1.82
$Ni^{2+} + 2e^- \longrightarrow Ni(s)$	−0.25	$F_2(g) + 2e^- \longrightarrow 2 F^-$	2.87

Note: Unless otherwise stated, assume that for all questions involving solutions and/or chemical equations, the system is in water and at room temperature.

Directions: Each group of lettered answer choices below refers to the numbered statements or questions that immediately follow. For each question or statement, select the one lettered choice that is the best answer and fill in the corresponding circle on the answer sheet. An answer choice may be used once, more than once, or not at all in each set of questions.

Questions 1–3

A. F

B. Co

C. Sr

D. Be

E. O

1. Which has the lowest ionization energy?

2. Of those elements with negative oxidation states, which has the fewest such states?

3. Which has the smallest ionic radius?

Questions 4–8

A. SO_2

B. SiH_4

C. CO_2

D. Be_2

E. NO

4. In which of the choices is there polar double bonding in a nonpolar molecule?

5. In which molecule(s) does resonance occur?

6. In which molecule is the bond order $2\frac{1}{2}$?

7. Which of the molecules has four sp^3 hybrid bonds?

8. Which molecule would you expect to be unstable on the basis of molecular orbital theory?

Questions 9–11

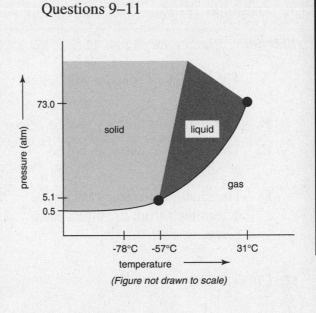

(Figure not drawn to scale)

A. −78°C

B. −57°C, 5.1 atm

C. 31°C, 73 atm

D. 31°C

E. None of the above.

9. What does the phase diagram above show to be the normal boiling point of carbon dioxide?

10. Which point represents the critical point?

11. Which point represents the triple point?

GO ON TO THE NEXT PAGE

Questions 12–16

Directions: *Predict the change in entropy.*

A. The change in entropy will be positive.

B. The change in entropy will be zero.

C. The change in entropy will be negative.

D. The change in entropy cannot be determined from the information given.

12. $Cl_2 \rightarrow 2\ Cl(g)$

13. $H_2(g)$ at 5.0 atm $\rightarrow H_2(g)$ at 1.0 atm

14. Sublimation of solid CO_2

15. $2\ H_2(g) + O_2(g) \rightarrow 2\ H_2O(g)$

16. $PCl_5(g) \leftrightarrow PCl_3(g) + Cl_2(g)$

17. Unknown element X combines with oxygen to form the compound XO_2. If 44.0 grams of element X combine with 8.00 grams of oxygen, what is the atomic mass of element X?

A. 16 amu

B. 44 amu

C. 88 amu

D. 176 amu

E. 352 amu

18. When water evaporates at constant pressure, the sign of the change in enthalpy

A. is negative

B. is positive

C. depends on the temperature

D. depends on the volume of the container

E. does not exist; that is, the enthalpy change is zero

19. Which of the following statements is NOT correct?

A. At constant temperature, the pressure of a certain amount of gas increases with increasing volume.

B. At constant volume, the pressure of a certain amount of gas increases with increasing temperature.

C. At constant pressure, the volume of a certain amount of gas increases with increasing temperature.

D. In dealing with gas laws, the most convenient scale of temperature to use is the Kelvin temperature scale.

E. Equal numbers of molecules of all gases exert about the same pressure at a certain temperature and volume.

20. Within a period, an increase in atomic number is usually accompanied by

A. a decrease in atomic radius and an increase in electronegativity

B. an increase in atomic radius and an increase in electronegativity

C. a decrease in atomic radius and a decrease in electronegativity

D. an increase in atomic radius and a decrease in electronegativity

E. None of these answer choices is correct.

21. A molecule exhibits sp^3d^2 hybridization in its bonding structure. The most probable geometric shape of this molecule is

A. triangular bipyramidal

B. T-shaped

C. octahedral

D. linear

E. hexagonal

22. What is the proper name of $[Co(NH_3)_5Br]Cl_2$?

A. Cobaltpentaamine bromo-dichloride

B. Pentaamminecobalt(III) bromo-dichloride

C. Dichlorocobalt(V) bromodichloride

D. Dichloropentaaminecobalt(III) bromide

E. Pentaaminebromocobalt(III) chloride

23. Which one of the following does NOT show hydrogen bonding?

A. Ammonia, NH_3

B. Hydrazine, N_2H_4

C. Hydrogen peroxide, H_2O_2

D. Dimethyl ether, CH_3OCH_3

E. Methyl alcohol, CH_3OH

24. How many moles of solid $Ca(NO_3)_2$ should be added to 450 milliliters of 0.35 M $Al(NO_3)_3$ to increase the concentration of the NO_3^- ion to 1.7 M? (Assume that the volume of the solution remains constant.)

A. 0.07 mole

B. 0.15 mole

C. 0.29 mole

D. 0.45 mole

E. 0.77 mole

Questions 25–29

$$A(g) + 2B(g) + 3C(g) \rightarrow 4D(g) + 5E(g)$$

rate of formation of $E = \dfrac{d[E]}{dt} = k[A]^2[B]$

25. If one were to double the concentration of B, the rate of the reaction shown above would increase by a factor of

A. ½

B. 1

C. 2

D. 4

E. 8

GO ON TO THE NEXT PAGE

26. $\dfrac{-d[B]}{dt}$ is equal to

   A. $\dfrac{-d[A]}{dt}$

   B. $\dfrac{-d[C]}{dt}$

   C. $\dfrac{-d\frac{1}{2}[D]}{dt}$

   D. $\dfrac{d\frac{1}{5}[E]}{dt}$

   E. none of these

27. To decrease the rate constant k, one could

   A. increase [E]

   B. decrease [B]

   C. decrease the temperature

   D. increase the volume

   E. increase the pressure

28. If one were to reduce the volume of a container by ⅓, the rate of the reaction would increase by a factor of

   A. 3

   B. 9

   C. 16

   D. 27

   E. Reducing the volume of the container has no effect on the rate.

29. Sulfur trioxide gas dissociates into sulfur dioxide gas and oxygen gas at 1250°C. In an experiment, 3.60 moles of sulfur trioxide were placed into an evacuated 3.0-liter flask. The concentration of sulfur dioxide gas measured at equilibrium was found to be 0.20 M. What is the equilibrium constant, K_c, for the reaction?

   A. 1.6×10^{-4}

   B. 1.0×10^{-3}

   C. 2.0×10^{-3}

   D. 4.0×10^{-3}

   E. 8.0×10^{-3}

30. A solution has a pH of 11.0. What is the hydrogen ion concentration?

   A. 1.0×10^{-11} M

   B. 1.0×10^{-3} M

   C. 0.0 M

   D. 1.0×10^{3} M

   E. 1.0×10^{11} M

31.

Species	$\Delta H_f°$ (kJ/mole) at 25°C and 1 atm	$\Delta S°$ (J/mole · K)
$BaCO_3(s)$	−1170.	100.00
$BaO(s)$	−600.	70.00
$CO_2(g)$	−400.	200.00

At what temperature does $\Delta G°$ become zero for the following reaction:

$$BaCO_3(s) \rightarrow BaO(s) + CO_2(g)$$

A. 0 K

B. 1.0×10^1 K

C. 1.0×10^2 K

D. 1.0×10^3 K

E. 1.0×10^4 K

32. Calculate the rate constant for the radioactive disintegration of an isotope that has a half-life of 6930 years.

A. 1.00×10^{-5} yr^{-1}

B. 1.00×10^{-4} yr^{-1}

C. 1.00×10^{-3} yr^{-1}

D. 1.00×10^{3} yr^{-1}

E. 1.00×10^{4} yr^{-1}

Questions 33–37

A. Amide

B. Amine

C. Ketone

D. Thiol

E. Salt

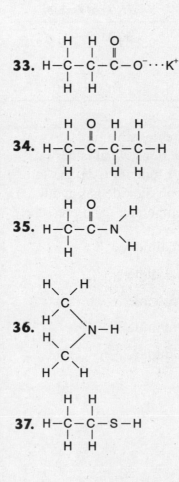

GO ON TO THE NEXT PAGE

38. A mining company supplies an ore that is 15.915% chalcocite, Cu_2S, by weight. How many metric tons of ore should be purchased in order to produce 6.0×10^2 metric tons of an alloy containing 12.709% Cu?

A. 5.0×10^{-1} metric tons

B. 1.0×10^1 metric tons

C. 2.0×10^2 metric tons

D. 3.0×10^2 metric tons

E. 6.0×10^2 metric tons

39. At constant temperature and pressure, the heats of formation of $H_2O(g)$, $CO_2(g)$, and $C_2H_6(g)$ (in kilocalories per mole) are as follows:

Species	ΔH_f (kcal /mole)
$H_2O(g)$	−60.0
$CO_2(g)$	−94.0
$C_2H_6(g)$	−20.0

If ΔH values are negative for exothermic reactions, what is ΔH (in kcal/mole) for 1 mole of C_2H_6 gas to oxidize to carbon dioxide gas and water vapor (same temperature and pressure)?

A. −1340.0 kcal/mole

B. −696.0 kcal/mole

C. −348.0 kcal/mole

D. −134.0 kcal/mole

E. 348.0 kcal/mole

40. Which one of the following is NOT an assumption of the kinetic theory of gases?

A. Gas particles are negligibly small.

B. Gas particles undergo a decrease in kinetic energy when passed from a region of high pressure to a region of low pressure.

C. Gas particles are in constant motion.

D. Gas particles don't attract each other.

E. Gas particles undergo elastic collisions.

41. How many electrons can be accommodated in all the atomic orbitals that correspond to the principal quantum number 4?

A. 2

B. 8

C. 18

D. 32

E. 40

42. A linear molecule can have the general formulas AA, AB, or AB_2. Given a molecule with the general formula AB_2, which one of the following would be the most useful in determining whether the molecule was bent or linear?

A. Ionization energies

B. Electron affinities

C. Dipole moments

D. Electronegativities

E. Bond energies

43. What is the charge of Zn in $Zn(H_2O)_3(OH)^+$?

A. 0

B. +1

C. +2

D. +3

E. +5

44. A certain organic compound has a vapor pressure of 132 mm Hg at 54°C. To determine the vapor pressure of 2.00 moles of the compound at 37°C, taking the heat of vaporization for the compound to be 4.33×10^4 J/mole, you would use

A. the Arrhenius equation

B. the Clausius-Clapeyron equation

C. the combined gas laws

D. the ideal gas law

E. Raoult's law

45. What is the molality of a 10.% (by weight) C_6H_2O (MW = 90.) solution?

A. 0.012 m

B. 0.12 m

C. 1.2 m

D. 12 m

E. Not enough information is provided.

46. If a 10. cm^3 sample of unknown contains 1 cm^3 of 0.1 M $AlCl_3$, then the concentration of Al^{3+} in the unknown is about

A. 0.001 M

B. 0.01 M

C. 0.1 M

D. 1 M

E. 10. M

47. If a reactant concentration is doubled, and the reaction rate increases by a factor of 8, the exponent for that reactant in the rate law should be

A. ¼

B. ½

C. 2

D. 3

E. 4

48. For the reaction

$$2 NO(g) + O_2(g) \rightarrow 2 NO_2(g)$$

which two of the following possible intermediate mechanisms would support this reaction?

1. $2 NO(g) \rightarrow N_2O_2(g)$

2. $NO(g) + O_2(g) \rightarrow NO_3(g)$

3. $2 NO_2(g) \rightarrow N_2O_2(g) + O_2(g)$

4. $NO_3(g) + NO(g) \rightarrow 2 NO_2(g)$

5. $NO(g) \rightarrow NO(g) + O_2(g)$

A. 1 and 2

B. 2 and 3

C. 3 and 4

D. 2 and 4

E. 1 and 4

GO ON TO THE NEXT PAGE

49. The value of K_a for lactic acid, HLac, is 1.5×10^{-5}. What is the value of K_b for the lactate anion, Lac^-?

A. 1.0×10^{-14}

B. 8.5×10^{-10}

C. 6.7×10^{-10}

D. 8.5×10^{10}

E. It cannot be determined from the information provided.

50. How many grams of NaOH are required to neutralize 700 mL of 3.0 N HCl?

A. 2.1 grams

B. 21 grams

C. 42 grams

D. 84 grams

E. 102 grams

51. Solid calcium carbonate decomposes to produce solid calcium oxide and carbon dioxide gas. The value of $\Delta G°$ for this reaction is 130.24 kJ/mole. Calculate ΔG at 100°C for this reaction if the pressure of the carbon dioxide gas is 1.00 atm.

A. −998.56 kJ/mole

B. −604.2 kJ/mole

C. 56.31 kJ/mole

D. 130.24 kJ/mole

E. 256.24 kJ/mole

52. Which of the following choices represents $^{239}_{94}Pu$ producing a positron?

A. $^{239}_{94}Pu \rightarrow\ ^{235}_{94}Pu +\ ^{4}_{2}He$

B. $^{239}_{94}Pu \rightarrow\ ^{0}_{-1}e +\ ^{239}_{93}Np$

C. $^{239}_{94}Pu +\ ^{0}_{-1}e \rightarrow\ ^{239}_{93}Np$

D. $^{239}_{94}Pu +\ ^{0}_{1}e \rightarrow\ ^{239}_{93}Np$

E. $^{239}_{94}Pu +\ ^{4}_{2}He \rightarrow\ ^{235}_{92}U$

53. $Ni(s)\big|Ni^{2}(aq)\big\|Ag^+(aq)\big|Ag(s)$

$Ni^{2+} + 2e^- \rightarrow Ni(s)\ E°_{red} = -0.25\,volt$

$Ag^+ + e^- \rightarrow Ag(s)\ E°_{red} = 0.80\,volt$

Which of the following statements is true of the above reaction?

A. The reaction is spontaneous, $E° = 1.05$ volts.

B. The reaction is nonspontaneous, $E° = -1.05$ volts.

C. The reaction is spontaneous, $E° = -1.05$ volts.

D. The reaction is spontaneous, $E° = 0.55$ volt.

E. The reaction is nonspontaneous, $E° = -0.55$ volt.

54. An excess of $S_8(s)$ is heated with a metallic element until the metal reacts completely. All excess sulfur is combusted to a gaseous compound and escapes from the crucible. Given the information that follows, determine the most probable formula for the residue.

mass of crucible, lid, and metal = 55.00 grams

mass of crucible and lid = 41.00 grams

mass of crucible, lid, and residue = 62.00 grams

A. CuS

B. Cu_2S

C. FeS

D. Fe_2S_3

E. Not enough information is given to solve the problem.

55. When 3.00 grams of a certain metal are completely oxidized, 3.80 grams of its oxide are produced. The specific heat of the metal is 0.052 cal/g · °C. What is the approximate atomic weight of this metal?

A. 35 g/mole

B. 65 g/mole

C. 124 g/mole

D. 150 g/mole

E. 180 g/mole

56. The density of a gas is directly proportional to its

A. pressure

B. volume

C. kinetic energy

D. temperature

E. molecular mass

57. The valence electron configuration of element A is $3s^2 3p^1$ and that of B is $3s^2 3p^4$. What is the probable empirical formula for a compound of the two elements?

A. A_2B

B. AB_2

C. A_3B_2

D. A_2B_3

E. AB

58. Which formula correctly represents the diamminediaquadibromochromium(III) ion?

A. $\left[Cr(H_2O)_2(NH_3)_2Br_2\right]^+$

B. $\left[(NH_3)_2(H_2O)Br_2Cr\right]^{3+}$

C. $\left[Cr(H_2O)_2(NH_3)_2Br_2\right]^{3+}$

D. $\left[(NH_3)_2(H_2O)_2Br_2+Cr\right]^+$

E. $\left[Cr(H_2O)_2(NH_3)Br_2\right]^{2+}$

GO ON TO THE NEXT PAGE

59. According to Raoult's law, which statement is false?

A. The vapor pressure of a solvent over a solution is less than that of the pure solvent.

B. Ionic solids ionize in water, increasing the effect of all colligative properties.

C. The vapor pressure of a pure solvent when measured over a solution decreases as the mole fraction increases.

D. The solubility of a gas increases as the temperature decreases.

E. The solubility of a gas in solution increases as the pressure of the gas increases.

60. Calculate the volume of a 36.45% solution of hydrochloric acid (density = 1.50 g/mL) required to prepare 9.0 liters of a 5.0-molar solution.

A. 0.5 liter

B. 1.0 liter

C. 2.0 liters

D. 2.5 liters

E. 3.0 liters

61. Dinitrogen pentoxide decomposes according to the following balanced equation:

$$N_2O_5(g) \rightarrow 2\ NO_2(g) + \tfrac{1}{2}\ O_2(g)$$

The rate of decomposition was found to be 0.80 mole $\cdot$ liter^{-1} $\cdot$ sec^{-1} at a given concentration and temperature. What would the rate be for the formation of oxygen gas under the same conditions?

A. 0.20 mole $\cdot$ liter^{-1} $\cdot$ sec^{-1}

B. 0.40 mole $\cdot$ liter^{-1} $\cdot$ sec^{-1}

C. 0.80 mole $\cdot$ liter^{-1} $\cdot$ sec^{-1}

D. 1.60 moles $\cdot$ liter^{-1} $\cdot$ sec^{-1}

E. 3.20 moles $\cdot$ liter^{-1} $\cdot$ sec^{-1}

62. The K_{sp} of lead (II) chloride is 2.4×10^{-4}. What conclusion can be made about the concentration of [Cl$^-$] in a solution of lead chloride if [Pb^{2+}] = 1.0 M?

A. [Cl$^-$] can have any value.

B. [Cl$^-$] cannot be greater than $\tfrac{1}{2} K_{sp}$.

C. [Cl$^-$] cannot be less than $\tfrac{1}{2} K_{sp}$.

D. [Cl$^-$] cannot be equal to $\tfrac{1}{2} K_{sp}$.

E. [Cl$^-$] must also be equal to 1.0 M.

Directions: For Questions 63–65, use the following information:

25.0 mL of a sample of vinegar (a solution of $HC_2H_3O_2$, MW = 60.00 g · mole^{-1}) is neutralized by 50.0 mL of a 0.50 N NaOH solution.

63. What is the normality of the acid?

    A. 0.25 N

    B. 0.50 N

    C. 0.75 N

    D. 1.0 N

    E. 2.0 N

64. Calculate the number of grams of acetic acid per liter of the vinegar.

    A. 5.0 grams

    B. 25.0 grams

    C. 30.0 grams

    D. 50.0 grams

    E. 60.0 grams

65. Calculate the weight percentage of acetic acid in the vinegar. The vinegar has a density of 1.0 g/mL.

    A. 1.0%

    B. 1.5%

    C. 2.5%

    D. 6.0%

    E. 6.5%

66. When a solid melts, which of the following is true?

    A. $\Delta H > 0, \Delta S > 0$

    B. $\Delta H < 0, \Delta S < 0$

    C. $\Delta H > 0, \Delta S < 0$

    D. $\Delta H < 0, \Delta S > 0$

    E. More information is required before we can specify the signs of ΔH and ΔS.

67. For the following reaction

$$Zn(s) + 2\,Ag^+(aq) \rightarrow Zn^{2+}(aq) + 2\,Ag(s)$$

the standard voltage E°_{cell} has been calculated to be 1.56 volts. To decrease the voltage from the cell to 1.00 volt, one could

    A. increase the size of the zinc electrode

    B. reduce the coefficients of the reactions so that it reads $1/2\,Zn(s) + Ag^+(aq) \rightarrow 1/2\,Zn^{2+}(aq) + Ag(s)$

    C. decrease the concentration of the silver ion in solution

    D. increase the concentration of the silver ion in solution

    E. decrease the concentration of the zinc ion in solution

GO ON TO THE NEXT PAGE

68. A radioactive isotope has a half-life of 6.93 years and decays by beta emission. Determine the approximate fraction of the sample that is left undecayed at the end of 11.5 years.

   A. 1%

   B. 5%

   C. 30%

   D. 75%

   E. 99%

69. The solubility product constant at 25°C for AgCl is 1.6×10^{-10} mol$^2 \cdot$ L^{-2} and that for AgI is 8.0×10^{-17} mol$^2 \cdot$ L^{-2}. Determine the equilibrium constant for the reaction of silver chloride with I^-(aq).

   A. 1.3×10^{-26} mol$^2 \cdot$ L^{-2}

   B. 5.0×10^{-7} mol$^2 \cdot$ L^{-2}

   C. 1.0×10^{3} mol$^2 \cdot$ L^{-2}

   D. 2.0×10^{6} mol$^2 \cdot$ L^{-2}

   E. 1.3×10^{16} mol$^2 \cdot$ L^{-2}

70.

Species	Bond Energy (kcal/mole)
F–F	33
H–H	103
H–F	135

Calculate the value of ΔH for the following reaction:

$$H_2(g) + F_2(g) \rightarrow 2\ HF(g)$$

   A. −406 kcal/mole

   B. −320 kcal/mole

   C. −271 kcal/mole

   D. −134 kcal/mole

   E. −1.00 kcal/mole

71. What happens to the velocities of different molecules as the temperature of a gas increases?

   A. The velocities of all component molecules increase equally.

   B. The velocity range among different molecules at higher temperatures is smaller than that at lower temperatures.

   C. The effect on the velocities of the molecules depends on whether the pressure remains constant.

   D. The velocity range among different molecules at higher temperatures is wider than the range at lower temperatures.

   E. None of these answer choices are correct.

72. For the isoelectronic series S^{2-}, Cl^-, Ar, K^+, and Sc^{3+}, which species requires the least energy to remove an outer electron?

    A. S^{2-}

    B. Cl^-

    C. Ar

    D. K^+

    E. Sc^{3+}

73. The silver ion in the complex $[Ag(CN)_2]^-$ has a coordination number of

    A. 2

    B. 3

    C. 4

    D. 5

    E. 6

74. Cesium (atomic radius = 0.255 nm) crystallizes with a body-centered cubic unit cell. What is the approximate length of a side of the cell? $\left(\sqrt{3} = 1.73\right)$

    A. 0.4 nm

    B. 0.5 nm

    C. 0.6 nm

    D. 0.8 nm

    E. 0.9 nm

75. How many grams of H_3PO_4 are required to make 100.0 mL of a 0.100 N H_3PO_4 solution (Assume complete or 100% ionization)?

    A. 0.0164 g

    B. 0.164 g

    C. 0.327 g

    D. 0.654 g

    E. 1.31 g

IF YOU FINISH BEFORE TIME IS CALLED, CHECK YOUR WORK ON THIS SECTION ONLY. DO NOT WORK ON ANY OTHER SECTION IN THE TEST.

PERIODIC TABLE OF THE ELEMENTS

1 **H** Hydrogen 1.00797							
3 **Li** Lithium 6.939	**4** **Be** Beryllium 9.0122						
11 **Na** Sodium 22.9898	**12** **Mg** Magnesium 24.312						

19 **K** Potassium 39.102	**20** **Ca** Calcium 40.08	**21** **Sc** Scandium 44.956	**22** **Ti** Titanium 47.90	**23** **V** Vanadium 50.942	**24** **Cr** Chromium 51.996	**25** **Mn** Manganese 54.9380	**26** **Fe** Iron 55.847	**27** **Co** Cobalt 58.9332
37 **Rb** Rubidium 85.47	**38** **Sr** Strontium 87.62	**39** **Y** Yttrium 88.905	**40** **Zr** Zirconium 91.22	**41** **Nb** Niobium 92.906	**42** **Mo** Molybdenum 95.94	**43** **Tc** Technetium (99)	**44** **Ru** Ruthenium 101.07	**45** **Rh** Rhodium 102.905
55 **Cs** Cesium 132.905	**56** **Ba** Barium 137.34	**57** **La** Lanthanum 138.91	**72** **Hf** Hafnium 179.49	**73** **Ta** Tantalum 180.948	**74** **W** Tungsten 183.85	**75** **Re** Rhenium 186.2	**76** **Os** Osmium 190.2	**77** **Ir** Iridium 192.2
87 **Fr** Francium (223)	**88** **Ra** Radium (226)	**89** **Ac** Actinium (227)	**104** **Rf** Rutherfordium (261)	**105** **Db** Dubnium (262)	**106** **Sg** Seaborgium (266)	**107** **Bh** Bohrium (264)	**108** **Hs** Hassium (269)	**109** **Mt** Meitnerium (268)

Lanthanide Series

58 **Ce** Cerium 140.12	**59** **Pr** Praseodymium 140.907	**60** **Nd** Neodymium 144.24	**61** **Pm** Promethium (145)	**62** **Sm** Samarium 150.35	**63** **Eu** Europium 151.96

Actinide Series

90 **Th** Thorium 232.038	**91** **Pa** Protactinium (231)	**92** **U** Uranium 238.03	**93** **Np** Neptunium (237)	**94** **Pu** Plutonium (242)	**95** **Am** Americium (243)

							He Helium 4.0026

		B Boron 10.811	C Carbon 12.01115	N Nitrogen 14.0067	O Oxygen 15.9994	F Flourine 18.9984	10 Ne Neon 20.183
		13 Al Aluminum 26.9815	14 Si Silicon 28.086	15 P Phosphorus 30.9738	16 S Sulfur 32.064	17 Cl Chlorine 35.453	18 Ar Argon 39.948

28 Ni Nickel 58.71	29 Cu Copper 63.546	30 Zn Zinc 65.37	31 Ga Gallium 69.72	32 Ge Germanium 72.59	33 As Arsenic 74.9216	34 Se Selenium 78.96	35 Br Bromine 79.904	36 Kr Krypton 83.80
46 Pd Palladium 106.4	47 Ag Silver 107.868	48 Cd Cadmium 112.40	49 In Indium 114.82	50 Sn Tin 118.69	51 Sb Antimony 121.75	52 Te Tellurium 127.60	53 I Iodine 126.9044	54 Xe Xenon 131.30
78 Pt Platinum 195.09	79 Au Gold 196.967	80 Hg Mercury 200.59	81 Tl Thallium 204.37	82 Pb Lead 207.19	83 Bi Bismuth 208.980	84 Po Polonium (210)	85 At Astatine (210)	86 Rn Radon (222)
110 Uun Ununnilium (269)	111 Uuu Unununium (272)	112 Uub Ununbium (277)	113 Uut §	114 Uuq Ununquadium (285)	115 Uup §	116 Uuh Ununhexium (289)	117 Uus §	118 Uuo Ununoctium (293)

64 Gd Gadolinium 157.25	65 Tb Terbium 158.924	66 Dy Dysprosium 162.50	67 Ho Holmium 164.930	68 Er Erbium 167.26	69 Tm Thulium 168.934	70 Yb Ytterbium 173.04	71 Lu Lutetium 174.97
96 Cm Curium (247)	97 Bk Berkelium (247)	98 Cf Californium (251)	99 Es Einsteinium (254)	100 Fm Fermium (257)	101 Md Mendelevium (258)	102 No Nobelium (259)	103 Lr Lawrencium (260)

§ Note: Elements 113, 115, and 117 are not known at this time, but are included in the table to show their expected positions.

Section ▮ ▮ Free-Response Questions

Equations and Constants

Atomic Structure

$$\Delta E = h\nu \quad c = \lambda\nu \quad \lambda = \frac{h}{mv} \quad p = mv$$

E = energy

ν = frequency (Greek nu)

λ = wavelength

n = principal quantum number

p = momentum

v = velocity (Italic v)

m = mass

speed of light: $c = 3.00 \times 10^8$ m/s

Planck's constant: $h = 6.63 \times 10^{-34}$ joule $\cdot$ s

Boltzmann's constant: $k = 1.38 \times 10^{-23}$ joule/K

Avogadro's number $= 6.022 \times 10^{23}$/mole

electron charge: $e = -1.602 \times 10^{-19}$ coulomb

1 electron volt/atom = 96.5 kilojoules/mole

Equilibrium

$$K_a = \frac{[H^+][A^-]}{[HA]} \quad K_b = \frac{[OH^-][HB^+]}{[B]}$$

$$K_w = [OH^-][H^+] = 10^{-14} \; K_w, \text{ at } 25°C \, (= K_a \cdot K_b)$$

$$pH = -\log[H^+] \quad pOH = -\log[OH^-] \quad 14 = pH + pOH$$

$$pH = pK_a + \log\frac{[A^-]}{[HA]} \quad pOH = pK_b + \log\frac{[HB^+]}{[B]}$$

$$pK_a = -\log K_a \quad pK_b = -\log K_b$$

$$K_p = K_c(RT)^{\Delta n}, \text{ where } \Delta n = \text{moles product gas} - \text{moles reactant gas.}$$

K_a (weak acid) K_p (gas pressure)

K_b (weak base) K_c (molar concentration)

K_w (water)

Thermochemistry

$\Delta S^\circ = \Sigma S^\circ_{f\ products} - \Sigma S^\circ_{f\ reactants}$

$\Delta H^\circ = \Sigma \Delta H^\circ_{f\ products} - \Sigma \Delta H^\circ_{f\ reactants}$

$\Delta G^\circ = \Sigma \Delta G^\circ_{f\ products} - \Sigma \Delta G^\circ_{f\ reactants}$

$\Delta G^\circ = \Delta H^\circ - T\Delta S^\circ = -RT \ln K = -2.303\ RT \log K = -n\mathscr{F}E^\circ$

$\Delta G = \Delta G^\circ + RT \ln Q = \Delta G^\circ + 2.303\ RT \log Q$

$q = m \cdot c \cdot \Delta T$ $\qquad$ $C_p = \Delta H / \Delta T$

S° = standard entropy $\qquad$ n = moles

H° = standard enthalpy $\qquad$ m = mass

G° = standard free energy $\qquad$ q = heat

E° = standard voltage $\qquad$ c = specific heat capacity

T = absolute temperature $\qquad$ C_p = molar heat capacity at constant pressure

Gases, Liquids, and Solutions

$P_A = P_{total} \cdot X_A$, where $X_A = \dfrac{\text{moles A}}{\text{total moles}}$

$P_{total} = P_a + P_b + P_c + \ldots$

$n = \dfrac{m}{M}$

$K = {}^\circ C + 273$

$\dfrac{P_1 V_1}{T_1} = \dfrac{P_2 V_2}{T_2}$

$D = \dfrac{m}{V}$

$u_{rms} = \sqrt{\dfrac{3kT}{m}} = \sqrt{\dfrac{3RT}{M}}$

KE per molecule = $\frac{1}{2}mv^2$ $\qquad$ KE per mole = $\frac{3}{2}RTn$

$\dfrac{r_1}{r_2} = \sqrt{\dfrac{M_2}{M_1}}$

molarity: M = moles solute/liter solution

molality = moles solute/kilogram solvent

$\Delta T_f = i \cdot k_f \cdot \text{molality}$ $\qquad$ $\Delta T_b = i \cdot k_b \cdot \text{molality}$

$\pi = \dfrac{nRT}{V}\, i$

P = pressure

V = volume

T = absolute temperature

n = number of moles

D = density

m = mass

v = velocity

u_{rms} = root mean square velocity

KE = kinetic energy

r = rate of effusion

M = molar mass

π = osmotic pressure

i = van't Hoff factor

k_f = molal freezing-point depression constant

k_b = molal boiling-point elevation constant

Q = reaction quotient

Gas constant: R = 8.31 joules/(mole $\cdot$ K)

$\qquad$ = 0.0821 (liter $\cdot$ atm)/(mole $\cdot$ K)

$\qquad$ = 8.31 (volt $\cdot$ coulomb)/(mole $\cdot$ K)

Boltzmann's constant: $k = 1.38 \times 10^{-23}$ joule/K

k_f for H_2O = 1.86 (K $\cdot$ kg)/mole

k_b for H_2O = 0.512 (K $\cdot$ kg)/mole

STP = 273 K and 1.000 atmospheres

Oxidation-Reduction and Electrochemistry

$$Q = \frac{[C]^c [D]^d}{[A]^a [B]^b} \text{ where } aA + bB \rightarrow cC + dD$$

$$I = \frac{q}{t}$$

$$E_{cell} = E_{cell}^\circ - \frac{RT}{n\mathscr{F}} \ln Q = E_{cell}^\circ - \frac{0.0592}{n} \log Q \text{ at } 25^\circ C$$

$$\log K = \frac{nE^\circ}{0.0592}$$

I = current (amperes)

Q = charge (coulombs)

t = time (seconds)

E° = standard potential

K = equilibrium constant

1 Faraday ($\mathscr{F}$)= 96,500 coulombs/mole

E°_{red} Potentials in Water Solution at 25°C

$Li^+ + e^-$	$\rightarrow$	$Li(s)$	−3.05
$Cs^+ + e^-$	$\rightarrow$	$Cs(s)$	−2.92
$K^+ + e^-$	$\rightarrow$	$K(s)$	−2.92
$Sr^{2+} + 2e^-$	$\rightarrow$	$Sr(s)$	−2.89
$Ca^{2+} + 2e^-$	$\rightarrow$	$Ca(s)$	−2.87
$Na^+ + e^-$	$\rightarrow$	$Na(s)$	−2.71
$Mg^{2+} + 2e^-$	$\rightarrow$	$Mg(s)$	−2.37
$Al^{3+} + 3e^-$	$\rightarrow$	$Al(s)$	−1.66
$Mn^{2+} + 2e^-$	$\rightarrow$	$Mn(s)$	−1.18
$Zn^{2+} + 2e^-$	$\rightarrow$	$Zn(s)$	−0.76
$Fe^{2+} + 2e^-$	$\rightarrow$	$Fe(s)$	−0.44
$Cr^{3+} + e^-$	$\rightarrow$	Cr^{2+}	−0.41
$Ni^{2+} + 2e^-$	$\rightarrow$	$Ni(s)$	−0.25
$Sn^{2+} + 2e^-$	$\rightarrow$	$Sn(s)$	−0.14
$Pb^{2+} + 2e^-$	$\rightarrow$	$Pb(s)$	−0.13
$2 H^+ + 2e^-$	$\rightarrow$	$H_2(g)$	0.00
$S(s) + 2H^+ + 2e^-$	$\rightarrow$	H_2S	0.14
$Sn^{4+} + 2e^-$	$\rightarrow$	Sn^{2+}	0.15
$Cu^{2+} + e^-$	$\rightarrow$	Cu^+	0.16
$Cu^{2+} + 2e^-$	$\rightarrow$	$Cu(s)$	0.34
$Cu^+ + e^-$	$\rightarrow$	$Cu(s)$	0.52
$Fe^{3+} + e^-$	$\rightarrow$	Fe^{2+}	0.77
$Ag^+ + e^-$	$\rightarrow$	$Ag(s)$	0.80
$Hg^{2+} + 2e^-$	$\rightarrow$	$Hg(l)$	0.85
$Br_2(l) + 2e^-$	$\rightarrow$	$2Br^-$	1.07
$O_2(g) + 4H^+ + 4e^-$	$\rightarrow$	$2H_2O$	1.23
$Cl_2(g) + 2e^-$	$\rightarrow$	$2Cl^-$	1.36
$F_2(g) + 2e^-$	$\rightarrow$	$2F^-$	2.87

Water Vapor Pressure

Temperature (°C)	Water Vapor Pressure (mm Hg)
0	4.6
10	9.2
15	12.7
20	17.4
30	31.5
35	41.8
40	55.0
50	92.2
60	149.2

Section II (Free-Response Questions)

Part A: Question 1

Directions: Clearly show the methods used and steps involved in arriving at your answers. It is to your advantage to do this, because you may earn partial credit if you do, and you will receive little or no credit if you do not. Attention should be paid to significant figures. Be sure to write all your answers to the questions on the lined pages following each question the booklet. The Section II score weighting for this question is 20%.

Solve the following problem.

1. Ethylamine reacts with water as follows:

$$C_2H_5NH_2\ (aq) + H_2O(\ell) \rightarrow C_2H_5NH_3^+(aq) + OH^-(aq)$$

The base-dissociation constant, K_b, for the ethylamine ion is 5.6×10^{-4}.

(a) A student carefully measures out 65.987 mL of a 0.250 M solution of ethylamine. Calculate the OH^- ion concentration.

(b) Calculate the pOH of the solution.

(c) Calculate the % ionization of the ethylamine in the solution in part (a).

(d) What would be the pH of a solution made by adding 15.000 grams of ethylammonium bromide ($C_2H_5NH_3Br$) to 250.00 ml of a 0.100-molar solution of ethylamine?

(e) If a student adds 0.125 grams of solid silver nitrate to the solution in part (a), will silver hydroxide form as a precipitate? The value of K_{sp} for silver hydroxide is 1.52×10^{-8}.

Part A: Question 2 or 3

Directions: Answer either Question 2 or 3. Only one of these two questions will be graded. If you start both questions, be sure to cross out the question you do not want graded. The Section II score weighting for the question you choose is 20%.

2. Water is introduced into a test tube that contains 2.51 grams of $SbCl_3$. The products of the reaction are collected, analyzed, and found to be

- 1.906 gram of a solid containing only antimony, oxygen, and chlorine.

- 0.802 grams of a single gas that is found to be 97.20% by weight chlorine and 2.75% by weight hydrogen.

  (a) Determine the simplest formula for the gas.

  (b) What fraction of the chlorine atoms are found in the solid compound, and what fraction are found in the gas phase, after the reaction?

(c) What is the formula of the solid product?

(d) Write a balanced equation for the reaction. Assume that the empirical formula of the gas is the true formula.

3. Methyl alcohol oxidizes to produce methanoic (formic) acid and water according to the following reaction and structural diagram:

$$CH_3OH(aq) + O_2(g) \rightarrow HCOOH(aq) + H_2O(l)$$

$$
\begin{array}{c}
\quad\ \ H \\
\quad\ \ | \\
H-C-O-H+O_2 \rightarrow
\end{array}
\begin{array}{c}
O \\
\parallel \\
H-C-O-H + H_2O \\
\end{array}
$$

Given the following data:

Substance	ΔH_f° (kJ/mole)	S° (J/K · mole)
CH₃OH(aq)	−238.6	129
O₂(g)	0	205.0
HCOOH(aq)	−409	127.0
H₂O(ℓ)	−285.84	69.94

(a) Calculate ΔH° for the oxidation of methyl alcohol.

(b) Calculate ΔS° for the oxidation of methyl alcohol.

(c) (1) Is the reaction spontaneous at 25°C? Explain.

  (2) If the temperature were increased to 100°C, would the reaction be spontaneous? Explain.

(d) The heat of fusion of methanoic acid is 12.7 kJ/mole, and its freezing point is 8.3°C. Calculate ΔS° for the reaction

$$HCOOH(\ell) \rightarrow HCOOH(s)$$

(e) (1) What is the standard molar entropy of HCOOH(s)?

  S° HCOOH(ℓ) = 109.1 J/mole · K.

  (2) Is the magnitude of S° for HCOOH(s) in agreement with the magnitude of S° for HCOOH(ℓ)? Explain.

(f) Calculate ΔG° for the ionization of methanoic acid at 25°C. $K_a = 1.9 \times 10^{-4}$.

GO ON TO THE NEXT PAGE

Section II Free-Response Questions

Part B: Question 4

Directions: No Calculators may be used with Part B. Write the formulas to show the reactants and the products for any FIVE of the laboratory situations described below. Answers to more than five choices will not be graded. In all cases a reaction occurs. Assume that solutions are aqueous unless otherwise indicated. Represent substances in solution as ions if the substances are extensively ionized. Omit formulas for any ions or molecules that are unchanged by the reaction. You need not balance the equations. The Section II score weighting for this question is 15%.

4. *Example:* A strip of magnesium is added to a solution of silver nitrate.

$$Mg + Ag^+ \rightarrow Mg^{2+} + Ag$$

(a) A piece of solid tin is heated in the presence of chlorine gas.

(b) Ethane is burned completely in air.

(c) Solid copper shavings are added to a hot, dilute nitric acid solution.

(d) Dilute sulfuric acid is added to a solution of mercuric nitrate.

(e) Sulfur trioxide gas is heated in the presence of solid calcium oxide.

(f) Copper sulfate pentahydrate is strongly heated.

(g) A strong ammonia solution is added to a suspension of zinc hydroxide.

(h) Ethane gas is heated in the presence of bromine gas to yield a monobrominated product.

Part B: Question 5 and 6

Directions: No Calculators may be used with Part B. Your responses to the rest of the questions in this part of the examination will be graded on the basis of the accuracy and relevance of the information cited. Explanations should be clear and well organized. Examples and equations may be included in your responses where appropriate. Specific answers are preferable to brad, diffuse responses. Answer both Question 5 and 6. Both of these questions will be graded. The Section II score weighting for this question is 30% (15% each).

5. Give a brief explanation for each of the following:

   (a) Water can act either as an acid or as a base.

   (b) HF is a weaker acid than HC1.

   (c) For the triprotic acid H_3PO_4, K_{a1} is 7.5×10^{-3} whereas K_{a2} is 6.2×10^{-8}.

   (d) Pure HCl is not an acid.

   (e) $HClO_4$ is a stronger acid than $HClO_3$, HSO_3^-, or H_2SO_3.

6. Interpret each of the following four examples using modern bonding principles.

   (a) C_2H_2 and C_2H_6 both contain two carbon atoms. However, the bond between the two carbons in C_2H_2 is significantly shorter than that between the two carbons in C_2H_6.

   (b) The bond angle in the hydronium ion, H_3O^+, is less than 109.5°, the angle of a tetrahedron.

   (c) The lengths of the bonds between the carbon and the oxygens in the carbonate ion, CO_3^{2-}, are all equal and are longer than one might expect to find in the carbon monoxide molecule, CO.

   (d) The CNO$^-$ ion is linear.

GO ON TO THE NEXT PAGE

Section II Free-Response Questions

Part B: Question 7 or 8

Directions: No Calculators may be used with Part B. Answer either Question 7 or 8. Only one of these two questions will be graded. If you start both questions, be sure to cross out the question you do not want graded. The Section II score weighting for the question you choose is 15%.

7. If one completely vaporizes a measured amount of a volatile liquid, the molecular weight of the liquid can be determined by measuring the volume, temperature, and pressure of the resulting gas. When using this procedure, one must use the ideal gas equation and assume that the gas behaves ideally. However, if the temperature of the gas is only slightly above the boiling point of the liquid, the gas deviates from ideal behavior. Explain the postulates of the ideal gas equation and explain why, if measured just above the boiling point, the molecular weight deviates from the true value.

8. The boiling points of the following compounds increase in the order in which they are listed below:

   $F_2 < PH_3 < H_2O$

   Discuss the theoretical considerations involved, and use them to account for this order.

IF YOU FINISH BEFORE TIME IS CALLED, CHECK YOUR WORK ON THIS SECTION ONLY. DO NOT WORK ON ANY OTHER SECTION IN THE TEST.

Answers and Explanations for the Practice Test

Answer Key for the Practice Test

Section I (Multiple-Choice Questions)

1. C	**26.** C	**51.** D
2. A	**27.** C	**52.** D
3. D	**28.** D	**53.** A
4. C	**29.** D	**54.** A
5. A or C or E	**30.** A	**55.** C
6. E	**31.** D	**56.** A
7. B	**32.** B	**57.** D
8. D	**33.** E	**58.** A
9. E	**34.** C	**59.** C
10. C	**35.** A	**60.** E
11. B	**36.** B	**61.** B
12. A	**37.** D	**62.** B
13. A	**38.** E	**63.** D
14. A	**39.** C	**64.** E
15. C	**40.** B	**65.** D
16. B	**41.** D	**66.** A
17. D	**42.** C	**67.** C
18. B	**43.** C	**68.** C
19. A	**44.** B	**69.** D
20. A	**45.** C	**70.** D
21. C	**46.** B	**71.** D
22. E	**47.** D	**72.** A
23. D	**48.** D	**73.** A
24. B	**49.** C	**74.** C
25. C	**50.** D	**75.** C

Predicting Your AP Score

The table below shows historical statistical relationships between students' results on the multiple-choice portion (Section I) of the AP chemistry exam and their overall AP score. The AP score ranges from 1 to 5, with 3, 4, or 5 generally considered to be passing. Over the years, around 60% of the students who take the AP chemistry exam receive a 3, 4, or 5.

After you've taken the multiple-choice practice exam under timed conditions, count the number of questions you got correct. From this number, subtract the number of wrong answers $\times \frac{1}{4}$. Do *not* count items left blank as wrong. Then refer to this table to find your "probable" overall AP score. For example, if you get 39 questions correct, based on historical statistics you have a 25% chance of receiving an overall score of 3, a 63% chance of receiving an overall score of 4, and a 12% chance of receiving an overall score of 5. Note that your actual results may be different from the score this table predicts. Also, remember that the free-response section represents 55% of your AP score.

No attempt is made here to combine your specific results on the practice AP chemistry free-response questions (Section II) with your multiple-choice results (which is beyond the scope of this book). However, you should have your AP chemistry instructor review your essays before you take the AP exam so that he or she can give you additional pointers.

Number of Multiple-Choice Questions Correct*	Overall AP Score				
	1	2	3	4	5
47 to 75	0%	0%	1%	21%	78%
37 to 46	0%	0%	25%	63%	12%
24 to 36	0%	19%	69%	12%	0%
13 to 23	15%	70%	15%	0%	0%
0 to 12	86%	14%	0%	0%	0%
Percent of Test Takers Receiving Score	21%	22%	25%	15%	17%

*Corrected for wrong answers

Answers and Explanations for the Practice Test

Section I (Multiple-Choice Questions)

1. (C) Shielding and large ionic radius minimize electrostatic attraction.

2. (A) F has only one negative oxidation state (−1).

3. (D) Be^{2+} now has electrons in the first energy level only.

4. (C) $\ddot{O}=C=\ddot{O}$

Oxygen is more electronegative than carbon, resulting in polar bonding. Because there are no unshared pairs of electrons for carbon, a linear molecule results.

5. (A) $:\!\ddot{O}\!\cdot\!\diagup\!\overset{\ddot{S}}{\diagdown}\!\ddot{O}\cdot \longrightarrow \cdot\ddot{O}\!\diagup\!\overset{\ddot{S}}{\diagdown}\!\ddot{O}\!:$

There are three molecules listed that exhibit resonance:

$$:\!\ddot{O}\!\cdot\!\diagup\!\overset{\ddot{S}}{\diagdown}\!\ddot{O}\!: \longleftrightarrow :\!\ddot{O}\!\cdot\!\diagup\!\overset{\ddot{S}}{\diagdown}\!\ddot{O}\!: \qquad \ddot{O}=C=\ddot{O} \longleftrightarrow :\ddot{O}-C\equiv O: \longleftrightarrow :O\equiv C-\ddot{O}: \qquad \cdot\dot{N}=\ddot{O}: \longleftrightarrow :\dot{N}=\ddot{O}\cdot$$

6. (E) $\text{bond order} = \dfrac{\substack{\text{number of}\\ \text{bonding electrons}} - \substack{\text{number of}\\ \text{antibonding electrons}}}{2}$

or bond order $= (\text{total valence electrons} - \text{non-bonding electrons}) \times \frac{1}{2}$

$$= (11 - 6) \times \frac{1}{2} = 2\frac{1}{2}$$

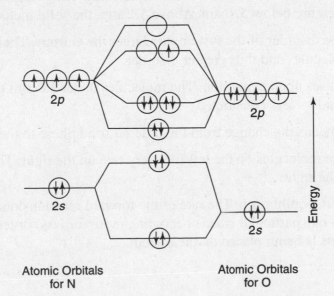

Atomic Orbitals for N   Atomic Orbitals for O

7. (B)

H
|
Si
H H H
|
H

Silicon, in order to bond four hydrogen atoms to itself, must exhibit sp^3 hybridization.

8. (D) Two of the four valence electrons would go to the σ_{2s} bonding orbital, and the other two would go to the σ_{2s}^* antibonding orbital. The electron configuration would be $\left(\sigma_{2s}\right)^2\left(\sigma_{2s}^*\right)^2$.

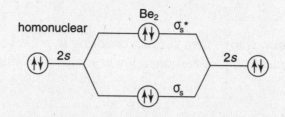

9. (E) "Normal" means 1 atm (760 mm Hg) pressure. Boiling occurs at a temperature at which the substance's vapor pressure becomes equal to the pressure above its surface. On this phase diagram, at 1 atm pressure, there is no intercept on a line separating the liquid phase from the gas phase. In other words, carbon dioxide cannot be liquefied at 1 atm pressure. It is in the liquid form only under very high pressures. At 1.0 atm pressure, solid CO_2 will sublime — that is, go directly to the gas phase.

10. (C) The critical point is the point at which the liquid — gas curve ends at a point at which the temperature and pressure have their critical values. Critical temperature is the temperature above which the liquid state of a substance no longer exists. Critical pressure is the pressure at the critical temperature.

11. (B) All three phases are in equilibrium at the triple point. The solid CO_2 sublimes if warmed at any pressure below 5.1 atm. Above 5.1 atm, the solid melts if warmed.

12. (A) The greater the disorder of the system, the larger the entropy. There is an increase in the number of molecules and thus greater disorder.

13. (A) Entropy increases upon expansion. The molecules under 1.0 atm of pressure are more free to move around — less constricted.

14. (A) Sublimation means the change from the ordered solid phase to the random gas phase.

15. (C) There are three molecules on the left for every two on the right. Things are becoming more ordered on the right.

16. (B) The system is in equilibrium. The rate of the forward reaction equals the rate of the reverse reaction. No one particular side is becoming more (or less) ordered than the other. No additional stress is being placed on the system.

17. (D) 8.00 g of oxygen *atoms* represent 0.500 mole.

$$\frac{8.00 \,\text{g O}}{0.500 \,\text{mole O}} \times \frac{2 \,\text{moles O}}{1 \,\text{mole X}} \times \frac{44.0 \,\text{g X}}{8.00 \,\text{g O}} = 176 \,\text{g X} / \text{mole X}$$

18. (B) Remember that in an endothermic process, energy is being absorbed. All endothermic changes are defined with a + sign. Going from the liquid to the gaseous phase requires energy and thus is endothermic.

19. (A) Remember Boyle's law: As the volume decreases (at constant temperature), the pressure increases.

20. (A) The atomic radius decreases because of increasing effective nuclear charge and electrostatic attraction. There are more protons and electrons, so electrons are needed to create a complete shell; thus, there is an increase in electronegativity.

21. (C) Refer to the table entitled "Geometry and Hybridization Patterns," page 82.

22. (E) NH_3 is a neutral ligand; the bromide and the chloride ion both have a −1 charge. Cobalt would have to have a +3 charge in this compound for the complex compound to be electrically neutral.

Naming Complex Compounds

1. Name the cation before the anion.

2. Name the ligands before the metal ion.

3. For the ligands:

    a. Add −o to the root name of the anion — in this case, bromo.

    b. Use aquo for H_2O, amine for NH_3, carbonyl for CO, nitrosyl for NO.

    c. Use the name of the ligand for other neutral ligands.

    d. Name the ligands alphabetically.

4. Use prefixes to denote the number of ligands of the same kind present.

    a. mono-, di-, tri-, tetra-, penta-, hexa- (here, pentaamine).

    b. bis-, tris-, tetrakis for complex ligands.

5. Use Roman numerals to designate the oxidation state of the metal ion [cobalt(III)].

6. Add -ate as a suffix to the metal if the complex ion is an anion.

23. (D) Hydrogen bonding is a very strong *inter*molecular force that occurs between an H atom of one molecule that is bonded to either a fluorine, an oxygen, or a nitrogen atom. In choice (D), the hydrogens are bonded to carbon, not to F, O, or N.

24. (B) The molarity of a solution multiplied by its volume equals the number of moles of solute. In this case, 450 mL of 0.35 M $Al(NO_3)_3$ can be shown as

$$\frac{0.35\,\text{mole}\,Al(NO_3)_3}{1\,\cancel{\text{liter solution}}} \times \frac{0.45\,\cancel{\text{liter solution}}}{1} = 0.16\,\text{mole}\,Al(NO_3)_3$$

$Al(NO_3)_3$ is completely soluble, so there would be three times the number of moles of nitrate ions present in the solution because

$$Al(NO_3)_3 \rightarrow Al^{3+}(aq) + 3NO_3^-(aq)$$

Therefore, the number of moles of nitrate ions in the original solution would be 0.16 x 3 = 0.48.

The number of moles of nitrate ions needs to be brought up to 0.77 because the volume did not change (it remained at 0.45 liter).

$$\frac{1.7\,\text{moles}\,NO_3^-}{1\,\cancel{\text{liter solution}}} \times \frac{0.45\,\cancel{\text{liter solution}}}{1} = 0.77\,\text{mole of}\,NO_3^-\,\text{in final solution}$$

The solution begins with 0.48 mole of nitrate ions and must end up with 0.77 moles of nitrate ions; therefore, the solution needs an additional 0.29 mole of nitrate ions:

$$(0.77 - 0.48) = 0.29\,\text{mole}\,NO_3^-\,\text{needed}$$

Calcium nitrate, $Ca(NO_3)_2$, produces 2 moles of nitrate ions in solution for each mole of solid calcium nitrate added to the solution. Therefore, because 0.29 mole of NO_3^- is needed, you will need 0.29 / 2 = 0.15 mole of solid $Ca(NO_3)_2$.

25. (C) In examining the rate expression, note that B is first-order, so the rate is directly proportional to the concentration of the reactant. Holding [A] constant, and doubling [B] would double the rate.

26. (C) The term $-d[B]/dt$ represents the rate of decrease in the concentration of B as time elapses. For every mole of B that is lost on the reactant side, ½ x 4, or 2, moles of D are gained on the product side over the same amount of time (dt).

27. (C) The rate constant is independent of the concentration of the reactants. However, k depends on two factors:

- The nature of the reaction. "Fast" reactions typically have large rate constants.

- The temperature. Usually k increases with an increase in temperature.

With all other variables held constant, choice (D) would reduce the rate of molecular collisions, but increasing the volume is analogous to decreasing the concentration.

28. (D) The overall order of the reaction is the sum of the orders of the individual reactants. Here, $[A]^2[B]^1 = 2 + 1 = 3$. For reactions with an overall order of 3, the rate is proportional to the cube of the concentration of the reactants. Reducing the volume by 1/3 effectively triples their concentration: $3^{(\text{conc})3(\text{order})} = 27$.

29. (D)

Step 1: Write the balanced equation in equilibrium:

$$2 SO_3(g) \leftrightarrow 2 SO_2(g) + O_2(g)$$

Step 2: Write the equilibrium expression:

$$K_c = \frac{[SO_2]^2[O_2]}{[SO_3]^2}$$

Step 3: Create a chart showing initial and final concentrations.

Species	Initial Concentration	Final Concentration
SO_3	1.20 M	1.20 M – 0.20 M = 1.00 M
SO_2	0 M	0.20 M
O_2	0 M	0.10 M

Step 4: Substitute the final equilibrium concentrations into the equilibrium expression.

$$K_c = \frac{[SO_2]^2[O_2]}{[SO_3]^2} = \frac{(0.20)^2(0.10)}{(1.00)^2} = 4.0 \times 10^{-3}$$

30. (A) Remember, log $[H^+] = -pH$, so $[H^+] = 10^{-pH}$.

31. (D) When $\Delta G° = 0$,

$$T = \frac{\Delta H°}{\Delta S°} = \frac{[(-600.) + (-400.) - [1170.]]}{[(0.0700) + (0.2000)] + [0.10000]}$$

$$= 1.00 \times 10^3 \, K$$

32. (B)

$$k = \frac{0.693}{t_{1/2}} = \frac{0.693}{6930 \, yr} = 1.00 \times 10^{-4} \, yr^{-1}$$

33. (E) The functional group of a salt is

$$\overset{O}{\underset{}{\overset{\|}{-C-O^- \cdots M^+}}}$$
M = metal

The name of this compound is potassium propionate.

34. (C) The functional group of a ketone is

$$\overset{O}{\overset{\|}{-C-}}$$

The name of this ketone is methyl ethyl ketone.

35. (A) The functional group of an amide is

$$\begin{matrix} & O \\ & \parallel \\ -C & -N \diagup \\ & \diagdown \end{matrix}$$

The name of this amide is acetamide.

36. (B) The functional group of an amine is $-\overset{|}{N}-$

The name of this amine is dimethylamine.

37. (D) The functional group of a thiol is

—S–H

The name of this thiol is ethanethiol.

38. (E) Do this problem by using the factor-label method (m.t. stands for metric tons, 1 m.t. = 10^3 kg).

$$\frac{6.0 \times 10^2 \; \text{m.t.alloy}}{1} \times \frac{12.709 \; \text{m.t.Cu}}{100. \; \text{m.t.alloy}} \times \frac{159.15 \; \text{m.t.Cu}_2\text{S}}{127.09 \; \text{m.t.Cu}}$$

$$\times \frac{100. \, \text{m.t.ore}}{15.915 \; \text{m.t.Cu}_2\text{S}} = 6.0 \times 10^2 \, \text{m.t.ore}$$

39. (C) Begin this problem by balancing the reaction.

$$2\,C_2H_6(g) + 7\,O_2(g) \rightarrow 4\,CO_2(g) + 6\,H_2O(g)$$

Because $\Delta H^\circ = \Sigma\Delta H^\circ_f \text{ products} - \Sigma\Delta H^\circ_f \text{ reactants}$, you can substitute at this point.

$\Delta H^\circ = [4(-94.0) + 6(-60.0)] - 2(20.0) = -696.0$

However, remember that the question calls for the answer per mole of C_2H_6. Thus, because the balanced equation is written for 2 moles of C_2H_6, simply divide −696.0 by 2 and you get the answer −348.0.

40. (B) Review the postulates of the kinetic molecular theory of gases, which are listed on page 58.

41. (D) A principal quantum number of 4 tells you that you are in the fourth energy level. The fourth energy level contains electrons in the *s*, *p*, *d*, and *f* orbitals. Counting the maximum numbers of electrons available in each of the four types of sublevels — 2 in the *s*, 6 in the *p*, 10 in the *d*, and 14 in the *f* — yields a total of 32.

42. (C) When presented with a generic formula, such as AB_2, the best way to answer the question is to use familiar examples that satisfy the conditions of the question. H_2O would satisfy AB_2 as a bent molecule, and CO_2 would satisfy AB_2 as a linear molecule. In CO_2, a linear molecule, the two dipoles cancel each other, resulting in a nonpolar molecule. However, in H_2O, which also satisfies the AB_2 requirement, the two dipoles do not cancel each other out and result in a net dipole moment and a bent molecule. For both CO_2 and H_2O, we have data on ionization energy, electron affinity, electronegativity, and bond energy, but these are of no use, by themselves, in determining the geometry of the species.

43. (C) Water molecules are neutral. Hydroxide ions (OH$^-$) have a -1 charge. The overall charge of the complex is $+1$. Zinc would have to have a $+2$ charge in order for the complex to end up with a $+1$ charge. If you let $x =$ the charge of the zinc ion, then

$$+1 = x + 3(0) + 1(-1)$$
$$x = +2$$

44. (B) To do this problem, you would use the Clausius-Clapeyron equation:

$$\log \frac{P_2}{P_1} = \frac{\Delta H_{vap}}{2.303\,R}\left(\frac{T_2 - T_1}{T_2 T_1}\right)$$

where

P_1 (132 mm Hg) is the vapor pressure of the liquid at T_1 (327 K).

P_2 (x) is the vapor pressure of the liquid at T_2 (310 K).

R is a universal gas constant: 8.314 joules/(mole $\cdot$ K).

Although the problem does not require you to solve the equation, it is presented below. Substituting the values of the problem into the equation gives

$$\log \frac{x}{132\,\text{mm Hg}} = \frac{4.33 \times 10^4 \text{ J/mole}}{2.303 \times 8.314 \text{ J/(mole·K)}}\left(\frac{310\,\text{K} - 327\,\text{K}}{310\,\text{K} \cdot 327\,\text{K}}\right)$$

Simplifying this problem gives you

$$\log \frac{x}{132} = -0.379$$

Solving for x yields $x = 55.2$ mm Hg.

Note that the question tells you there are 2.00 moles of the compound. This information is irrelevant to solving the problem since equilibrium vapor pressure is independent of the amount of compound.

45. (C) This problem can be solved by using the factor-label method.
100. g solution $-$ 10. g solute $= 90$ g solvent (H_2O)

$$\frac{10.\text{ g } C_6H_2O}{90.\text{ g } H_2O} \times \frac{1000\text{ g } H_2O}{1\text{ kg } H_2O} \times \frac{1\text{ mole } C_6H_2O}{90.\text{ g } C_6H_2O} \approx 1.2m$$

46. (B) The Al^{3+} has been diluted tenfold:

$$\frac{1.0\text{ M} \times 1\text{ cm}^3}{10.\text{ cm}^3} = 0.10\text{ M}$$

47. (D)

$$\frac{\text{rate}_2}{\text{rate}_1} = \frac{(\text{new conc.})^x}{(\text{old conc.})^x} = \frac{8}{1} = \left(\frac{2 \cdot \text{old conc.}}{\text{old conc.}}\right) = 2^x$$

$$8 = 2^x$$
$$x = 3$$

48. (D) All intermediate mechanisms must add up to yield the original, overall balanced equation.

$$NO(g) + O_2(g) \rightarrow NO_3(g)$$

$$\frac{NO_3(g) + NO(g) \rightarrow 2NO_2(g)}{2NO(g) + O_2(g) \rightarrow 2NO_2(g)}$$

49. (C) Remember that $K_a \times K_b = 10^{-14}$. Therefore,

$$K_b \, Lac^- = \frac{10^{-14}}{1.5 \times 10^{-5}} = 6.7 \times 10^{-10}$$

50. (D) Use the relationship gram-equivalents acid = gram-equivalents base. Solve for the gram-equivalents of acid.

$$\frac{0.700 \, \cancel{\text{liter acid}}}{1} \times \frac{3.0 \, \text{gram} - \text{equiv. acid}}{1 \, \cancel{\text{liter}}} = 2.1 \, \text{gram} - \text{equivalents of acid}$$

At neutralization, the gram-equivalents of acid = the gram-equivalents of base. Therefore,

$$\frac{2.1 \, \cancel{\text{gram equiv. NaOH}}}{1} \times \frac{40.00 \, \text{g NaOH}}{1 \, \cancel{\text{gram equiv. NaOH}}} = 84 \, \text{g NaOH}$$

51. (D) $\Delta G°$ represents the free energy at standard conditions: 25°C and 1 atm pressure. ΔG represents the free energy at nonstandard conditions. In this problem, we have the non-standard condition of 100°C. In order to solve for the free energy of this reaction, you must use the following equation:

$$\Delta G = \Delta G° + 2.303 \, RT \log Q_p$$

where the constant $R = 8.314 \, \text{J} \cdot \text{K}^{-1} \cdot \text{mole}^{-1}$ and Q_p is called the reaction quotient. The reaction quotient has the same form as the equilibrium constant K_p.

Step 1: Write a balanced equation.

$$CaCO_3(s) \rightarrow CaO(s) + CO_2(g)$$

Step 2: Determine the value of Q_p, the reaction quotient.

$$Q_p = [CO_2(g)] = 1.00$$

Step 3: Substitute into the equation.

$$\Delta G = \Delta G° + 2.303 \, RT \log Q_p$$
$$= 130,240 \, \text{J/mole} + 2.303(8.314 \, \text{J} \cdot \text{K}^{-1} \cdot \text{mole}^{-1})(373 \, \text{K})(\log 1.00)$$
$$= 130,240 \, \text{J/mole}$$
$$= 130.240 \, \text{kJ/mole}$$

52. (D) The positron is a particle with the same mass as the electron but the opposite charge. The net effect is to change a proton to a neutron. Begin by writing the nuclear equation.

$$^{239}_{94}\text{Pu} \rightarrow {}^{0}_{1}\text{e} + {}^{A}_{Z}\text{X}$$

Remember that the total of the A and Z values must be the same on both sides of the equation.

Solve for the Z value of X: $Z + 1 = 94$, so $Z = 93$.

Solve for the A value of X: $A + 0 = 239$, so $A = 239$.

Therefore, you have $^{239}_{93}\text{X}$, or $^{239}_{93}\text{Np}$.

53. (A)

$$\text{Ni}(s) \mid \text{Ni}^{2+}(aq) \parallel \text{Ag}^{+}(aq) \mid \text{Ag}(s)$$
$$\text{anode (oxidation)} \parallel \text{cathode (reduction)}$$
$$\text{Ni}(s) \rightarrow \text{Ni}^{2+}(aq) + 2\text{e}^{-} \parallel 2\,\text{Ag}^{+}(aq) + 2\text{e}^{-} \rightarrow 2\,\text{Ag}(s)$$

By convention, in the representation of the cell, the anode is represented on the left and the cathode on the right. The anode is the electrode at which oxidation occurs (AN OX), and the cathode is the electrode at which reduction takes place (RED CAT). The single vertical lines ($\mid$) indicate contact between the electrode and solution. The double vertical lines ($\parallel$) represent the porous partition, or salt bridge, between the two solutions in the two half-cells. The ion concentration or pressures of a gas are enclosed in parentheses.

Take the two equations that decoded the standard cell notation and include the $E°$ reduction and the $E°$ oxidation voltages:

(Change the sign!)

ox: $\text{Ni}(s) \rightarrow \text{Ni}^{2+}(aq) + 2\text{e}^{-}$

red: $2\,\text{Ag}^{+}(aq) + 2\text{e}^{-} \rightarrow 2\,\text{Ag}\,(s)$

$E°_{ox} = +0.25$ volt

$\underline{E°_{red} + 0.80 \text{ volt}}$

$E°_{cell} = +1.05$ volts

By definition, $E°_{cell}$ voltages that are positive indicate a spontaneous reaction.

54. (A) Begin by writing as much of an equation as you can:

$$\text{S}_8(s) + \text{M}(s) \rightarrow \text{M}_a\text{S}_b(s)$$

From the information provided, you can determine that the residue, $\text{M}_a\text{S}_b(s)$, weighed 21.00 grams (62.00 − 41.00) and that the metal M weighed 14.00 grams (55.00 − 41.00). According to the Law of Conservation of Mass, the sulfur that reacted with the metal must have weighed 7.00 grams (21.00 − 14.00). You can now set up a proportion that relates the grams of S_8 and M to their respective equivalent weights.

$$\frac{7.00\,\text{grams sulfur}\,(\text{S}_8)}{16.0\,\text{grams/equiv. sulfur}} = \frac{14.00\,\text{grams M}}{x\,\text{grams/equiv.}}$$

Solving for x, you obtain 32.00 grams/equiv. for metal M. From this information, it would seem reasonable that the unknown metal is copper, forming the compound CuS. Copper, with a +2 valence, has an equivalent weight of 31.78.

55. (C) The law of Dulong and Petit states that

(molar mass) × (specific heat) ≈ 25 J/mole · °C

Substituting the given information into this relationship yields

(x g/mole) × 0.052 cal/g · °C ≈ 25 J/mole · °C

Because you know that 4.184 joules = 1 calorie, convert the 25 joules to calories so that units can cancel.

(x ~~g~~/mole) × 0.052 cal/~~g~~ · °C = 6.4 cal/mole · °C

Solving for x yields ~124. You can also learn something more about the metal from the concept of equivalent weights.

0.80 gram oxygen/equivalent #8 = 3.00 grams x/equivalent #?

? = 30 is the gram-equivalent weight for the metal. However, this assumes that the charge of the metal is +1. Because equivalent weight = atomic weight/valence number,

30 gram-equivalent weight = 120 g · mole^{-1}/x

x = +4, which means that the metal, whose atomic weight is approximately 120, has a valence of +4. Thus, the metal is probably tin.

56. (A)

$$\text{Density} = \frac{\text{mass}}{V} = \frac{P \cdot MM}{R \cdot T} = \frac{\text{atm} \cdot \text{g/mole}}{L \cdot \text{atm} \cdot \text{mole}^{-1} K^{-1} \cdot K} = \frac{g}{L}$$

Many students would choose answer choice (E). The reason molecular mass is the wrong answer is the question did not state the conditions when comparing one gas to another. For example, 1 mole of hydrogen gas at 50 K in a 1.0-liter container might have a higher density than 0.01 mole of uranium hexafluoride (UF$_6$ at 200 K in a 20-liter container, even though the UF$_6$ has a greater molecular mass.

57. (D) Element A keys out to be Al, which, being a metal in Group IIIA, would have a +3 charge. Element B would key out as sulfur, a nonmetal with a charge of −2, giving the formula Al$_2$S$_3$, or A$_2$B$_3$.

58. (A) If you missed this question, go back to the rules for naming coordination compounds found in the answer to question 22 (page 357).

The (III) indicates that the central positive ion should have a +3 charge. In adding up the charges for the ligand you get:

2 H$_2$O = 0

2 NH$_3$ = 0

2(Br$^-$) = −2

$\underline{1 (Cr^{+?}) = ?}$

+1 (overall charge of compound)

Therefore, the Cr ion must have a charge of +3.

Choice (D) also yields a +3 charge for Cr, and yet does not conform to standard methods of writing complex ion formulas.

59. (C) Raoult's law states that the partial pressure of a solvent over a solution, P_1, is given by the vapor pressure of the pure solvent, P_1°, times the mole fraction of the solvent in the solution, X_1.

$$P_1 = X_1 P_1^\circ$$

A decrease in vapor pressure is directly proportional to the concentration (measured as mole fraction) of the solute present.

60. (E) In dilution problems, we use the formula $M_1V_1 = M_2V_2$; therefore, it is necessary to determine the molarity of the initial solution first.

$$\frac{1.50 \text{ g solution}}{1 \text{ mL solution}} \times \frac{1000 \text{ mL solution}}{1 \text{ liter solution}} \times \frac{36.45 \text{ g HCl}}{100 \text{ g solution}} \times \frac{1 \text{ mole HCl}}{36.45 \text{ g HCl}} = 15.0 \text{ M}$$

Next we use the relationship $M_1V_1 = M_2V_2$:

$(15.0 \text{ M})(x \text{ liters}) = (5.0 \text{ M})(9.0 \text{ liters})$

$$x = 3.0 \text{ liters}$$

61. (B) In examining the balanced equation, note that for each mole of N_2O_5 gas that decomposes, ½ mole of O_2 gas is formed. Therefore, the rate of formation of oxygen gas should be half the rate of decomposition of the N_2O_5.

62. (B) Begin by writing the equilibrium equation.

$PbCl_2 \rightleftharpoons Pb^{2+}(aq) + 2Cl^-(aq)$

Next, write the equilibrium expression.

$K_{sp} = \left[Pb^{2+}\right]\left[Cl^-\right]^2$

In reference to the chloride ion concentration, rewrite the expression for $[Cl^-]$:

$$\left[Cl^-\right] = \left(\frac{K_{sp}}{\left[Pb^{2+}\right]}\right)^{1/2}$$

At any value greater than this expression, $PbCl_2(s)$ will precipitate, removing $Cl^-(aq)$ from solution.

63. (D) Use the equation $N_aV_a = N_bV_b$. Solve the equation for N_a.

$$N_a = \frac{N_bV_b}{V_a} = \frac{0.50 \text{ N} \times 50.0 \text{ mL}}{25.0 \text{ mL}} = 1.0 \text{ N}$$

64. (E) This problem can be done using the factor-label method.

$$\frac{1.0 \text{ gram equiv. } HC_2H_3O_2}{1 \text{ liter vinegar}} \times \frac{60.00 \text{ g } C_2H_3O_2}{1 \text{ gram equiv. } HC_2H_3O_2 / \text{liter of vinegar}}$$

65. (D) This problem can be solved using the factor-label method.

$$\text{wt.\%} = \frac{\text{parts } HC_2H_3O_2}{\text{solution}} \times 100\%$$

$$= \frac{60. \text{ g } HC_2H_3O_2}{1000. \text{ mL solution} \cdot 1.0 \text{ g/mL}} \times 100\% = 6.0\%$$

66. (A) Heat needs to be absorbed when a solid melts; therefore, the reaction is endothermic, $\Delta H > 0$. When a solid melts and becomes a liquid, it is becoming more disordered, $\Delta S > 0$.

67. (C) The question concerns the effect of changing standard conditions of a cell to nonstandard conditions. To calculate the voltage of a cell under nonstandard conditions, use the Nernst equation

$$E = E° - \frac{0.0591}{n} \log Q = E° - \frac{0.0591}{2} \log \frac{\left[Zn^{2+}\right]}{\left[Ag^+\right]^2}$$

where $E°$ represents the cell voltage under standard conditions, E represents the cell voltage under nonstandard conditions, n represents the number of moles of electrons passing through the cell, and Q represents the reaction quotient.

Choices (D) and (E) would have the effect of increasing the cell voltage. Choices (A) and (B) would have no effect on the cell voltage.

68. (C) To solve this problem, use the equation

$$\log \frac{x_0}{x} = \frac{k \cdot t}{2.30}$$

with the corresponding half-life $t_{1/2} = 0.693/k$, where x_0 is the number of original radioactive nuclei and x represents the number of radioactive nuclei at time t. k represents the first-order rate constant. Substituting into the equation yields

$$\log \frac{x_0}{x} = \frac{(0.693/6.93 \text{ years})(11.5 \text{ years})}{2.30} = 0.5$$

$$\frac{x_0}{x} \approx 3$$

$$\frac{x}{x_0} \approx \frac{1}{3} \times 100\% \approx 33\% \text{ that remain unreacted}$$

69. (D) Begin by writing the equations which define the equilibrium constants.

$$AgCl(s) \rightarrow Ag^+(aq) + Cl^-(aq) \qquad K_{sp1} = 1.6 \times 10^{-10} \text{ mol}^2 \cdot L^{-2}$$

$$AgI(s) \rightarrow Ag^+(aq) + I^-(aq) \qquad K_{sp2} = 8.0 \times 10^{-17} \text{ mol}^2 \cdot L^{-2}$$

The K_{eq} is needed for the following equation:

$$AgCl(s) + I^-(aq) \rightarrow AgI(s) + Cl^-(aq) \qquad K_{eq} = \frac{[Cl^-]}{[I^-]}$$

$$\frac{1.6 \times 10^{-10} \text{ mole}^2}{L^2} \times \frac{L^2}{8.0 \times 10^{-17} \text{ mole}^2} = 2.0 \times 10^6$$

70. (D)

Bond breaking $(\Delta H_1) = H\!\!-\!\!H + F\!\!-\!\!F = 103 \text{ kcal} \cdot \text{mole}^{-1} + 33 \text{ kcal} \cdot \text{mole}^{-1}$
$= 136 \text{ kcal} \cdot \text{mole}^{-1}$

Bond forming $(\Delta H_2) = 2 \text{ H}\!\!-\!\!F = 2(-135 \text{ kcal} \cdot \text{mole}^{-1}) = -270 \text{ kcal} \cdot \text{mole}^{-1}$

$\Delta H° = \Delta H_1 + \Delta H_2 = 136 \text{ kcal} \cdot \text{mole}^{-1} + (-270 \text{ kcal} \cdot \text{mole}^{-1}) = -134 \text{ kcal/mole}$

71. (D) Whether you can answer this question depends on whether you are acquainted with what is known as the Maxwell-Boltzmann distribution. This distribution describes the way that molecular speeds or energies are shared among the molecules of a gas. If you missed this question, examine the following figure and refer to your textbook for a complete description of the Maxwell-Boltzmann distribution.

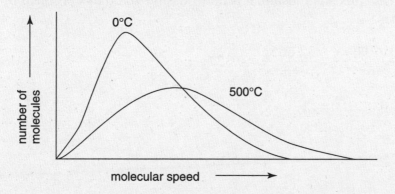

72. (A) Because all choices have 18 electrons in their valence shell, you should pick the species with the fewest protons in the nucleus; this would result in the weakest electrostatic attraction. That species is sulfur.

73. (A) The central metal ion forms only two bonds to ligands, so the coordination number is 2.

74. (C) The body-centered cubic cell looks like this:

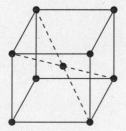

The formula that relates the atomic radius (r) to the length of one edge of the cube (s) for a body-centered cubic cell is $4r = s\sqrt{3}$.

$$s = \frac{4(0.255\,\text{nm})}{\sqrt{3}} = 0.600\,\text{nm}$$

75. (C) H_3PO_4 is a triprotic acid; that is, there are 3 moles of H^+ ions produced for each mole of H_3PO_4 that completely ionizes. Normality is the number of equivalents per liter. Assuming complete or 100% ionization, a 1-molar HCl solution is 1 normal. A 1-molar H_2SO_4 solution is 2 normal, and a 1-molar solution of H_3PO_4 is 3 normal.

In order to use the concept of normality, one must know the reaction involved. Phosphoric acid, H_3PO_4 undergoes three simultaneous ionizations:

$$H_3PO_4 + H_2O \rightleftharpoons H_3O^+ + H_2PO_4^- \quad K_{a1} = 5.9 \times 10^{-3}$$
$$H_2PO_4^- + H_2O \rightleftharpoons H_3O^+ + HPO_4^{2-} \quad K_{a2} = 6.2 \times 10^{-8}$$
$$HPO_4^{2-} + H_2O \rightleftharpoons H_3O^+ + PO_4^{3-} \quad K_{a3} = 4.8 \times 10^{-13}$$

Because 1 mole of H_3PO_4 weighs 97.995 grams, 1 equivalent of H_3PO_4 would weigh ⅓ as much, or 32.665 grams. Given this relationship, it is now possible to do this problem by using factor-label techniques.

$$\frac{100.0\,\text{mL}}{1} \times \frac{1\,\text{liter}}{1000\,\text{mL}} \times \frac{0.100\,\text{equiv}}{1\,\text{liter}} \times \frac{32.665\,\text{g}}{1\,\text{equiv}} = 0.327\,\text{g}$$

Section II (Free-Response Questions)

Scoring Guidelines

One point deduction for mathematical error (maximum once per question)

One point deduction for error in significant figures* (maximum once per question)

*number of significant figures must be correct within +/– one digit

Part A: Question 1

1. Given: $C_2H_5NH_2(aq) + H_2O(\ell) \leftrightharpoons C_2H_5NH_3^+(aq) + OH^-(aq)$

   K_b for $C_2H_3NH_2 = 5.6 \times 10^{-4}$

(a) Given: 65.987 mL of 0.250 M $C_2H_5NH_2$

Restatement: Find $[OH^-]$

Step 1: Rewrite the balanced equation for the ionization of ethylamine.

$C_2H_5NH_2 + H_2O \leftrightharpoons C_2N_5NH_3^+ + OH^-$

Step 2: Write the expression for the base-dissociation constant.

$$K_b = \frac{[C_2H_5NH_3^+][OH^-]}{[C_2H_5NH_2]} = 5.6 \times 10^{-4}$$

Step 3: Create a chart showing initial and final concentrations (at equilibrium) of the involved species. Let x be the amount of $C_2H_5NH_3^+$ that forms from $C_2H_5NH_2$. Because $C_2H_5NH_3^+$ is in a 1:1 molar ratio with OH^-, $[OH^-]$ also equals x.

Species	Initial Concentration	Final Concentration (at equilibrium)
$C_2H_5NH_2$	0.250 M	$0.250 - x$
$C_2H_5NH_3^+$	0 M	x
OH^-	0 M	x

Step 4: Substitute the equilibrium concentrations from the chart into the equilibrium expression and solve for x.

$$K_b = \frac{[C_2H_5NH_3^+][OH^-]}{C_2H_5NH_2} = 5.6 \times 10^{-4} = \frac{(x)(x)}{0.250 - x}$$

You have a choice in solving for x. The first method would require the quadratic equation—not a good idea because compared to the magnitude of 0.250, the value of x is negligible. If you used the quadratic, you would be wasting time. The second method would assume that $[C_2H_5NH_2]$ remains constant at 0.250 M; $5.6 \times 10^{-4} = x^2/0.250$.

$$x = [OH^-] = 0.012 \text{ M}$$

By the way, the 65.987 mL is not needed because concentration is independent of the amount of solution measured.

(b) Restatement: Find pOH of solution.

$$pOH = -\log [OH^-]$$
$$pOH = -\log [0.012] = 1.92$$

(c) Restatement: Find % ionization of ethylamine

$$\% = \frac{\text{part}}{\text{whole}} \times 100\% = \frac{0.012}{0.250} \times 100\% = 4.8\%$$

(d) Given: 15.000 g $C_2H_5NH_3Br$ + 250.00 mL 0.100 M $C_2H_5NH_2$

Restatement: Find pH of solution.

Step 1: Note that when $C_2H_5NH_3Br$ dissolves in water, it dissociates into $C_2H_5NH_3^+$ and Br^-. Furthermore, $C_2H_5NH_3^+$ is a weak acid.

Step 2: Rewrite the balanced equation at equilibrium for the reaction.

$$C_2H_5NH_3^+ \rightleftharpoons C_2H_5NH_2 + H^+$$

Step 3: Write the equilibrium expression.

$$K_a = K_w/K_b = \frac{[C_2H_5NH_2][H^+]}{[C_2H_5NH_3^+]} = 10^{-14}/5.6 \times 10^{-4}$$

$$= 1.8 \times 10^{-11}$$

Step 4: Calculate the initial concentrations of the species of interest.

$$[C_2H_5NH_2] = 0.100 \text{ M (given)}$$

$$[C_2H_5NH_3^+] = \frac{15.000 \, g \, C_2H_5NH_3Br}{1} \times \frac{1 \, mole \, C_2H_5NH_3^+}{126.05 \, g \, C_2H_5NH_3Br}$$

$$\times \frac{1}{0.250 \, liter} = 0.476 \text{ M}$$

$$[H^+] = 0$$

Step 5:

Species	Initial Concentration	Final Concentration (at equilibrium)
$C_2H_5NH_2$	0.100 M	0.100 + x
$C_2H_5NH_3^+$	0.476 M	0.476 − x
H^+	0 M	x

$$\frac{(0.100 + x)(x)}{(0.476 - x)} = 1.8 \times 10^{-4}$$

$$x = (H^+) = 8.57 \times 10^{-11}$$

$$pH = -\log(8.57 \times 10^{-11}) = 10.07$$

(e) Given: 0.125 g AgBr(*s*) with $AgNO_3$(s).

K_{sp} AgOH = 1.52×10^{-8}

Restatement: Will AgOH precipitate?

Step 1: Write the equation in equilibrium for the dissociation of AgOH.

$$AgOH(s) \leftrightharpoons Ag(aq) + OH^-(aq)$$

Step 2: Calculate the concentration of the ions present.

$$[Ag^+] = \frac{0.125 \text{ g AgBr}}{0.065987 \text{ liter sol'n}} \times \frac{1 \text{ mole AgBr}}{187.772 \text{ g AgBr}} \times \frac{1 \text{ mole Ag}^+}{1 \text{ mole AgBr}} = 0.0101 M$$

$$[OH^-] = 0.012 M$$

Step 3: Solve for the ion product, *Q*.

Q=[Ag$^+$][OH$^-$] = (0.0101)(0.012) = 1.32×10^{-4}

K_{sp} AgOH = 1.52×10^{-8}

Since Q > K_{sp}, AgOH will precipitate

Part A: Question 2

2. Given: H_2O + 2.51 g $SbCl_3 \rightarrow$ 1.906 g $Sb_xO_yCl_z$

0.802 g gas, 97.20% Cl, and 2.75% H

(a) Restatement: Simplest formula for gas.

If you had 100. grams of the gas, 97.20 grams would be due to the weight of chlorine atoms, and 2.75 grams would be due to the weight of hydrogen atoms.

97.20 grams Cl / (1 mole Cl / 35.453 g / mole) = 2.742 moles Cl

2.75 grams H / (1 mole H / 1.00794 g / mole) = 2.73 moles H

Because this is essentially a 1:1 molar ratio, the empirical formula of the gas is HCl.

(b) Restatement: Fraction of chlorine in the solid product and in the gas phase.

1. The mass of chlorine in the original compound:

2.51 g of $SbCl_3 \times$ 106.36 g Cl / 228.10 g $SbCl_3$ = 1.17 g Cl

2. Fraction of chlorine in the gas.

According to the Law of Conservation of Mass, if you have 1.17 grams of chlorine in the original compound, you must account for 1.17 grams of chlorine on the product side.

Because the gas is 97.2% by mass chlorine, the fraction of chlorine in the gas can be found as follows:

$$\frac{part}{whole} \times 100\% = \frac{0.802\,g\,gas\,(0.972)}{1.17\,g\,total\,chlorine} \times 100\% = 66.6\%\ gas$$

fraction of chlorine in the solid product = 100.00% − 66.6% = 33.4% solid

(c) Restatement: Formula of solid product.

You know from the question that the solid product contains Sb, Cl, and O atoms. The weight of Sb can be found by taking the weight of antimony chloride (which has all of the antimony atoms in it) and getting rid of the weight of chlorine atoms, which you have determined to be 1.17 g. Therefore, 2.51 g $SbCl_3$ − 1.17 grams of chlorine atoms = 1.34 grams of antimony.

The weight of chlorine in the solid product can be determined by taking the weight of chlorine in the original compound (which has all of the chlorine atoms in it) and multiplying it by the percent of chlorine found in the solid product. This becomes 1.17 grams of chlorine × 0.334 = 0.391 g chlorine in the solid product.

The weight of oxygen in the solid product can be found by taking the total weight of the solid product and subtracting the amount of antimony and chlorine previously determined. This becomes 1.906 g solid product − 1.34 g antimony − 0.391 g chlorine = 0.175 g oxygen atoms in the solid product.

Expressing these weights as moles yields.

1.34 g Sb → 0.0110 mole Sb

0.175 g O → 0.0109 mole O

0.391 g Cl → 0.0110 mole Cl

Thus, they are in essentially a 1:1:1 molar ratio, which indicates the molecular formula SbOCl.

(d) Given: Empirical formula is true formula.

Restatement: Balanced equation.

$SbCl_3(s) + H_2O(\ell) \rightarrow SbOCl(s) + 2\ HCl(g)$

Part A: Question 3

3. Given: $CH_3OH(aq) + O_2(g) \rightarrow HCOOH(aq) + H_2O(\ell)$. Data shown in the table

(a) Restatement: $\Delta H°$ for the oxidation of methyl alcohol.

$\Delta H°_f \, HCOOH(aq) = -409$ kJ/mole

$\Delta H°_f \, H_2O(\ell) = -285.84$ kJ/mole

$\Delta H°_f \, CH_3OH(aq) = -238.6$ kJ/mole

$\Delta H° = \Sigma\Delta H°_{f \text{ products}} - \Sigma\Delta H°_{f \text{ reactants}}$

$= (\Delta H°_f \, HCOOH + \Delta H°_f \, H_2O) - (\Delta H°_f \, CH_3OH)$

$= [(-409 \text{ kJ/mole}) + (-285.84 \text{ kJ/mole})] - (238.6 \text{ kJ/mole})$

$= -456$ kJ/mole

(b) Restatement: $\Delta S°$ for the oxidation of methyl alcohol.

$S° \, HCOOH(aq) = 129$ J/mole $\cdot$ K

$S° \, H_2O(\ell) = 69.94$ J/mole $\cdot$ K

$S° \, CH_3OH(aq) = 127.0$ J/mole $\cdot$ K

$S° \, O_2(g) = 205.0$ J/mole $\cdot$ K

$\Delta S° = \Sigma S°_{\text{products}} - \Sigma S°_{\text{reactants}}$

$= (S° \, HCOOH + S° \, H_2O) - (S° \, CH_3OH + S° \, O_2)$

$= (129 \text{ J/K} \cdot \text{mole} + 69.94 \text{ J/K} \cdot \text{mole}) - (127.0 \text{ J/K} \cdot \text{mole} + 205.0 \text{ J/K} \cdot \text{mole})$

$= -133$ J/K $\cdot$ mole

(c) (1) Restatement: Is the reaction spontaneous at 25°C?

Explain.

$\Delta G° = \Delta H° - T\Delta S$

$\quad = -456 \text{ kJ/mole} - 298 \text{ K}(-0.133 \text{ kJ/mole} \cdot \text{K})$

$\quad = -416$ kJ/mole

Reaction is spontaneous because $\Delta G°$ is negative.

(2) Restatement: If the temperature were increased to 100°C, would the reaction be spontaneous? Explain.

$\Delta G = \Delta H - T\Delta S$

$\quad = -456 \text{ kJ/mole} - 373 \text{ K}(-0.133 \text{ kJ/mole} \cdot \text{K})$

$\quad = -406$ kJ/mole

Reaction is spontaneous because ΔG is still negative.

(d) Given: ΔH_{fus} HCOOH = 12.71 kJ/mole at 8.3°C

Restatement: Calculate ΔS for the reaction

$$HCOOH(\ell) \rightarrow HCOOH(s)$$

The temperature at which liquid HCOOH converts to solid HCOOH is known as the freezing point; it is also the melting point. Because at this particular temperature a state of equilibrium exists — that is, $HCOOH(\ell) \leftrightarrow HCOOH(s)$ — you can set $\Delta G = 0$. Substituting into the Gibbs-Helmholtz equation yields

$$\Delta G = \Delta H - T\Delta S$$
$$0 = -12.71 \text{ kJ / mole} - 281.3 \text{ K}(\Delta S)$$

(Did you remember to make 12.71 negative, because you want ΔH for freezing, which is an exothermic process?)

$$\Delta S = \frac{-12.71 \text{ kJ / mole}}{281.3 \text{ K}}$$

$$= -0.04518 \text{ kJ / mole} \cdot \text{K} = -45.18 \text{ J / mole} \cdot \text{K}$$

(e) (1) Given: $S°$ HCOOH(ℓ) = 109.1 J / mole $\cdot$ K

Restatement: What is the standard molar entropy of HCOOH(s)?

$$HCOOH(\ell) \rightarrow HCOOH(s)$$
$$\Delta S° = \Sigma \Delta S°_{products} - \Sigma \Delta S°_{reactants}$$
$$= S° \text{ HCOOH}(s) - S° \text{ HCOOH}(\ell)$$
$$-45.18 \text{ J/mole} \cdot \text{K} = S° \text{ HCOOH}(s) - 109.1 \text{ J/mole} \cdot \text{K}$$
$$S° \text{ HCOOH}(s) = 63.9 \text{ J/mole} \cdot \text{K}$$

(2) Restatement: Is magnitude of $S°$ HCOOH(s) in agreement with magnitude of $S°$ HCOOH(ℓ)?

The magnitude of S° HCOOH(s) is in agreement with the magnitude of S° HCOOH(ℓ) because the greater the value of $S°$, the greater the disorder; the liquid phase has higher entropy than the solid phase.

(f) Given: ($K_a = 1.9 \times 10^{-4}$)

Restatement: $\Delta G°$ for the ionization of methanoic (formic) acid at 25°C.

$$\Delta G° = -2.303 \, R \cdot T \log K_a \qquad\qquad R = 8.314 \text{ J/K}$$
$$= -2.303(8.314 \text{ J/K}) \cdot 298 \text{ K}(-3.72)$$
$$= 2.1 \times 10^4 \text{ J} = 21 \text{ kJ}$$

Scoring Guidelines

Students choose five of the eight reactions. Only the answers in the boxes are graded (unless clearly marked otherwise). Each correct answer earns 3 points, 1 point for reactants and 2 points for products. All products must be correct to earn both product points. Equations do not need to be balanced and phases need not be indicated. Any spectator ions on the reactant side nullify the 1 possible reactant point, but if they appear again on the product side, there is no product-point penalty. A fully molecular equation (when it should be ionic) earns a maximum of 1 point. Ion charges must be correct.

Part B: Question 4

The roman numeral in each answer refers to the section in the chapter entitled "Writing and Predicting Chemical Reactions." For example, I is found on page 216.

4. Restatement: Give a formula for each reaction, showing the reactants and products.

(a) I. $Sn + Cl_2 \rightarrow SnCl_4$

(Usually pick the higher oxidation state of the metal ion.)

(b) II. $C_2H_6 + O_2 \rightarrow CO_2 + H_2O$

All hydrocarbons burn in oxygen gas to produce CO_2 and H_2O. ("Air" almost always means oxygen gas.) Note the use of the word "completely". Unless this word was in the problem, a mixture of CO and CO_2 gases would result.

(c) IX. $Cu + H^+ + NO_3^{3-} \rightarrow Cu^{2+} + H_2O + NO$

This reaction is well known and is covered quite extensively in textbooks. Note how it departs from the rubric. Copper metal does not react directly with H^+ ions, since it has a negative standard oxidation voltage. However, it will react with 6M HNO_3 because the ion is a much stronger oxidizing agent than H^+ The fact that copper metal is difficult to oxidize indicates that is readily reduced. This fact allows one to qualitatively test for the presence of by reacting it with dithionite (hydrosulfite) ion, as the reducing agent to produce copper:

$Cu^{2+}_{(aq)} + S_2O_4^{2-} + 2H_2O \rightarrow Cu + 2SO_3^{2-} + 4H^+$

(d) XIX. $SO_4^{2-} + Hg^{2+} \rightarrow HgSO_4$

(e) XXI. $SO_3 + CaO \rightarrow CaSO_4$

(f) XXIII. $CuSO_4 \cdot 5\,H_2O \rightarrow CuSO_4 + 5\,H_2O$

(g) XXIV. $Zn(OH)_2 + NH_3 \rightarrow Zn(NH_3)_4^{2+} + OH^-$

(h) XXV. $C_2H_6 + Br_2 \rightarrow C_2H_5Br + HBr$

Note: Unless the word "monobrominated" had been used in the problem, a whole host of products would have been possible; i.e. polybrominated ethanes, polymers, and so on.

Part B: Question 5

For the following answer, try using the bullet format.

5. Restatement: Explain each of the following:

(a) Water can act as either an acid or a base.

- Water can provide both H^+ and OH^-.

$$H_2O \rightleftharpoons H^+ + OH^-$$

- According to Brønsted-Lowry theory, a water molecule can accept a proton, thereby becoming a hydronium ion. In this case, water is acting as a base (proton acceptor).

$$H_2O + H^+ \rightarrow H_3O^+$$

- When water acts as a Brønsted-Lowry acid, it donates a proton to another species, thereby converting to the hydroxide ion.

$$\underset{\text{base}}{H_2O} + \underset{\text{acid}}{H_2O} \rightleftharpoons \underset{\substack{\text{conjugate} \\ \text{base}}}{OH^-} + \underset{\substack{\text{conjugate} \\ \text{acid}}}{H_3O^+}$$

- According to Lewis theory, water can act as a Lewis base (electron pair donor). Water contains an unshared pair of electrons that is utilized in accepting a proton to form the hydronium ion.

$$\overset{H}{\underset{}{:\ddot{O}:}}H + H^+ \rightarrow \left[H:\overset{H}{\underset{H}{\ddot{O}:}} \right]^+$$

(b) HF is a weaker acid than HCl.

- Fluorine is more electronegative than Cl.

- The bond between H and F is therefore stronger than the bond between H and Cl.

- Acid strength is measured in terms of how easy it is for the H to ionize. The stronger the acid, the weaker the bond between the H atom and the rest of the acid molecule; measured as K_a or, if the acid is polyprotic, K_{a1}, K_{a2}, K_{a3}, . . .

(c) For the triprotic acid H_3PO_4, K_{a1} is 7.5×10^{-3}, whereas K_{a2} is 6.2×10^{-8}.

- K_{a1} represents the first hydrogen to depart the H_3PO_4 molecule, leaving the conjugate base, $H_2PO_4^-$.

- The conjugate base, $H_2PO_4^-$, has an overall negative charge.

- The overall negative charge of the $H_2PO_4^-$ species increases the attraction of its own conjugate base HPO_4^{2-} to the departing proton. This creates a *stronger* bond, which indicates that it is a weaker acid.

(d) Pure HCl is not an acid.

- An acid is measured by its concentration of H^+ (its pH).

- Pure HCl would not ionize; the sample would remain as molecular HCl (a gas).

- In order to ionize, a water solution of HCl is required:

$$HCl(aq) \rightarrow H^+(aq) + Cl^-(aq)$$

(e) $HClO_4$ is a stronger acid than $HClO_3$, HSO_3^-, or H_2SO_3.

- As the number of lone oxygen atoms (those not bonded to H) increases, the strength of the acid increases. Thus, $HClO_4$ is a stronger acid than $HClO_3$.

- As electronegativity of central atom increases, acid strength increases. Thus, Cl is more electronegative than S.

- Loss of H^+ by a neutral acid molecule (H_2SO_3) reduces acid strength. Thus, H_2SO_3 is a stronger acid than HSO_3^-.

- As effective nuclear charge (Z_{eff}) on the central atom increases, acid strength is likewise increased. Thus, a larger nuclear charge draws the electrons closer to the nucleus and binds them more tightly.

Part B: Question 6

Question 6 might best be answered in the bullet format.

6. Restatement: Interpret using bonding principles.

(a) Restatement: Compare carbon-to-carbon bond lengths in C_2H_2 and C_2H_6.

- Lewis structure of C_2H_2:

$$H-C\equiv C-H$$

- Lewis structure of C_2H_6:

```
    H  H
    |  |
H−C−C−H
    |  |
    H  H
```

- C_2H_2 has a triple bond, whereas C_2H_6 consists only of single bonds.

- Triple bonds are shorter than single bonds since bond energy is larger for a multiple bond. The extra electron pairs strengthen the bond, making it more difficult to separate the bonded atoms from each other.

(b) Restatement: The O—H bond angle of H_3O^+ is less than 109.5°.

- Lewis structure of H_3O^+:

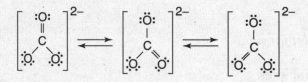

- H_3O^+ is pyramidal in geometry due to a single pair of unshared electrons.

- Angle of tetrahedron is 109.5°; this exists only if there are no unshared electrons.

- Repulsion between shared pairs of electrons is less than repulsion between an unshared pair and a shared pair. This stronger repulsion found in the shared-unshared pair condition, as seen in H_3O^+, decreases the bond angle of the pure tetrahedron (109.5°).

(c) Restatement: Compare C—O bond lengths as found in CO and CO_3^{2-}

- Lewis structure of CO:

:C≡O:

- Lewis structures of CO_3^{2-}

- CNO^- exists in three "resonance" forms. CO bond length is considered to be the average of the lengths of all single and double bonds.

(d) Restatement: CNO^- ion is linear.

- Lewis structure of CNO^-:

- There are no unshared pairs of electrons around the central atom N, resulting in a linear molecule.

- The molecule is polar because O is more electronegative than C.

Part B: Question 7

7. Restatement: Explain how MW measured just above boiling point deviates from its ideal value in terms of the ideal gas law.

The ideal gas equation, $PV = nRT$, stems from three relationships known to be true for gases:

   i) The volume is directly proportional to the number of moles: $V \sim n$

   ii) The volume is directly proportional to the absolute temperature: $V \sim T$

   iii) The volume is inversely proportional to the pressure: $V \sim 1/P$

n, the symbol used for the moles of gas, can be obtained by dividing the mass of the gas by the molecular weight. In effect, $n = $ mass / molecular weight ($n = m$ / MW). Substituting this relationship into the ideal gas law gives

$$PV = \frac{m \cdot R \cdot T}{MW}$$

Solving this equation for the molecular weight yields

$$MW = \frac{m \cdot R \cdot T}{PV}$$

Real gas behavior deviates from the values obtained using the ideal gas equation because the ideal equation assumes that (1) the molecules do not occupy space and (2) there is no attractive force between the individual molecules. However, at low temperatures (just above the boiling point of the liquid), these two postulates are not true and one must use an alternative equation known as the van der Waals equation, which accounts for these factors.

Because the attraction between the molecules becomes more significant at the lower temperatures, the compressibility of the gas is increased. This causes the product $P \cdot V$ to be smaller than predicted. PV is found in the denominator in the equation listed above, so the molecular weight tends to be higher than its ideal value.

Part B: Question 8

Question 8 could be answered by using the bullet format. You should try to arrange your points in logical order, but because time is a consideration, you may not be able to organize all of your bullets in perfect sequence.

8. Given: $F_2 < PH_3 < H_2O$

Restatement: Discuss boiling point (BP) order.

General Trends

   • BP is a result of the strength of intermolecular forces—the forces between *molecules.*

   • A direct relationship exists between the strength of intermolecular forces and the BP: The stronger the intermolecular force, the higher the BP.

   • Relative strength of intermolecular forces: H bonds > dipole forces > dispersion forces.

- BP is directly proportional to increasing MW-dispersion force (van der Waals force).

- Greater MW results in greater dispersion forces.

- Strength of dispersion force depends on how readily electrons can be polarized.

- Large molecules are easier to polarize than small, compact molecules. Hence, for comparable MW, compact molecules have lower BP.

- Polar compounds have slightly higher BP than nonpolar compounds of comparable MW.

- Hydrogen bonds are very strong intermolecular forces, causing very high BP.

Lowest BP: F_2

- F_2 is nonpolar; the only intermolecular attraction present is due to dispersion forces.

- F_2 has a MW of 38 g/mole.

- F_2 is covalently bonded.

Intermediate BP: PH_3

- PH_3 is polar; geometry is trigonal pyramidal; presence of lone pair of electrons.

- PH_3 is primarily covalently bonded; two nonmetals.

- There are dipole forces present between PH3 molecules because PH3 is polar.

- PH_3 has a MW of 34 g/mole (even though PH3 has a lower MW than F2 and might be expected to have a lower BP, the effect of the polarity outweighs any effect of MW).

Highest BP: H_2O

- H_2O is covalently bonded.

- H_2O is a bent molecule; hence, it is polar.

- H_2O has a MW of 18 g/mole.

- Between H_2O molecules there exist hydrogen bonds.

- Even though H_2O has the lowest MW of all three compounds, the hydrogen bonds outweigh any effects of MW or polarity.

The Final Touches

1. Spend your last week of preparation on a general review of key concepts, test-taking strategies, and techniques. Be sure to review the key terms and key concepts for each subject-matter chapter in this preparation guide.

2. Don't cram the night before the test! It's a waste of time.

3. Remember to bring the proper materials: three or four sharpened #2 pencils, an eraser, several ballpoint pens, a calculator, and a watch.

4. Start off crisply, answering the questions you know first and then coming back to the harder ones.

5. If you can eliminate one or more of the answers in a multiple-choice question, make an educated guess.

6. On the test, underline key words and decide which questions you want to do first. Remember, you do not have to do the essay questions in order in your answer booklet. Just be sure to number all of your essays properly.

7. Make sure that you're answering the question that is being asked and that your answers are legible.

8. Pace yourself; don't run out of time.

PART V

APPENDIXES

Appendix A: Commonly Used Abbreviations, Signs, and Symbols

Å	angstrom (unit of wavelength measure)
abs	absolute
ac	alternating current
amor or amorph	amorphous
A	ampere
anhyd	anhydrous
aq	aqueous
atm	atmosphere
at no	atomic number
at wt	atomic weight
bp	boiling point
Btu	British thermal unit
°C	degree Celsius
ca.	approximately
Cal	large calorie (kilogram calorie or kilocalorie)
cal	small calorie (gram calorie)
cg	centigram
cgs	centimeter-gram-second
cm	centimeter
cm^2	square centimeter
cm^3	cubic centimeter
conc	concentrated
cos	cosine
cp	candlepower
cu	cubic
D	density diopter
dB	decibel
dc	direct current

deg or °	degree
dg	decigram
dil	dilute
dr	dilute
E	electric tension: electromotive force
e.g.	for example
emf	electromotive force
esu	electrostatic unit
etc.	and so forth
et seq.	and the following
eV	electronvolt
° F	Fahrenheit
F	Frictional loss
f or v	frequency
ft	foot
ft^2	square foot
ft^3	cubic foot
ft·c	foot-candle
ft·lb	foot-pound
g	acceleration due to gravity
g	gram
g•cal	gram-calorie
gal	gallon
gr	grain
h	Planck's constant
h	hour
hp	horsepower
hp·h	horsepower-hour
hyg	hygroscopic
Hz	hertz (formerly cycles per second, cps)
I	electric current
^A_ZI	symbol for isotope with atomic number Z and atomic number A
ibid.	in the same place

i.e.	that is
in	inch
in^2	square inch
in^3	cubic inch
insol	insoluble
iso	isotropic
J	joule; mechanical equivalent of heat
k	kilo (1000)
K	Kelvin
kc	kilocycle
kcal	kilogram-calorie
kg	kilogram
kW	kilowatt
kWh	kilowatt-hour
L	liter
l	lumen
l	length
λ	lambda; wavelength, coefficient of linear expansion
lb	pound
lb/ft^3	pound per cubic foot
ln	natural logarithm
log	logarithm
M	molecular weight; mass
M	molar, as 1 M
m	meter
m^2	square meter
m^3	cubic meter
μ	micro-(10^{-6})
μm	micrometer (micron)
meq	milliequivalent
MeV	million (or mega-) electronvolt
mg	milligram
min	minute

mks	meter-kilogram-second
mL	milliliter
mm	millimeter
mm^2	square millimeter
mm^3	cubic millimeter
mp	melting point
mph	miles per hour
N	normality as $1N$
n	index of refraction; neutron (component of atomic nucleus)
oz	ounce
ppt	precipitate
p sol	partly soluble
Q	energy of nuclear reaction
qt	quart
q.v.	which see
R	roentgen (international unit for x-rays)
satd	saturated
s	second
sin	sine
sol'n	solution
sp	specific
sp gr	specific
sp ht	specific heat, C_p
sq	square
T	temperature
t	time
tan	tangent
V	volt
W	watt
Wh	watt-hour
yr	year
+	plus; add; positive
−	minus; subtract; negative

$\pm$	plus or minus; positive or negative
$\times$ or $\cdot$	times. multiplied by
$\div$ or $/$	is divided by
$=$ or $::$	is equals; as
$\approx$	is identical to; congruent with
$\neq$	is not identical
$>$	is greater than
$<$	is less than
$\geq$	is equal to or greater than
$\leq$	is equal to or less than
$:$	is ratio of
∞	varies as (is proportional to)
$\sqrt{a}$ or $a^{\frac{1}{2}}$	is square root of a
a^2	a squared, the second power of a; $a \times a$
a^3	a cubed, the third power of a; $a \times a \times a$
a^{-1}	is $1/a$ (reciprocal of a)
a^{-2}	is $1/a^2$ (reciprocal of a^2)
Σ	summation of
Δ	difference in
k	a mathematical constant
π	the ratio of the circumference of a circle to its diameter; roughly equal to 3.1416

Appendix B: Acid-Base Indicators

Range	Indicator	Lower Color	Upper Color
0.0–2.5	Methyl violet	Yellow-green	Violet
0.5–2.0	Malachite green HCl	Yellow	Blue
1.0–2.8	Thymol blue	Red	Yellow
1.2–4.0	Benzopurpurin	Violet	Red
1.3–3.0	Orange IV	Red	Yellow
1.5–2.6	Naphthol yellow S	Colorless	Yellow
2.1–2.8	p-Phenylazoaniline	Orange	Yellow
2.5–4.4	Methyl orange	Red	Yellow
3.0–4.7	Bromophenol blue	Orange-yellow	Violet
3.0–5.0	Congo red	Blue	Red
3.5–6.3	Gallein	Orange	Red
3.8–5.4	Bromcresol green	Yellow	Blue
4.0–5.8	2,5- Dinitrophenol	Colorless	Yellow
4.2–4.6	Ethyl orange	Colorless	Yellow
4.4–6.2	Methyl red	Salmon	Orange
4.5–8.3	Litmus	Red	Blue
4.8–6.2	Chlorphenol red	Yellow	Red
5.1–6.5	Propyl red	Pink	Yellow
5.4–6.8	Bromocresol purple	Green-yellow	Violet
5.6–7.2	Alizarin	Yellow	Red
6.0–7.6	Bromthymol blue	Yellow	Blue
6.0–7.6	Bromoxylenol blue	Orange-yellow	Blue
6.4–8.2	Phenol red	Yellow	Red-violet
7.1–8.8	Cresol red	Yellow	Violet
7.5–9.0	m-cresol purple	Yellow	Violet
8.1–9.5	Thymol blue	Yellow	Blue
8.3–10.0	Phenolphthalein	Colorless	Dark pink
8.6–9.8	o-Cresolphthalein	Colorless	Pink

Range	Indicator	Lower Color	Upper Color
9.5–10.4	Thymolphthalein	Colorless	Blue
9.9–11.8	Alizarin yellow R	Yellow	Dark orange
10.6–13.4	Methyl blue	Blue	Pale violet
11.1–12.8	Acid fuchsin	Red	Colorless
11.4–13.0	Indigo carmine	Blue	Yellow
11.7–12.8	2,4,6-Trinitrotoluene	Colorless	Orange
12.0–14.0	Trinitrobenzene	Colorless	Orange

Appendix C: Flame Tests for Elements

BLUE	
Azure	Lead, selenium, $CuCl_2$ (and other copper compounds when moistened with HCl); $CuBr_2$ appears azure blue, then is followed by green.
Light Blue	Arsenic and some of its compounds; selenium.
Blue-green	Phosphates moistened with sulfuric acid; B_2O_3.
GREEN	
Emerald Green	Copper compounds other than halides (when not moistened with HCl), thallium compounds.
Greenish-blue	$CuBr_2$; arsenic; lead; antimony.
Pure green	Thallium and tellurium compounds.
Yellow-green	Barium; possibly molybdenum; borates with (H_2SO_4).
Faint green	Antimony and ammonium compounds.
Whitish green	Zinc.
RED	
Carmine	Lithium compounds (masked by barium or sodium), are invisible when viewed through green glass, appear violet through cobalt glass.
Scarlet	Calcium compounds (masked by barium), appear greenish when viewed through cobalt glass and green through green glass.
Crimson	Strontium compounds (masked by barium), appear violet through cobalt glass, yellowish through green glass.
VIOLET	Potassium compounds other than silicates, phosphates and borates; rubidium and cesium are similar. Color is masked by lithium and/or sodium, appears purple-red through cobalt glass and bluish-green glass.
YELLOW	Sodium, even the smallest amount, invisible when viewed through cobalt glass.

Appendix D: Qualitative Analysis of Cations and Anions

**Qualitative Analysis Of Cations
Flow Chart**

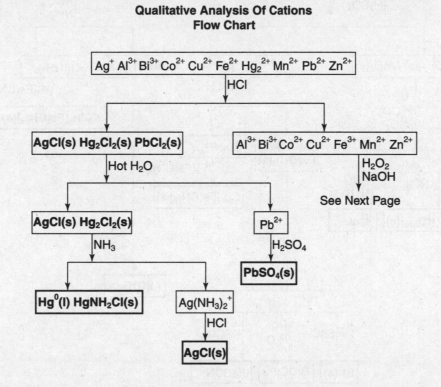

Note: Boldface type indicates that a solid, liquid or gas forms.

Flow Chart continued

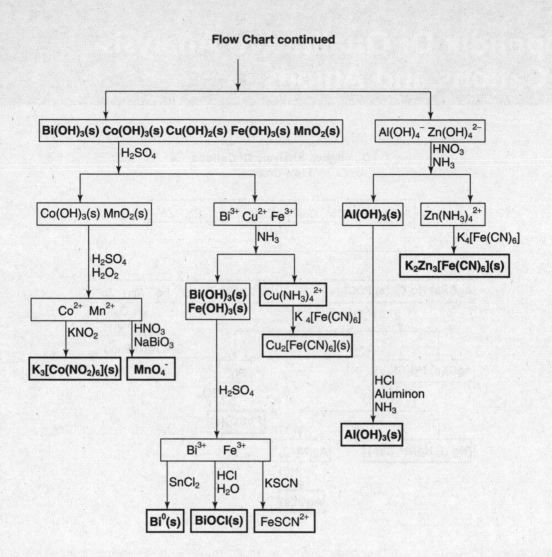

Qualitative Analysis Of Anions
Flow Chart

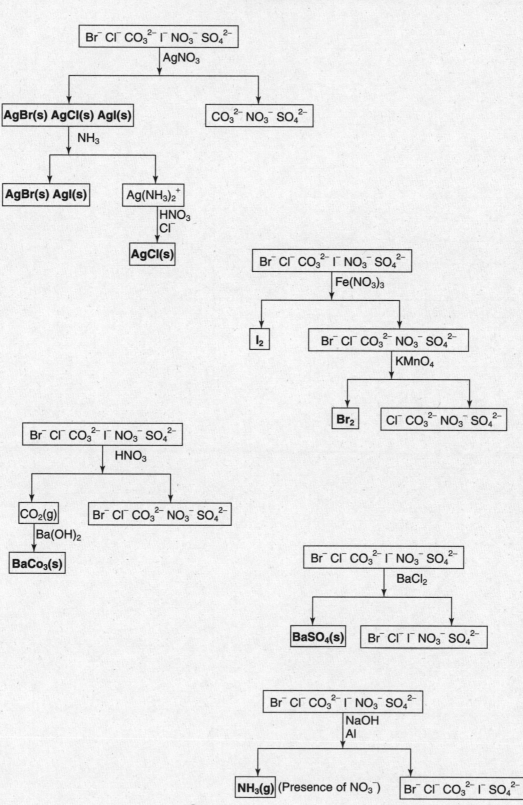

Notes

CliffsNotes

LITERATURE NOTES

Absalom, Absalom!
The Aeneid
Agamemnon
Alice in Wonderland
All the King's Men
All the Pretty Horses
All Quiet on Western Front
All's Well & Merry Wives
American Poets of the 20th Century
American Tragedy
Animal Farm
Anna Karenina
Anthem
Antony and Cleopatra
Aristotle's Ethics
As I Lay Dying
The Assistant
As You Like It
Atlas Shrugged
Autobiography of Ben Franklin
Autobiography of Malcolm X
The Awakening
Babbit
Bartleby & Benito Cereno
The Bean Trees
The Bear
The Bell Jar
Beloved
Beowulf
The Bible
Billy Budd & Typee
Black Boy
Black Like Me
Bleak House
Bless Me, Ultima
The Bluest Eye & Sula
Brave New World
Brothers Karamazov
The Call of the Wild & White Fang
Candide
The Canterbury Tales
Catch-22
Catcher in the Rye
The Chosen
The Color Purple
Comedy of Errors...
Connecticut Yankee
The Contender
The Count of Monte Cristo
Crime and Punishment
The Crucible
Cry, the Beloved Country
Cyrano de Bergerac
Daisy Miller & Turn...Screw
David Copperfield
Death of a Salesman
The Deerslayer
Diary of Anne Frank
Divine Comedy-I. Inferno
Divine Comedy-II. Purgatorio
Divine Comedy-III. Paradiso
Doctor Faustus

Dr. Jekyll and Mr. Hyde
Don Juan
Don Quixote
Dracula
Electra & Medea
Emerson's Essays
Emily Dickinson Poems
Emma
Ethan Frome
The Faerie Queene
Fahrenheit 451
Far from the Madding Crowd
A Farewell to Arms
Farewell to Manzanar
Fathers and Sons
Faulkner's Short Stories
Faust Pt. I & Pt. II
The Federalist
Flowers for Algernon
For Whom the Bell Tolls
The Fountainhead
Frankenstein
The French Lieutenant's Woman
The Giver
Glass Menagerie & Streetcar
Go Down, Moses
The Good Earth
Grapes of Wrath
Great Expectations
The Great Gatsby
Greek Classics
Gulliver's Travels
Hamlet
The Handmaid's Tale
Hard Times
Heart of Darkness & Secret Sharer
Hemingway's Short Stories
Henry IV Part 1
Henry IV Part 2
Henry V
House Made of Dawn
The House of the Seven Gables
Huckleberry Finn
I Know Why the Caged Bird Sings
Ibsen's Plays I
Ibsen's Plays II
The Idiot
Idylls of the King
The Iliad
Incidents in the Life of a Slave Girl
Inherit the Wind
Invisible Man
Ivanhoe
Jane Eyre
Joseph Andrews
The Joy Luck Club
Jude the Obscure
Julius Caesar
The Jungle
Kafka's Short Stories
Keats & Shelley
The Killer Angels
King Lear
The Kitchen God's Wife
The Last of the Mohicans

Le Morte Darthur
Leaves of Grass
Les Miserables
A Lesson Before Dying
Light in August
The Light in the Forest
Lord Jim
Lord of the Flies
Lord of the Rings
Lost Horizon
Lysistrata & Other Comedies
Macbeth
Madame Bovary
Main Street
The Mayor of Casterbridge
Measure for Measure
The Merchant of Venice
Middlemarch
A Midsummer-Night's Dream
The Mill on the Floss
Moby-Dick
Moll Flanders
Mrs. Dalloway
Much Ado About Nothing
My Ántonia
Mythology
Narr. ...Frederick Douglass
Native Son
New Testament
Night
1984
Notes from Underground
The Odyssey
Oedipus Trilogy
Of Human Bondage
Of Mice and Men
The Old Man and the Sea
Old Testament
Oliver Twist
The Once and Future King
One Day in the Life of Ivan Denisovich
One Flew Over Cuckoo's Nest
100 Years of Solitude
O'Neill's Plays
Othello
Our Town
The Outsiders
The Ox-Bow Incident
Paradise Lost
A Passage to India
The Pearl
The Pickwick Papers
The Picture of Dorian Gray
Pilgrim's Progress
The Plague
Plato's Dialogues
Plato's The Republic
Poe's Short Stories
A Portrait of Artist...
The Portrait of a Lady
The Power and the Glory
Pride and Prejudice
The Prince
The Prince and the Pauper
A Raisin in the Sun

The Red Badge of Courage
The Red Pony
The Return of the Native
Richard II
Richard III
The Rise of Silas Lapham
Robinson Crusoe
Roman Classics
Romeo and Juliet
The Scarlet Letter
A Separate Peace
Shakespeare's Comedies
Shakespeare's Histories
Shakespeare's Minor Plays
Shakespeare's Sonnets
Shakespeare's Tragedies
Shaw's Pygmalion & Arms...
Silas Marner
Sir Gawain...Green Knight
Sister Carrie
Slaughterhouse-Five
Snow Falling on Cedars
Song of Solomon
Sons and Lovers
The Sound and the Fury
Steppenwolf & Siddhartha
The Stranger
The Sun Also Rises
T.S. Eliot's Poems & Plays
A Tale of Two Cities
The Taming of the Shrew
Tartuffe, Misanthrope...
The Tempest
Tender Is the Night
Tess of the D'Urbervilles
Their Eyes Were Watching God
Things Fall Apart
The Three Musketeers
To Kill a Mockingbird
Tom Jones
Tom Sawyer
Treasure Island & Kidnapped
The Trial
Tristram Shandy
Troilus and Cressida
Twelfth Night
Ulysses
Uncle Tom's Cabin
The Unvanquished
Utopia
Vanity Fair
Vonnegut's Works
Waiting for Godot
Walden
Walden Two
War and Peace
Who's Afraid of Virginia...
Winesburg, Ohio
The Winter's Tale
The Woman Warrior
Worldly Philosophers
Wuthering Heights
A Yellow Raft in Blue Water